T0252740

Series on

Reproduction and Development in Aquatic Invertebrates

Volume 1

Reproduction and Development in Crustacea

Series on

Reproduction and Development in Aquatic Invertebrates

Volume 1

Reproduction and Development in Crustacea

T. J. Pandian

INSA Senior Scientist
Central Marine Fisheries Research Institute
Kochi-682018, Kerala
India

CRC Press
Taylor & Francis Group
Boca Raton London New York

CRC Press is an imprint of the
Taylor & Francis Group, an **informa** business

A SCIENCE PUBLISHERS BOOK

Cover Illustrations: Peripheral figures show representatives of over 96% crustaceans brooding eggs on one or the other parts of their body. Brooding eggs is a hallmark strategy of crustaceans. The remaining 4% crustaceans (in the centre of the picture) shed their eggs (for more details see Fig. 2.10).

CRC Press
Taylor & Francis Group
6000 Broken Sound Parkway NW, Suite 300
Boca Raton, FL 33487-2742

First issued in paperback 2021

© 2016 by Taylor & Francis Group, LLC
CRC Press is an imprint of Taylor & Francis Group, an Informa business

No claim to original U.S. Government works

ISBN-13: 978-0-367-78302-0 (pbk)
ISBN-13: 978-1-4987-4828-5 (hbk)(Volume 1)

Library of Congress Cataloging-in-Publication Data

Names: Pandian, T. J., author.
Title: Reproduction and development in crustacea / T.J. Pandian.
Description: Boca Raton : Taylor & Francis, 2016. | Series: Series on reproduction and development in aquatic invertebrates ; volume 1 | Includes bibliographical references and index.
Identifiers: LCCN 2015044150 | ISBN 9781498748285 (alk. paper)
Subjects: LCSH: Crustacea--Reproduction. | Crustacea--Development.
Classification: LCC QL445.2 .P36 2016 | DDC 595.3--dc23
LC record available at http://lccn.loc.gov/2015044150

Visit the Taylor & Francis Web site at
http://www.taylorandfrancis.com

and the CRC Press Web site at
http://www.crcpress.com

Preface to the Series

Invertebrates surpass vertebrates not only in species number but also in diversity of sexuality, modes of reproduction and development. Yet, we know much less of them than we know of vertebrates. During the 1950s, the multi-volume series by L. H. Hyman accumulated bits and pieces of information on reproduction and development of aquatic invertebrates. Through a couple of volumes published during 1960s, A. C. Giese and A. S. Pearse provided a shape to the subject of Aquatic Invertebrate Reproduction. Approaching from the angle of structure and function in their multi-volume series on Reproductive Biology of Invertebrates during 1990s, K. G. Adiyodi and R. G. Adiyodi elevated the subject to a visible and recognizable status.

Reproduction is central to all biological events. The life cycle of most aquatic invertebrates involves one or more larval stage(s). Hence, an account on reproduction without considering development shall remain incomplete. With passage of time, publications are pouring through a large number of newly established journals on invertebrate reproduction and development. The time is ripe to update the subject. This treatise series proposes to (i) update and comprehensively elucidate the subject in the context of cytogenetics and molecular biology, (ii) view modes of reproduction in relation to Embryonic Stem Cells (ESCs) and Primordial Germ Cells (PGCs) and (iii) consider cysts and vectors as biological resources.

Hence, the first chapter on Reproduction and Development of Crustacea opens with a survey of sexuality and modes of reproduction in aquatic invertebrates and bridges the gaps between zoological and stem cell research. With capacity for no or slow motility, the aquatic invertebrates have opted for hermaphroditism or parthenogenesis/polyembryony. In many of them, asexual reproduction is interspersed within sexual reproduction. Acoelomates and eucoelomates have retained ESCs and reproduce asexually also. However, pseudocoelomates and hemocoelomates seem not to have retained ESCs and are unable to reproduce asexually. This series provides possible explanations for the exceptional pseudocoelomates and hemocoelomates that reproduce asexually. For posterity, this series intends to bring out six volumes.

August, 2015   **T. J. Pandian**
Madurai-625 014

Preface

Being arthropods, the crustaceans molt to grow and/or reproduce. Of 52,000 species, >96% crustaceans brood their eggs on their body. Consequently, they share the available resources among intensely competing processes (i) growth including molting, (ii) breeding and (iii) brooding. Hence, they are unique among aquatic invertebrates and render the study of reproduction and development a fascinating field of research. Not surprisingly, many books are written but they are limited to one or another taxon or aspect of reproduction and development. This book represents the first attempt to comprehensively elucidate almost all aspects of reproduction and development covering from anostracan *Artemia* to xanthid crabs.

Commencing with resource allocation to reproduction, the second and third chapters display that the investment on generation time as a fraction of life span averages to 28, 48 and 85% in tropical, temperate and Arctic crustaceans. Consequently, the tropical cladocerans, for example, produce 10 times more neonates than their temperate counterparts. That their hallmark strategy of brooding embryos costs considerable time and resource investment is demonstrated by the fact that the egg shedding calanoids and penaeids produce 3-20 times more eggs than the brooders. A second consequence of brooding is the liability of the brooders to reproductive senescence. For the first time, the book has adduced adequate evidence to establish the fact that brooding gonochorics, parthenogenics and hermaphrodites of all taxons across Crustacea do undergo reproductive senescence. With inclusion of many species characterized by sessile, sedentary or slow motile groups, gonochorism in crustaceans is limited to 92%; the remaining 8% employ alternate sexuality of parthenogenic and hermaphoditism. To escape from inbreeding, parthenogenics sporadically or regularly produce sexual males and females, or intertwine hybridization and polyploidization. On the other hand, hermaphrodites employ sequential hermaphroditism or androdioesic mating system.

The fourth chapter explains that the same tissue types are involved in regeneration of the lost part(s) of chelar appendages and asexual reproduction of the externae in the exceptional hemocoelomate of the parasitic colonial rhizocephalan females. They are limited to a few tissues derived from ectoderm and mesoderm alone. Hence, the so called asexual reproduction in these rhizocephalans simulates more of regeneration than of agametic cloning of a whole animal.

The fifth chapter is concerned with cyst and their equivalents like the ephihippia and diapausing embryos/larvae. In *Artemia* cyst, protein synthesis and degradation are ceased, trehalose is used as metabolite, metabolism is depressed down to 500,000 times and the cyst can survive anoxia for four years. A number of environmental factors are known to trigger cyst hatching. However, even the most potent cue does not initiate hatching in all cysts in an egg bank. Hatching in batches is a crustacean trait. In brooded embryos, hatching is initiated by the mother with the release of a chemical substance. However, almost nothing is known on the mechanism, which controls cyst hatching in different batches.

Interestingly, collections of widely scattered information reveal that (i) the sex in crustaceans is genetically determined at fertilization and (ii) the presence of both male and female heterogametic sex determination mechanisms. The ostracods have explored employing numerically more X (X_1-X_{11}) chromosomes to determine the female sex and anostracans used B chromosome to determine the male sex.

Chapter seven provides an account of sex differentiation by endocrines and its disruption by endocrine disruptors, parasites and food. Unlike vertebrates, endocrine and spermatogenic functions are clearly separated into distinct androgenic glands and testes, respectively. In crustaceans, sex differentiation process is irrevocably completed on or before metamorphosis or birth. Juvenile hormone or its analogs induce(s) parthenogenic male production in cladocerans. But the ability of these chemically induced males to fertilize eggs remains doubtful. Limitation of food induces production of males in cladocerans but females in copepods. Specific alga induces direct development of females in protrandric hippolytids. For the first time, cymathoid parasites are recognized and designated as protrandrics and the other bopyrids as protogynics. Briefly, parasitism in bopyrids has retained sexual plasticity until their cryptoniscus larvae settle on the definitive hosts and also eliminated the reproductive senescence.

This book is a comprehensive synthesis of over 972 publications carefully selected from widely scattered information from 249 journals and 101 other literature sources. The holistic approach and incisive analysis have led to harvest several new findings and ideas related to reproduction and development in crustaceans. Hopefully, this book serves as a launching pad to further advance our knowledge on reproduction and development of crustaceans.

August, 2015   **T. J. Pandian**
Madurai-625 014

Acknowledgements

It is with great pleasure, that I wish to thank Drs. N. Munuswamy and E. Vivekanandan for critically reading parts of manuscript of this book and offering valuable suggestion. I must thank profusely my student Prof. R. Gadagkar for persuading me to author this book series and for offering a token support by the Indian National Science Academy, New Delhi. Besides my earlier review, my students Drs. C. Balasundaram, S. Katre and S. Kumari (USA), whose publications from my laboratory, have helped me to author this book. The Central Marine Fisheries Research Institute, (CMFRI) Kochi provides the best library and excellent service to the visitors. I wish to place on record my sincere thanks to Dr. A. Gopalakrishnan and his library staff. Dr. A. Gopalakrishnan, Director, CMFRI has thoughtfully chosen to engage and support me to accomplish this book series, for which, I remain thankful. I also thank Drs. T. Balasubramanian R. Jeyabaskaran, P. Murugasen and Rajakumar for regularly helping me with publications. Thanks are due to Drs. J. A. Baeza, (USA) G. Kumarasen, S. M. Naqvi, K. K. Vijayan, S. S. S. Sarma (Mexico) and S. C. Weeks (USA) for readily providing me the requested publications. The manuscript of this book was prepared by Ms C. K. Chitra Prabha, M.Sc., M.Phil. and I wish to record my grateful appreciation for her competence, patience and excellent work.

To reproduce figures and tables from published domain I gratefully appreciate the permissions issued by Central Marine Fisheries Research Institute, CRC Press (*Ophelia* and others), International Ecology Institute (*Marine Ecology Progress Series, Diseases of Marine Organisms*), Oxford and IBH Publishers (Reproductive Biology of Invertebrates), University of New England and CSIRO Publishers (Crustacean Parasites: Marine Parasitology). For permission to reproduce their figures and tables, I remain indebted to Drs. K. Altaff, C. Balasundaram, CMFRI, Z. M. Gliwics, M. Ikhwanuddin, N. Munuswamy, K. K. C. Nair, N. Rabet, S. B. Sainath, Scripps Institute of Oceanography, K. Tande, G. Vogt and S. C. Weeks. For advancing our knowledge in this area by their rich contributions, I thank my fellow scientists, whose publications are cited in this book.

August, 2015   T. J. Pandian
Madurai-625 014

Contents

Section I

Aquatic Invertebrate Reproduction: Stem Cells

1

Reproduction and Stem Cells

1.1 Introduction

Based on available relevant publications, this account attempts to (i) survey the sexuality and modes of reproduction in aquatic invertebrates and (ii) bridge the gaps between zoological and stem cell research. Reproduction is central to all biological events. Organisms may reproduce asexually or sexually. Asexual reproduction requires only a single parent and costs no extra time and energy to produce offspring. All offspring produced asexually are genetically identical to one another and to the parent. Sexual reproduction generates offspring with different mix of alleles than their parents. The variations in traits among the offspring increase the chances that some members of a species will survive in a changing environment. Sex is a luxury and a system, where two individuals are required for reproduction. Hence it costs time and energy but ensures genetic diversity, one of the major sources for evolution. As benefits accruing from genetic recombination outweigh the costs of time and energy, sex is successful, has evolved as early as 1.6-2.0 billion years ago (Butlin, 2002) and is manifested in a wide range of microbes, plants and animals. The invertebrates are no exception to this dictum. In fact, no invertebrate species is thus far reported to exclusively reproduce asexually. A computer search revealed the existence of only asexual race within gonochoric species (e.g. *Schmidtea mediterranea*, Baguna et al., 1989, *Dugestia tigrina*, *D. gonocephala* and *D. tahitiensis*, Benazzi and Benazzi Lentati, 1993).

In terms of structural diversity, number, biomass, sexual plasticity and modes of reproduction, the invertebrates surpass vertebrates. The endless experimentation undertaken by them with form and function since the early Cambrian has generated about 1.6 million aquatic invertebrate species (of 2.62 million animal species, about one million are terrestrial invertebrates), which are systematically classified into eight major (Table 1.1) and 25 minor phyla (Table 1.2). Strikingly, acoelomates like the poriferans, cnidarians, platyhelminthes and eucoelomates, like the annelids and echinodermates are also capable of asexual reproduction. In them, totipotent or pluripotent adult stem cells have been shown to perform agametic cloning and asexual reproduction (Skold et al., 2009). However, pseudocoelomates, Nematoda, Gastrotricha, Rotifera and a few others, and hemocoelomates, Mollusca and

Arthropoda are not known to reproduce asexually. In them, regeneration is limited to one or more tissues (see Table 1.3) but not to a whole animal. Apparently, the presence of Pluripotent Stem Cells (PSCs) is linked to asexual reproduction but in their absence, regeneration is limited to a few tissues but not the whole animal.

TABLE 1.1

Agametic and gametic modes of reproduction in major phyla of aquatic invertebrates (compiled from Chapman, 2009)

Name of Phylum	Species (No.)	Remarks
Acoelomata		
Porifera*	6,000	Sexual, Hermaphrodite, Asexual
Cnidaria*	9,295	Sexual, Hermaphrodite, Asexual
Ctenophora***	166	Sexual, Asexual
Protostomia: Acoelomata		
Platyhelminthes*	20,000	Sexual, Hermaphrodite, Asexual
Schizocoelomata (Hemocoelomata)		
Mollusca†	85,000	Sexual, Hermaphrodite, Asexual in 1 species *Clio pyramidata*
Arthropoda†	1,170,000	Sexual, Hermaphrodites, Asexual in 3 rhizocephalan spp
Deuterostomia (Eucoelomates)		
Annelida**	16,763	Sexuals, Hermaphrodite, Asexual
Echinodermata**	7003	Sexual, Asexual

According to Skold et al. (2009), *totipotent adult cells perform agametic cloning, **Pluripotent cells perform agametic cloning, ***Stemness of agamatic cells is yet to be proved, †Regeneration is limited to one or two organs (see Table 1.3)

1.2 Ontogenetic Pathways

Figure 1.1 shows a phylogenetic tree of acoelomate, pseudocoelomate, hemocoelomate and eucoelomate phyla displaying different modes of reproduction. At this juncture, it has become necessary to define some terms invariably but inconsistently used to denote sexual and asexual modes of reproduction (e.g. Simon et al., 2003). Surprisingly, almost all invertebrates are sexualized; some of them express the most divergent expression of sex. For example, there are simultaneous and sequential hermaphrodites. Many of them also involve an array of the so called 'asexual reproduction', while passing through different ontogenetic pathways. There may be more pathways but Fig. 1.2 describes some of them, through which ontogenetic development of invertebrate's progresses. In organisms, manifestation of sex, gametogenesis and fertilization are critical events that characterize sexual reproduction. During the course of evolution of ontogenesis and life cycle, one or more of these events were eliminated or shifted to another location in these sequences. Consequently, 'asexual reproduction' was invariably interwoven within the ontogenetic pathways of sexual reproduction in many invertebrates. In fact, 'asexual reproduction' has emerged from their respective sexual ancestors during the course of evolution of different metazoan lineages repeatedly and independently. Hence, it has explored many ontogenetic pathways, remain flexible but is not conserved (see Isaeva, 2011).

TABLE 1.2

Agametic and gametic modes of reproduction in aquatic invertebrates belonging to minor phyla of (compiled from Chapman, 2009, Skold et al., 2009)

Name of Phylum	Species (No.)	Remarks
Protostomia: Acoelomata		
Mesozoa	106	Sexual, Parasitic, Polyembryony
Nemertina	1,200	Sexual, Asexual e.g. *Lineus*
Gnathostomulia	97	Sexual, Hermaphrodite
Pseudocoelomata		
Rotifera	2,180	Sexual, Parthenogenesis
Gastrotricha	400	Sexual, Hermaphrodite, Parthenogenesis
Kinorhyncha	130	Sexual, Hermaphrodite, Parthenogenesis
Nematoda	46,000†	Sexual, Hermaphrodite
Nematomorpha	331	Sexual, asexual in *Lineus*
Acanthocephala	1,150	Sexual, Endoparasites, No gut
Schizocoelomata: Hemocoelomata		
Priapulida	16	Sexual
Sipuncula	300*	Sexual, Asexual in *Sipunculus nodus, Aspidospihon brocki*
Echiura	176	Sexual, Induced parthenogenesis
Pogonophora		Sexual, no gut
Tardigrada	1,045	Sexual, Parthenogenesis
Onycophora	165	Sexual
Pentastomida	105	Sexual, Endoparasites
Schizocoelomata: Lophophorata		
Phoronida	10	Sexual, Hermaphrodite
Entoprocta**	170	Sexual, Asexual e.g. *Phoronis ovalis*
Bryozoa**	5,000†	Sexual, Hermaphrodite, Polyembryony
Brachiopoda	550	Sexual, Hermaphrodite
Cycllophora*	2	Asexual, Symbiotic sp
Eucoelomata		
Chaetognatha	121	Hermaphrodites
Xenoturbella		Asexual
Hemichordata	108	Sexual, Hermaphrodite, Asexual
Urochordata*	2,300*	Sexual, Hermaphrodite, Asexual

†Mean of reported values ranging from 12,000 to 80,000, *Totipotent cells perform agametic cloning, **Pleuripotent cells perform agametic cloning

TABLE 1.3

Regeneration in pseudocoelomate and hemocoelomate invertebrates (from K.G. Adiyodi and R.G. Adiyodi, 1993, 1994)

Name of Phylum	Remarks
	Major Phyla
	Hemocoelomata
Mollusca	Limited to organs like head as well as tentacles, inhalant/exhalant siphons
Arthropoda	Limited to appendages alone
	Minor Phyla
	Pseudocoelomata
Rotifera	Limited to coronal arms, which are syncytial extensions with no nuclei
Gastrotricha	No somatic cell division, no regeneration
Kinorhyncha	
Nematoda	
Nematomorpha	
Acanthocephala	No regeneration
	Hemocoelomata
Priapulida	
Sipuncula	Limited to tentacles, parts of brain, oesophagus, nerve cord. coelomocytes and nerve cells are responsible for regeneration
Echiura	
Pogonophora	
Tardigrada	No regeneration
Onycophora	
Pentastomida	No regeneration
Phoronida	
Entoprocta	Periodic regeneration of tentacles, stalks that are lost, calyx
Brachiopoda	
	Eucoelomata
Chaetognatha	Limited to caudal section and fin rays
Hemichordata	Limited powers to regenerate

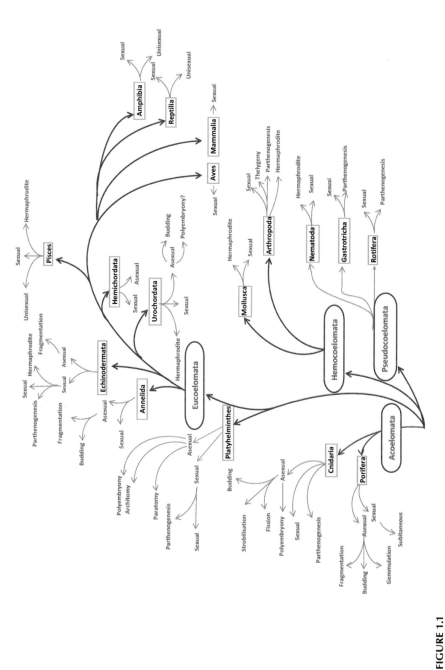

FIGURE 1.1

Phylogenetic tree showing the modes of reproduction in acoelomate, pseuodocoelomate, hemocoelomate and eucoelomate animals.

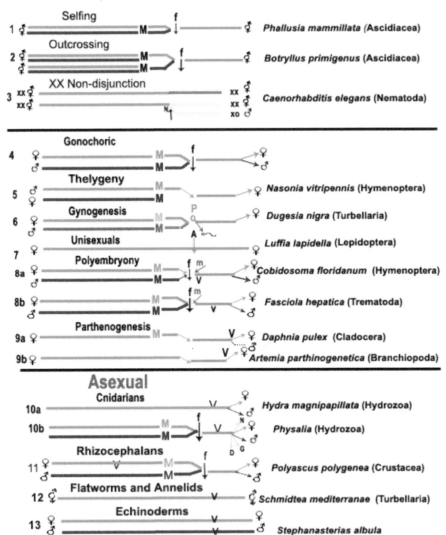

FIGURE 1.2
Ontogenetic pathways of sexual and asexual modes of reproduction in invertebrates.
M = meiosis, f = fertilization, V = location of asexual reproduction, m = morula,
A = activation, P = polar body, G = Gonozoid, T = Trophozoid, N = Pneumatophore, D =
Dactylozoid

In thelygenic insects like *Coccus hesperidium*, fertilized eggs produce both females and males but unfertilized eggs, following restoration of diploidy by automixis, produce females alone (see Muthukrishnan, 1994, p 135). In the thelygenic parasitoid wasp *Nasonia vitripennis*, fertilized eggs develop into diploid females but unfertilized eggs develop into haploid males. In other words "sons have no father and fathers have no sons" (Gadagkar, 1997, p 33). In unisexuals and ameiotic or apomictic parthenogenics, female sex is manifested but gametogenesis occurs, though limited to ameiotic oogenesis. In oogenesis of apomictic parthenogenesis, synapsis of homologous chromosomes does not occur; hence, there is no segregation or crossing over. In them, no fertilization is required, as ameiotic oogenesis generates diploid eggs only. In this mode of reproduction, 'activation' by sperm of a sympatric species is required to commence embryogenesis in unisexuals but it is not required in parthenogenesis (Fig. 1.2). In the nematode *Caenorhabditis elegans*, there are two natural sexes namely XX hermaphrodite and XO male. Hermaphrodites are somatic females and can reproduce by either self-fertilization or mating with the males. Matings generate broods of 50% XX and 50% XO offspring. Progenies arising from self fertilization are mostly XX but one out of 500 is an XO male, which is generated through non-disjunction of XX chromosomes (Zarkower, 2006).

Polyembryony is natural cloning at morula stage of an animal (Fig. 1.2). It occurs mostly in parasites like the flukes (Trematoda) and insects. In the parasitoid wasp *Copidosoma floridanum*, the zygote develops into a morula consisting of about 200 cells. Instead of developing into a blastula and then gastrula, the mitotically active cells of morula of *C. floridanum* divide repeatedly to produce more than 1,000 secondary morulae and from each of these secondary morulae, a progeny arises (Donell et al., 2004, Corley et al., 2005). In trematodes and eucestodes, polyembryony is shifted to larval stages. From every fertilized egg of 20 eucestode species belonging to six families, more than one offspring is produced (Whitefield and Evans, 1983).

In unisexuals as well as thelygenic and parthenogenic invertebrates, 'asexual reproduction' involves changes in ploidy levels and associated cytological mechanisms (see Simon et al., 2003). Parthenogenic females occasionally generate parthenogenically females and males by a not yet known cytological mechanism. However, active participation of Pluripotent Stem Cells (PSCs) is involved in budding, fission and fragmentation. From the foregone description, the following may be concluded: 1. No invertebrate is known to obligatorily or exclusively reproduce asexually. In known ontogenetic pathways, asexual reproduction occurs only within sexual reproduction. 2. In most of these ontogenetic pathways, females play a key role. Rarely and uniquely, do males play a role in it. For example, following its entry into an egg of a sturgeon, the sperm of narcomedusa *Polypodium hydriforme* develops into an inverted parasitic polyp with endoderm facing yolk (Raikova, 1973). 3. With the involvement of ploidy level changes alone and no participation by PSCs, parthenogenics (cf Simon et al., 2003), unisexuals and thelygenics can not be considered to reproduce asexually. However, the terms have long been in use and are continued here with the *sensu* restricted. True asexual reproduction occurs only, when an animal is budded, fissioned and fragmented.

1.3 Motility and Modes of Reproduction

Seawater covers 70% of the Earth with 97% of its water, while freshwater covers only about 1% of the Earth's surface and accounts for a little less than 0.01% of its water (Pandian, 2011, p 1). It contains 1 mg organic carbon/l. Free amino acids, comprising 5% of the total dissolved organic matter, occur at concentrations of 5×10^{-7} M/l in free water and 1.1×10^{-4} M/l in interstitial water of sediments (see Pandian, 1975, p 72). Indeed, sea water is an organic 'soup'. As early as in 1961, Stephens and Schinke reported the influx of amino acids from ambient seawater into representatives of 35 genera of marine invertebrates belonging to 11 phyla. Not surprisingly, the osmotrophic pogonophores, with no alimentary canal, flourish in the sea. It is not easy to starve aquatic animals like *Daphnia* to death, as "it is reluctant to die of starvation in laboratory cultures" (Gliwicz, 2003). The enormous volume of seawater with its nutrients and innumerable number of micro- and macro-habitats provide a vast scope for invertebrates to originate, evolve, speciate and flourish.

Motility of an animal plays an important role in acquiring food; a fraction of the acquired food, after its assimilation, is allocated for reproduction. Its importance can be understood from the following observation. Under comparable experimental conditions, the motile *Daphnia pulex* acquires 13% of available flagellate *Chlamydomonas reinhardi* but the sessile *Hydra oligactis* predates only 7% of the offered *Artemia salina* (see Pandian, 1975, p 93). Majority of aquatic invertebrates are suspension feeders. Flagella or cilia aid in filtering suspended particles from water. For example, *Leuconia aspera* possessing 2.25 million flagellated chambers filter 22.5 l/day (d) (see Pandian, 1975, p 76). Describing feeding bouts of the scyphozoan *Aurelia aurelia* at the polyp stage, Kamiyama (2011) has reported that the ciliary beats filter 0.76 m³/d. A single *Artemia* filters 240 ml/d. At the density of four individuals (indiv)/l, *Artemia* is capable of filtering the entire Great Salt Lake water once a day (Reeve, 1963, see also Gliwicz, 2003). To grow and reproduce, a filter-feeder has to filter suspended phytoplankton and/or detritus. The threshold levels of food density to initiate growth and the first wave of reproduction decrease with increasing body size in filter-feeding cladocerans (Gliwicz, 1990). Hence, the invertebrates are great movers of water column but have limited powers of motility through the denser aquatic medium. Sexual reproduction involves motility of mating partners to meet, recognize each other, positively respond and court, which culminates in the climax of impregnation of the female. Several filter-feeding invertebrates are sessile or sedentary (Table 1.4). Understandably, many aquatic invertebrates are hermaphroditic or parthenogenic and/or capable of asexual reproduction. With motility, aquatic arthropods and molluscs that reproduce sexually alone, are an exception to this generalization.

TABLE 1.4

Mode of living by members of minor phyla

Phylum	Mode of Life
Mesozoa	Parasitic
Nemertina	
Gnathostomulia	Sedentary
Rotifera	Sedentary/free living
Gastrotricha	Sedentary
Kinorhyncha	Sedentary
Nematoda	Sedentary
Nematomorpha	Sedentary
Acanthocephala	Endoparasitic
Priapulida	Sedentary
Sipuncula	Sedentary
Echiura	
Pogonophora	Sedentary in tubes
Tardigrada	
Onycophora	
Pentastomida	Endoparasitic
Phoronida	Sedentary
Entoprocta	Sessile
Bryozoa	Sessile
Brachiopoda	Sessile
Cycllophora	
Chaetognatha	Free living, pelagic drifters
Xenoturbella	
Hemichordata	Sessile/sedentary
Urochordata	Sessile/pelagic but motile

TABLE 1.5

Mobility of some invertebrates.

Taxonomic Group/Species	Motility (mm/s)	Speed (l/s)	Remarks	Reference
Turbellarians	1.7	–	Ciliary beats	Rompolas et al. (2013)
Polychaeta		–		
Nereis saccunea	85		Muscle contraction	Ram et al. (2008)
Gastropoda	3.3	–	Gliding	Lai et al. (2010)
Echinodermata		–		
Pteraster tesselatus	0.17		10-d old larva	Kelman and Emelt (1999)
Asterias rubens	0.8			Montgomery and
Archaster typicus	12.7			Palmer (2012)
Crustacea Spinicaudata				
Tony cypris	0.13		Jumping speed	Matzke-Karasz et al. (2014)
Copepoda				
Acartia tonsa	0.22			Matzke-Karasz et al. (2014)
Somatopoda				
Odontodactylus havanensis	1.42†			Campos et al. (2012)
Decopoda				
Acanthepyra exemia	57†	20.1	Tail dip jump	see Campos et al. (2012)
Crangon crangon	57†	40.5		see Campos et al. (2012)
Pandalus danae	70†	~ 41.2		see Campos et al. (2012)
Sergestes similis		1.8	Pleopod driven	see Campos et al. (2012)
Grapsus teniucrustatus	0.35†		Surfing crab	Martinez (2001)
Cephalopoda	1.6†		Pumping water	Zeidberg (2004)
Loligo vulgaris	65†	19.7	Jet propulsion	see Campos et al. (2012)
Esox sp	65†	21.0		see Campos et al. (2012)

†(m/s)

During the checkered history of evolution of invertebrates, motility levels of aquatic invertebrates may have played a decisive role to select hermaphroditism rather than gonochorism and intersperse asexual reproduction within the ontogenetic pathways of sexual reproduction. Sponges and cnidarians (exception: Scyphozoa and Siphonophora are pelagic drifters moving by pumping water) are sedentary or sessile. Aided by ciliary beats, turbellarians glide at ~ 1.7 mm/second(s) (Rompolas et al., 2013) (Table 1.5). With the appearance of segmentation and associated musculature, motility is improved a little more in annelids (e.g. *Nereis saccunea*, 85 mm/s, Ram et al., 2008) but the larvae of crustaceans (e.g. balanoid: nauplius 3 mm/s, cypris 10 mm/s, Walker, 2004) and echinoderms (10-d old larva of *Pteraster tesselatus*: 0.17 mm/s, Kelman and Emelt, 1999) move a little slowly. The gastropod slug also glides at 3.3 mm/s only, despite the presence of musculature in its foot (Lai et al., 2010). However, the surfing crabs *Grapsus tenuicrustatus* moves faster at 0.1-0.4 m/s (Martinez, 2001). Expectedly, the motility of powerful pumping by myopsid cephalopod is 1.6 m/s (Zeidberg, 2004). The speed of many decapod crustaceans is as fast as those of fishes. No information is yet available on the distance covered during spawning migration by the portunid crabs from pelagic realms to the coastal waters. But the estuarine crab *Callinectes sapidus* covers a distance of 1.9-10.6 km at the speed of 0.02-0.11 m/s (Carr et al., 2004). From an unique and perhaps the only investigation available on tagging and recapture of aquatic invertebrates undertaking spawning migration, the Central Marine Fisheries Research Institute (CMFRI, 2013) reported that the Indian prawn *Penaeus indicus* undertook spawning migration covering a distance of 380 km at the speed of 5.8 km/d and coastal migration covering a distance of 30 km at the speed of 8 km/d. These may be compared with those of Pacific salmon (430 km) and Atlantic eel (1,300 km, Aarestrup et al., 2009).

Figure 1.3 represents a model for motility of invertebrates based on the described data. Sedentary or sessile mode of life in Porifera, Cnidaria (except Scyphozoa) and low motility in Platyhelminths and Annelida have necessitated the selection and interspersion of hermaphroditism and asexualism in their respective sexual ontogenetic pathways (see Ghiselin, 1974, Charnov et al., 1976). Hermaphroditism and parthenogenesis are known to have repeatedly originated from their respective sexual ancestors. In scalpellid crustaceans, gonochorism evolved secondarily from an ancestral hermaphroditism (Hoeg et al., 2009, Yusa et al., 2012). Hermaphrodites have achieved the simultaneous expression of Primordial Germ Cells (PGCs) responsible for manifestation of the ovary and testis. Asexual reproduction is manifested with retention of Embryonic Stem Cells (ESCs) and/or Pluripotent Stem Cells (PSCs). However, asexual reproduction has also originated from sexual lineages. With relatively greater motility, the higher invertebrates namely Arthropoda and Mollusca shifted to almost exclusive sexual reproduction and gonochorism (Fig. 1.3).

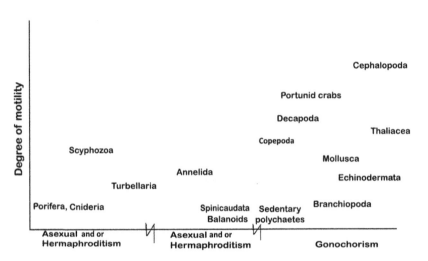

FIGURE 1.3

Motility levels of some invertebrates as functions of sexuality and modes of reproduction.

Most invertebrates belonging to minor phyla also do not have great powers of motility. In fact, most of them are sedentary or sessile (Table 1.3). Mesozoa, Nemertina and Acanthocephala are parasitic. Others belonging to a dozen of phyla are sedentary and 3 to 5 phyla are sessile. Members of Ptychoderidae in hemichordates are sessile. Barring the sessile Ascidiacea, the other urochordates Thaliacea and Larvacea are pelagic and motile. The all-hermaphroditic chaetognaths are pelagic drifters. Hence, the ontogenetic pathways in sexual mode of reproduction of these slowly/drifting invertebrates are invariably interspersed with hermaphroditism or parthenogenesis, but rarely asexualism.

1.4 Hermaphroditism

Hermaphrodite is a sex, in which female and male sexes are expressed and functional either simultaneously or sequentially in an individual, but in gonochoric male and female sexes are expressed and functional in different individuals. Surprisingly, hermaphroditism is overwhelmingly widespread in invertebrates. There are > 65,000 animal species that employ hermaphroditism (Jarne and Auld, 2006). In the cirripede *Chelonobia patula* male sex arises, when the barnacle attains a body size of ~ 2.5 cm and becomes a functional male at ~ 5 cm. Female sex also begins to develop at 4 mm size and becomes functional at ~ 7 mm. At this stage, *C. patula* is a fully functional simultaneous hermaphrodite (Crisp, 1983). The scleractinian colonial coral *Plesiastrea versipora* is dioecious, i.e. sexes are separate. However, as the colony grows to > 800 cm² in size, the female changes sex to male (Madsen et al., 2014). These two examples provide an idea (i) that PGCs and/or their derivatives Oogonial Stem Cells (OSCs)/Spermatogonial Stem Cells (SSCs) have bisexual potency and (ii) their simultaneous or sequential expression manifest simultaneous or protogynic hermaphroditism, respectively.

Hermaphroditism occurs in almost all taxa belonging to Porifera, Cnidaria, Ctenophora, Platyhelminths and Annelida (Table 1.1). All colonial species of cnidarians and ascidians are indeed hermaphrodites, as they possess zooids of both sexes. In fact, all zooids in a colony of ascidians like *Botryllus schlosseri* are protogynous and are synchronized in sexual development. When the colony is experimentally separated into two halves, the separated halves survive but sexual development of the zooids is no longer synchronized (Sabbadin, 1978), implying that the hermaphroditic colony is a single individual. Almost all polychaetes in Annelida (Schroeder, 1989) and a single, less speciose family Dioicocestidae in Eucestoda are gonochorics (Roberts and Kennedy, 1993). Arthropods and molluscs are gonochorics; however, hermaphrodites do exist in some taxa in these phyla (Table 1.1).

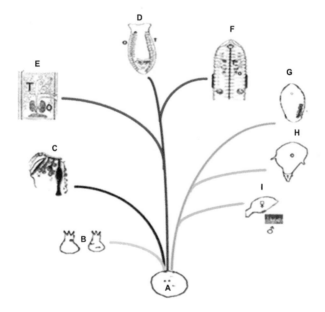

FIGURE 1.4
Diagrammatic representations of A = sponge, in which male and female gametes are diffusively distributed, B = male and female polyps of in *Hydra*, C = *Cormidium*, a hydroid colony, in which male and female zooids are held together, D = hermaphroditic *Planaria*, in which male and female reproductive systems are manifested (for clarity, male and female reproductive systems are shown for a single side only), E = A protoglottid of *Taenia*, in which male and female reproductive systems are brought together in each proglattid, F = hermaphroditic *Megascolex*, in which ovary and testis are condensed and located in a few segments, Protrandric hermaphroditic molluscs (G) and crustaceans (H), in which ovary and testis are brought together as ovotestis and I = *Verum brachiumcancri*, in which a female harbors many dwarf males; males are shown separately.

In hermaphroditic invertebrates, evolution seems to have proceeded by (i) separation of female and male germ cells, (ii) consolidation of these cells in solitary polyps but in separate zooids held together in colonies of cnidarians (however, sexes are separate in colonial coral *P. versipora*, Madsen et al., 2014) and (iii) manifestation of separate male and female reproductive systems within an individual of earthworms. Further, it seems to have selected the process of bringing the male and female zooids or ovary and testis closer together. Accordingly, female and male germ cells are diffusively distributed in the body of sponges (Fig. 1.4). They are separated and consolidated in female and male zooids but the zooids are held together in an individual cnidarian colony (e.g. *Cormidium*). In platyhelminths, separate female and male reproductive systems are manifested; however, the ovary and testis are diffusively spread over the entire body (e.g. *Planaria*). They are condensed and restricted to a few segments in annelids (e.g. *Megascolex*). The ovary and testis are held together as ovotestis in hermaphroditic crustaceans (e.g. Crustacea: *Lystmata,* Bauer, 2000) and molluscs. In some crustaceans, the female harbors one or more 'dwarf males' to ensure fertilization (e.g. *Verum brachiumcancri*, Buhl-Mortensen and Hoeg, 2013). To meet the demands of parasitic mode of life, each proglattid of cestodes harbors both ovaries and testis in their diffused reproductive systems (e.g. *Taenia*).

Despite their sessile/sedentary mode of life, the taxa belonging to 10 of 25 minor phyla, exclusively reproduce sexually. This may be one of the reasons for the presence of only a few species in them. To escape from this limitation, (i) hermaphroditism is employed and interspersed within sexual reproduction in nine minor phyla, (ii) parthenogenesis in four phyla (iii) polyembryony in two phyla (iv) asexual reproduction in six phyla. Exclusive gonochorism has perhaps limited Priapulida with 16 species, Onychophora with 165 species and Nematophora with 331 species. Adoption of hermaphroditism seems to be the best means to increase speciation, as Urochordata, Bryozoa and Nematoda have 2,300, 5,000 and ~ 46,000 species, respectively. Uniquely, chaetognaths are all hermaphrodites. To eliminate or minimize inbreeding by selfing, sequentials include either protandry (e.g. *Pandalus*) or protogyny. In simultaneous hermaphrodites too, 'sequentials' (protrandry, e.g. eucestodes, see Roberts and Kennedy, 1993; protogyny, e.g. ascidian colony *B. schlosseri*) exist. The others have evolved genetic incompatibility (e.g. *B. primigenus*) as well as structural and behavioral strategies. In effect, many hermaphroditic invertebrates are 'sexuals' at a given point of time (cf Pandian, 2013, p 213). However, parthenogenesis and polyembryony do not serve to enhance speciation. For meiosis is suppressed in parthenogenesis but not in hermaphroditism.

1.5 Asexual Reproduction

Budding, fission and/or fragmentation (autotomy) are some modes of asexual reproduction. They are ubiquitously prevalent in many sponges, cnidarians, flatworms, polychaetes and echinoderms (Fig. 1.1, 1.2). Budding and fission involve the extension of the existing two or three germ layers. Budding may lead to colony formation, as in cnidarians (e.g. *Obelia longissima*) and ascidians (e.g. *Botryllus schlosseri*). Cloning and coloniality have been recorded in cnidarians, entoprocts, bryozoans and ascidians (Chakraborthy and Agoramoorthy, 2012). With capacity for sexual and asexual modes of reproduction, some colonial ascidians (e.g. *Botryllus*) have become excellent models for studies on primordial and embryonic stem cells. Through asexual reproduction, the hub of cnidarian adaptation, (i) hermatypic corals generate monumental colony of 3 m or more (in diameter) containing 30 million polyps, (ii) siphonophores reach apotheosis of colonial complexity and (iii) sea feathers achieve their magnificent symmetry. "Shoals of *Aurelia* and blooms of *Craspedacusta* are produced asexually" (Shostak, 1993). However, autotomy requires regeneration of the lost parts of body to generate a complete offspring; regeneration includes (i) initial wound healing, (ii) blastema formation, (iii) organogenesis and (iv) restoration of normality (Runham, 1993, see also Chapter 4).

In the context of polarity, Skold et al. (2009) proposed four modes of asexual reproduction. The simplest is the longitudinal fission, which retains antero-posterier polarity in aceols and phoronids (Fig. 1.5). In turbellarians and polychaetes, horizontal fragmentation and paratomic fission also result in retention of antero-posterier polarity. Budding in anthozoans, acoels, annelids, bryozoans and ascidians branches laterally at any angle around 360° of the animal but retains the antero-posterier polarity. Budding in some acoels may also revert and establish opposing polarity.

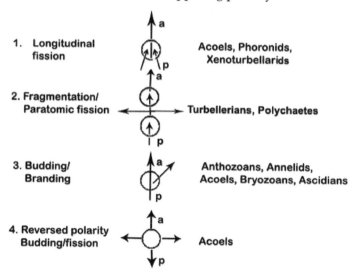

FIGURE 1.5
Schematic representation of different modes of asexual reproduction in the context of polarity. a-p represent the antero-posterior polarity (modified and redrawn from Skold et al., 2009).

Recently, evidence has been brought for the existence of asexual reproduction in the colonial parasitic hemocoelomate rhizocephalan crustaceans *Polyascus* (= *Sacculina*) *polygenea*, *Peltogasterella gracilis* and *Thylacoplethus isaevae* (Shukalyuk et al., 2007, Isaeva et al., 2009). In them, adult females reproduce asexually by budding prior to undertaking sexual reproduction (Fig. 1.2, see also Chapter 4).

Many scientists have endeavored to trace the ultimate progenitor cells, from which a complete offspring arises; the designated names are archeocytes and theocytes in sponges, interstitial cells and amoebocytes in cnidarians, neoblasts in turbellarians, blastocytes and eleocytes in annelids, coelomocytes in echinoderms and hemocytes in ascidians (Murugesan et al., 2010). From his famous experiment, Wilson (1907) found that following dissociation of cells of the sponge *Ephydatia fluviatilis*, homologous groups of cells got aggregated. The aggregate, composed of the PSCs namely archeocytes, developed into a whole sponge, while the aggregate groups of other somatic cell types inevitably died. Similar experimental studies on sea-urchin embryos showed that the aggregated embryonic cells *in vitro* formed 'embryoids' (Spiegel and Spiegel, 1986), which developed into normal larvae and after metamorphosis, became fertile sea-urchin (Hinegardner, 1975). Though each cell of a hydroid is totipotent (Frank at al., 2009), still a minimum mass of 2×10^4 cells were required to regenerate a whole animal in *Hydra attenuata* with 10^5 cells in it (Gierer et al., 1972). Notably, the aggregates, from which whole animals are generated, are indeed the ESCs or PSCs.

Using selective adhesive property of neoblasts, Franquinet (1976) eliminated all somatic cell types, which led to the culture of neoblasts with 'high purity'. Neoblasts are susceptible to irradiation. Following irradiation, the flatworm lost its ability to reproduce and died within 3-5 weeks. However, when an irradiated flatworm was injected with neoblasts, it not only regained its ability to regenerate but also survived long. Baguna et al. (1989) went a step further. They irradiated the asexual race of the flatworm *Schmidtea mediterranae* and subsequently injected neoblasts drawn from the sexual race of the worm into the irradiated ones. The restored asexual worm developed germ cells and copulatory apparatus as well as mated and produced cocoons. On the other hand, following the injection of neoblasts from asexual race into the irradiated sexual race of the flatworm, the formally sexual individuals began to reproduce asexually by fission. Clearly, the experiments of Baguna et al. have shown that the neoblasts or stem cells are of two entites: (i) Primordial Germ Cells (PGCs), that manifest sex followed by meiosis and fertilization in sexual flatworm and (ii) Embryonic Stem Cells (ESCs) with a capacity to regenerate a whole asexual flatworm. However, ESCs are capable of generating PGCs. Following differentiation, PCGs are capable of producing oogonia and/or spermatogonia.

1.6 Stem Cells

Investigations since early 1990s on regeneration of many invertebrates have shown that completely differentiated animals continue to harbor the unique 'Embryonic Stem Cells' . These cells have retained the ability to differentiate into offspring (Murugesan et al., 2010). They are slow-cycling undifferentiated cells. They divide asymmetrically into daughter cells, one of which is committed to differentiation and the other retains the capacity of original stem cell, which can differentiate all the cell types including germ line to generate an offspring (Knoblich, 2001). Because of their slow cycling, they can be identified by prolonged retention of nucleotide analog like bromodeoxyuridine (Borok et al., 2006). A basic difference between PGCs and ESCs is that PGCs are products of asymmetrically dividing progenitor and are capable of undergoing meiosis, whereas ESCs are produced by mitotic division of their progenitor and are capable of mitotic division alone. Hence, these two are different and are not to be confused with each other (cf Isaeva, 2011). Many reviewers do recognize these two entities of stem cells (e.g. B. Rinkevich, 2009).

Depending on the potency for differentiation, ESCs have been designated as (i) totipotent, (ii) pluripotent, (iii) multipotent, (iv) oligopotent and (v) unipotent (Skod et al., 2009), although the usage of this terminology is yet to be unified (e.g. Y. Rinkevich et al., 2009). Hence these terms are redefined. The stemness overlaps between multipotency and oligopotency. The multipotent stem cells can differentiate many cell types; hence they can regenerate more complicated organs like the head of the snail *Helix pomata*. But the oligopotent stem cells can differentiate only a few cell types and regenerate only the less complicated tentacles of snails and inhalant/exhalant siphons of bivalves (cf Runham, 1993). Information summarized in Table 1.1 reveals that asexual reproduction is limited to the retention of ESCs/Pluripotent Stem Cells (PSCs) in acoelomates and eucoelomates. Pseudocoelomates and hemocoelomates are unable to reproduce asexually, as they have not retained adequate mass of ESCs/PSCs.

Totipotency means the ability of a single cell to produce a whole organism. Only zygote and early blastomeres of some animals are totipotent. For example, halves and quarters of mature blastulae of the thecate hydroid *Clytia flavidula* developed into 'mini' but functional planula larvae (Zoja, 1895). In fact, each interstitial cell in *Hydractinia echinata* is totipotent (Frank et al., 2009, Chakraborthy and Agoramoorthy, 2012). When one of the first two blastomeres of the ascidian was destroyed, the remaining blastomere developed a half larva (Conklin, 1906). Naknauchi and Takeshita (1983), who extended the investigation of Conklin, found that the half larva of *Styela plicata*, produced from half embryo of one of the two blastomeres, hatched, metamorphosed and developed into a functional juvenile with completely normal morphology. The regulation of normal development did not occur during the period from cleavage to hatching but did it just prior to metamorphosis. Incidentally, a single stem cell of a trypsinized cysticercus larval stage of the cestode parasite *Taenia crassiceps* injected into the host mice developed into a whole larval cysticercus (Toledo et al, 1997) but is a rare exception from the common rule.

Pluripotent cells are characterized by two remarkable features: (i) their cell division rate is very high and (ii) they have extraordinary developmental plasticity (Geneviere et al., 2009). These cells are capable of agametic cloning, i.e. reproduce asexually in annelids, echinoderms and bryozoans (Skold et al., 2009). To regenerate a whole animal, a 'critical mass' of PSCs, i.e. what a totipotent cell, i.e. a zygote can do, can be made only by the 'mass effect' of PSCs. For example, Nikitin (1977) showed that a 'critical mass' of about 200 archaeocytes (= PSCs) is required to produce a whole animal in freshwater sponge *Ephydatia fluviatilis*. The number of PSCs required to produce a new individual in cnidarians, free-living flatworms and colonial ascidians is estimated to range between 100 and 300 (Y. Rinkevich et al., 2009). It can, however, be achieved with as small as 10-15 cells in the colonial parasitic rhizocephalans *Polyascus polygenea* and *Peltogasterella gracilis* (Isaeva, 2010). Incidentally, Skold et al. (2009) have suggested the presence of totipotent adult stem cells in Porifera, Cnideria, Platyhelminths, Urochordata and Cycllophora.

1.7 Primordial and Embryonic Stem Cells

In his study, Weismann (1883) was the first to conceive the idea of stem cells (*Stammzellen*) and germ cells (*UrKeimzellen*) in colonial hydroids, on the basis of which he formulated the germplasm theory. For two reasons, Weismann's germplasm theory could not initially be supported. Firstly, the germ cells are responsible for sexual reproduction and stem cells are for asexual reproduction (see below). Secondly, the somatic and germ cell lines have not yet been completely separated from each other in cnidarians. Selection of hydroid to bring experimental evidence in support of Weismann's theory proved to be a wrong choice. For example, a clone of single stem cells of *Hydra magnipapillata* differentiates into both somatic and germ cell lines (Bosch and David, 1987). Hence, germ cells are not exclusively sequestered in gonads. They may also develop in the total absence of gonadal tissue and even from gland cells derived from interstitial cells (Burnett et al., 1966). As germinal and somatic lines are unified through interstitial or amoebocytic stem cells, cnidarians fail to meet the criteria for persistently separated germ and somatic lines. Nevertheless, the germ and somatic cell lines are indeed consistently separated in higher invertebrates and vertebrates. According to Weismann, the stem cells retain germplasm (*Keimplasma*) containing the nuclear hereditary substances capable of differentiating into gametes and thereby provide continuity of the immortal germ line. Hence, both Weismann and many reviewers (e.g. Isaeva, 2011) wrongly considered the germ cells and stem cells as a single entity, as both these cells are morphologically similar and can be detected by *vasa*, a molecular marker. However, it is only the PGCs that have the ability to differentiate into gametes. Hence, the germ cells, i.e. PGCs differentiate into germ cell lineages, whereas ESCs into an organism. Of course, ESCs can produce gametes only after differentiation into PGCs.

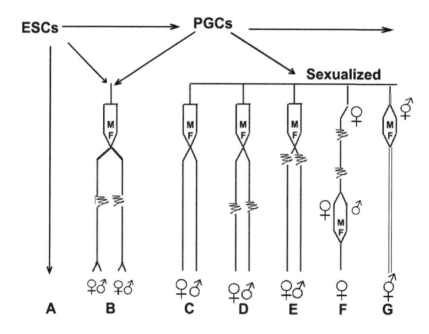

FIGURE 1.6
Origin of sex and evolution of ontogenetic pathways of aquatic invertebrates, in some of which polyembryonic, cyclic parthenogenic and asexual reproduction are interspersed within sexual mode of reproduction. A. Sexless microbes, B. Cnidarians, in which separation between ESCs and PGCs remains incomplete, C. Gonochorics, D. Gonochorics with asexual reproduction, E. Polyembryonics, F. Cyclic parthenogenics, G. Hermophrodites ≋ - indicates the occurrence of asexual reproduction/polyembryony.

Figure 1.6 shows the possible pathway, through which animals are sexualized and ontogenetic pathways through which polyembryonic, cyclic parthenogenic and agametic propagations are interspersed within the ontogenetic pathways of gametically reproducing sexual animals. Accordingly, ESCs differentiate into PGCs, which alone can generate germ line lineages and thereby establish reproductive systems in animals. With the origin of PCGs and sexual systems namely female, male and hermaphrodite have been established some 1.6-2.0 billion years ago (Butlin, 2002). Some cnidarians have retained ESCs, as every cell in them can agametically clone the whole cnidarian of either sex. Retaining adequate number (or mass) of PSCs, many acoelomates and eucoelomates clone progenies. Barring cladocerans, agametic cloning in these acoelomates and eucoelomates result in the same sex, as that of their parents. As evolution proceeded, the proportion of such agametically cloning sexual animals was progressively reduced. For example, only 0.4% of sea stars, brittle stars and sea cucumbers can agametically clone their progenies (Meladenov and Burke, 1994). This description is more of a suggestion. However, more research is required in this emerging area before any conclusive generalization can be made.

As more information is available for fishes, a brief account on the role of PGCs in manifestation of sex and bisexual potency in adults is summarized from Pandian (2011, 2012, 2013, 2014).

1. The gonadal primordium is formed by PGCs and surrounding Germ Cells Supporting Somatic Cells (GCSSCs). Following some structural changes, PGCs are transformed into oogonia or spermatogonia. The GCSSCs mitotically divide and differentiate into a few cell types including those of nursing and hormonal.

2. The PGCs are progenitors of germ cell lineage and possess the ability to differentiate into either oogonia or spermatogonia. Hence, they carry heritable information to the next generation. Their origin can histologically be traced somitogenic or late blastula at the earliest. They are roundish, relatively larger, cells with larger nuclei, clear envelope and granular chromatin. The use of *vasa* gene, a molecular marker of PGCs, has shown that the PGCs originate from maternally supplied mRNA. The ESCs share many of these characteristics (see Isaeva, 2011), which perhaps led Isaeva (2011) to consider ESCs and PGCs as a single entity.

3. In view of the small number of PGCs (25-100) and their visibility during the breeding season alone, Spermatogonial Stem Cells (SSCs) and Ovarian Stem Cells (OSCs) derived from PGCs with equal bisexual potency have also been used for transplantation studies.

4. Transplantation of PGCs drawn from dominant orange colored trout into early larvae (prior to gonadal differentiation) of recessive gray colored trout generated F_0 chimeras with morphology and anatomy (somatic tissues) of the recipients but with donor-derived gonads; crosses between these chimeras produced all donor-derived orange colored F_1 trouts.

5. Transplantation of SSCs drawn from the adult male (that have already milted) American iteroparous trout *Oncorhynchus mykiss*

into early larvae of the Japanese semelparous masu salmon *O. masou* produced both male and female progenies of the American trout. Similar transplantation of OSCs of the trout into masu also yielded male and female trout offspring. Clearly, PGCs, SSCs and OSCs retain bisexual potency of the donor adults that have earlier milted or spawned.

6. However, the SSCs transplantation into artificially sterilized mature testis of pejerrey and tilapia generated only male progenies, which produced only the donor-derived sperm. Other experimental studies from seemingly unrelated areas like the sex reversal by 17-α Methyltestosterone in female guppy and gonadectomy in many fishes have clearly shown that the GCSSCs lose bisexual potency at the time of gonadal differentiation.

7. The presence of PGCs and their derivatives SSCs/OSCs in the ovotestis facilitates the development and normal functioning of the ovary in protrandric *Acanthopagrus schlegeli* following castration. Contrastingly, *Thalassoma bifasciatum*, a protogynic diandric hermaphrodite, which do not retain PGCs in any organs other than its ovary, is unable to regenerate its gonad following the loss of the entire basket of its PGCs on ovariectomy.

8. Clearly, the presence of PGCs and their derivatives SSCs/OSCs is obligatorily required to establish and maintain sex in fishes. And the presence of ovotestis retaining PGCs and/or SSCs/OSCs in the heterologous zones facilitates normal development of the ovary following castration and presumably the testis after ovariectomy.

9. The presence of PGCs has been reported in many invertebrates (e.g. Metazoa: Xu et al., 2001, Nematoda: *Caenorhabditis elegans*, Saffman and Lasko, 1999, Chaetognatha, Alvarino, 1994, Crustacea: *Pandalus platyceros*, Hoffman, 1972 *Macrobrachium rosenbergii*, Damrongphol and Jaroensastraraks, 2001, Insecta: *Drosophila melanogaster*, Jaglars and Howard, 1995). Hence, PGCs may also manifest sexes namely female, hermaphrodite and male in invertebrates also. The simultaneous expression of bisexual potency of both PGCs and GCSSCs (or their equivalent) may manifest simultaneous hermaphroditism in many invertebrates.

10. This brief account of PGCs clearly shows that PGCs manifest sexes namely female, male and hermaphrodite in sexually reproducing animals. Notably, they carry heritable traits "including different mix of allele than their parents" to the next generation. On the other hand, the PSCs can agametically clone. In other words, the PSCs produce photocopies of the parent, whereas PGCs generate paintings, each of them differ from others and parents.

1.8 Pseudocoelom and Hemocoelom

Recently, a hypothesis was proposed suggesting that the presence of pseudocoelom in Nematoda, Gastrotrica, Rotifera and hemocoelom in Arthropoda and Mollusca has not facilitated the retention of PSGs, as these taxa do not reproduce asexually (Murugesan et al., 2010). Irrespective of free-living or endoparasitic (e.g. Acanthocephala), none of the invertebrate taxa belonging to the six pseudocoelomate minor phyla reproduce asexually (Table 1.2). Under schizocoelomata too, six taxa of seven hemocoelomate minor phyla and three of five lophophorate minor phyla do not reproduce asexually (Table 1.2). This comprehensive evidence goes to show that the Pandian's hypothesis of elimination of PSGs by pseudocoelom and hemocoelom may prove to be correct.

However, a single molluscan species (one of 85,000 species, i.e. 0.002%) the free-living pteropod *Clio pyramidata* is reported to reproduce asexually by a strobilization-like process (Van der Spoel, 1979). Three endoparasitic species belonging to rhizocephalan Crustacea (3 of 52,000 species, i.e. 0.006%), *Polyascus* (= *Sacculina*) *polygenea*, *Peltogasterella gracilis* and *Thylacoplethus isaevae* are also shown to reproduce asexually by budding (Shukalyuk et al., 2007, Isaeva et al., 2009). In minor phyla, two sipunculids (two of 300 species, i.e. 0.7%) *Aspidosiphon brocki* and *Sipunculus nodus* are known to naturally reproduce asexually by fission (see Rice and Pilger, 1993, Skold et al., 2009). In lophophorates, asexual reproduction occurs in *Phoronis ovalis* (one of 10 species) in Phoronida, and *Lophopus crystallinatus* (one of 5,000 species) in Bryozoa (Skold et al., 2009). In many Entoprocta, new zooids are formed from stolon or basal disk in colonial species but from the calyx in solitary species by budding (Nielson, 1990). Firstly, the number of species in these hemocelomatic major and minor phyla is very few and may be more an exception than a rule. It is not clear at present whether the presence of pseudocoelom and hemocoelom in these major and minor phyla eliminated PSCs *in toto* (Murugasen et al., 2010) or have not retained the critical mass of PSCs required to manifest asexual reproduction. As a few exceptional pseudocoelomates and hemocoelomates reproduce asexually, they may have secondarily retained the critical mass of PSCs to manifest asexual reproduction. The critical mass of PSCs required may be as few as 10-15 cells in the parasitic rhizocephalans *P. polygenea* and *P. gracilis* (Isaeva, 2010) with only a few tissues "an ovary, two male receptacles and ramified, nutrient-absorbing network of rootlets' (B. Rinkevich, 2009). Other exceptional pseudocoelomates and hemocoelomates are free-living and have many more tissues. Hence, asexual reproduction in them may require a larger number of PSCs. More publications in this area are required.

Briefly, this is perhaps the first taxonomic survey of sexuality, modes of reproduction and ontogenetic pathways of aquatic invertebrates. Sedentary and sessile modes of life and low motility seem to have driven the selection of hermaphroditism and interspersion of asexual reproduction within the ontogenetic pathways of sexual reproduction. This account has also attempted to interlink asexual reproduction with ESCs and the need for the critical mass of a smaller and larger number of PSCs to initiate asexual reproduction in the exceptional pseudocoelomates and hemocoelomates.

Section II

Crustacea

2

Introduction

2.1 Taxonomy and Diversity

The crustaceans comprise economically important shrimps, crabs, lobsters and foulers, the barnacles as well as 'fodder' animals, branchiopods including *Artemia*, cladocerans, copepods and euphausiids. They also include biologically important groups, branchiopods, copepods and ostracods, which produce resting eggs/cysts that hatch after a short or long period of diapause. Commencing from the early Cambrian, they had ample time to undertake endless experimentation with form and function; today, "no other group of plants or animals on the planet Earth exhibits the range of morphological diversity seen among the extant Crustacea" (Martin and Davis, 2001). For example, the diversity within decapods alone ranges from shrimps with an elongated laterally compressed body, muscular pleon and limbs mainly adapted for swimming to crabs displaying dorso-ventrally flattened broad body with a reduced pleon and uniramous walking limbs. Among large decapods, there are species with limbs specialized for digging, cracking molluscan shell and all kinds and number of spines, pincers and scissors (Scholtz et al., 2009). In terms of overall species richness, the crustaceans, with more than 52,000 species, are placed fourth behind insects, molluscs and chelicerates. Considering the diverse morphological features as the base for classification, the superclass Crustacea is divided among six classes, 13 subclasses, 46 orders and 802 families (Martin and Davis, 2001). This crustacean extant may be compared with more or less equally successful 32,000 species of teleostean fishes classified under six subclasses, 64 orders and 563 families (FishBase, 2010, www.fishbase.org). To familiarize the taxonomic status of the cited species, an introductory version of systematic resume of Martin and Davis (2001) is listed in Table 2.1.

TABLE 2.1

A condensed version of systematic resume of Crustacea (compiled from Martin and Davis, 2001)

Subphylum	:	Crustacea
Class	:	Branchiopoda
Sub class	:	Sarsostraca
Order	:	Anostraca, *Artemia, Streptocephalus*
Sub class	:	Phyllopoda
Order	:	Notostraca, *Triops, Lepidurus*
Sub order	:	Spinicaudata, *Limnadia, Eulimnadia*
Sub order	:	Cyclestherida, *Cyclestheria*
Sub order	:	Cladocera, *Daphnia*
Class	:	Cephalocaridae, *Sandersiella*
Class	:	Maxillopoda
Sub class	:	Thecostraca, *Ibla*
Infra class	:	Facetotecta
Infra class	:	Ascothoracida, *Lepas*
Infra class	:	Cirripedia, *Balanus*
Sub class	:	Tantulocarida
Sub class	:	Branchiyura, *Argulus*
Sub class	:	Copepoda
Order	:	Calanoida: *Calanus, Paracalanus*
Order	:	Cyclopoida: *Cyclops, Lernaea*
Order	:	Harpacticoida: *Harpacticus, Tisbe*
Class	:	Ostracoda, *Cypris*
Class	:	Malacastraca
Sub class	:	Hoplocarida
Sub class	:	Phyllocarida
Sub class	:	Entomostraca
Super order	:	Peracarida
Order	:	Mysida: *Mysis, Neomysis*
Order	:	Amphipoda: *Gammarus, Orchestia*
Order	:	Isopoda: *Oniscus, Ligia, Bopyrus*
Order	:	Tanaidacea: *Tanais*
Order	:	Cumacea: *Diastylis*
Order	:	Euphausiacea, *Euphausia*
Order	:	Decapoda
Family	:	Penaeidae: *Penaeus*
Infra order	:	Caridae
Super family	:	Palaemonidae: *Palaemon*
Super family	:	Crangonoidae: *Crangon*
Infra order	:	Palinura
Family	:	Palinuridae: *Palinurus*
Infra order	:	Anomura: *Emerita*
Infra order	:	Branchiyura: *Callinectes, Portunus*

Crustaceans inhabit every conceivable aquatic habitat. They float as plankton (e.g. 55 cladoceran *Alona pulchella*/ml, Nandini et al., 1998), swim as nekton (e.g. 22 caridean shrimp *Thor manningi*/m², Bauer, 1986) and thrive in sediments (e.g. 2-443 harpacticoid copepods/cm² at the Chennai coast, Sugumaran et al., 2009, 0.5 isopod and 1.5 amphipods/1275 cc core in sediment collected from 232 m depth, Weissberger et al., 2008, 0.5 *Pasiphaea multidentata*/ha, Cartes and Sarda, 1992, 3,200,000 resting eggs of calanoid copepod *Acartia clause*/m², Kashahara et al., 1975, 7.3 million floating cysts of *Artemia monica*/m², Anon, 1993). They also foul on almost all substrata including host animals (e.g. *Polyascus polygenea* on coastal crab *Hemigrapsus senguineus*, Shukalyuk et al., 2007, see also Fig. 2.5). They occur and flourish in permanent lakes and ephemeral puddles (e.g. clam shrimp *Limnadia badia*, Benvenuto et al., 2009, including rice fields *Triops longicaudatus*, Grigarick et al., 1961, 21 taxa belonging to Ctenopoda and Anomopoda, Maiphae et al., 2010) as well. The ability of branchiopods to tolerate, survive and thrive in a wide range of saline lakes of Australia is amazing and may pose a challenge to physiologists interested in osmoregulation. For example, Timms (2009) reported the occurrence of *Branchinella buchanensis* from 16 to 43 g/l, *T. australiensis* from 13 to 93 g/l and *Parartemia zietziana* from 74 to 353 g/l S. Equally interesting is the wide range of temperature tolerance of crustaceans. A few of them thrive at 0.8°C (e.g. *Weltnerium nymphocola*, see Buhl-Mortensen and Hoeg, 2013); a few others like calanoid copepods and euphausiids survive in super cool waters below ice sheet (Mesa and Eastman, 2012); but *A. franciscana* flourish at 36°C (e.g. Campos-Ramos et al., 2009). Resting eggs or cysts of many crustaceans can dry out or freeze, and survive anoxia in sediments for an extended time. *Daphina tibetana* and *D. dolichocephala* thrive in the world's highest lake of Nepal's Mount Everest region at 5,460 m above sea level (asl). Others survive in deep sea (up to 2,261 m below sea level [bsl], e.g. *P. multidentata*, Cartes and Sarda, 1992). Besides exploiting these wide ranges of aquatic habitats, many isopods (3,000 species of woodlouses) thrive in moist sediments on land and the robber crab *Birgus latro* on coconut trees.

Interestingly, freshwater, which covers only ~1% of the Earth's surface but accounts for a little less than 0.01% of its water (Pandian, 2010, p 1), has ~10%, i.e. 5,200 species of crustaceans (Dumont and Negrea, 2002), while the seawater, which covers ~70% of the globe with 97% of its water, has the remaining ~46,800 species. Hence, marine environment provides habitats for far more number of crustacean species than freshwater. Yet, 50, 90 and 95% of ostracods, branchiurans, and branchiopods occur only in freshwater. In fact, Cephalocaridae and Remipedia are found only in freshwater. Indeed, freshwater has also provided adequately diverse habitats to facilitate origin and evolution of many crustacean taxa.

In consonance with the wide morphological diversity, the size of crustaceans also ranges very widely; the smallest tantulocarid parasite on the antennules of the deep sea copepod *Stygotantulus stocki* measures just 74 μm in length

(Boxshall and Huys, 1989). Among the more known decapod crustaceans, it ranges from hermit crab *Pygmae pagurus* with the shield length of 0.76 mm and pinnothurid crab *Nannotheres moorei* with the carapace length of 1.5 mm to the Tasmanian crab *Pseudocarcinus gigas* with the carapace width of 46 cm and the giant Japanese spider crab *Macrocheirus kaempferi* with the leg span of ≈ 4 m (Martin and Davis, 2001). By weight, the crustaceans weigh from a few milligrams (e.g. 2 mg *Pareuchaeta norvegica*, Nemoto et al., 1976) to a few kilo grams (e.g. 20 kg *M. kaempferi*). With its standing biomass of 500 million tons, *Euphausia superba* surpasses the biomass of any other group of metazoans (Nicol and Endo, 1999). In terms of number too, the crustacean nauplia are called "the most abundant type of multi-cellular animals on earth" (Fryer, 1997).

Crustaceans are readily amenable to both field observations and laboratory experimentations. For example, Hopkins and Machin (1977) sampled 10,500 female *P. norvegica* to estimate the correct placement of spermatophores. For the same purpose, Crocos and Kerr (1983) sampled some 10,684 female *Penaeus merguiensis*. To describe secondary sexual characters, Villalobos-Rojas and Wehrtmann (2014) collected 5,985 female *Solenocera agassizii*. Banta (1939) reared several common cladocerans for 800 to 1,600 successive parthenogenic generations (for 27 years [y]) by frequently changing the culture medium. The clam shrimp *Lynceus brachyurus* was reared over 11 y; its translucent carapace and habit of laying on one side on the substratum facilitated observation and experiment on its life cycle and behavior (Patton, 2014). *Acartia tonsa* can be cultured in high quantities at the density of 1,000 indivi/l without any negative impact on development (Medina and Barata, 2004). The calanoid copepod *A. tonsa* was reared by Parrish and Wilson (1978) over 7 y extending for 32 filial generations. Voordouw et al. (2005) reared 18,094 copepod individuals of 608 families of *Tigriopus californicus* across three generations and sexed 15,754 adults. As early as in 1941, Heeley observed that sperm remained viable for > 2 y and females required only a single impregnation to produce a series of broods in *Oniscus asellus*. Not surprisingly, a voluminous literature is available on crustacean reproduction and development. Due to space limitation, this book provides a 'snap-shot' rather than an in depth and/or exhaustive description of each item mentioned in 'Contents'.

2.2 Life Span and Reproduction

Being a characteristic of arthropods, crustaceans molt to grow and/or reproduce at progressively increasing intervals during their life time. Molting and spawning are two major events that involve cyclic mobilization of nutrient reserves from storage sites to the epidermis and gonads, respectively. Hence, these events determine the reproductive patterns. In diecdysic crustaceans, characterized by long premolt and short intermolt periods, spawning is obligatorily preceded by a molt. Conversely, the anecdysic molt cycle is characterized by short premolt and long intermolt periods. Within the long intermolt period, clearly demarcated two antagonistic phases are programmed, i.e the initial reproductive and terminal somatic growth phases. Copepods and most brachyurans, especially xanthids display anecdysic molt cycle; on attaining sexual maturity, they do not molt any more.

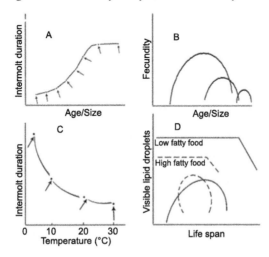

FIGURE 2.1
A. Molting duration as a function of age/body size in cladocerans, B. Fecundity as a function of age/body size in cladocerans, C. Intermolt duration as a function of temperature in cladocerans, D. Life span ⌐\ and fecundity ( ) in cladocerans fed on low fatty food (solid lines) and high fatty food (broken lines).

Availability of a large number of publications on life span, molting and their effects on spawning and brooding in cladocerans allows a few generalizations: 1. Intermolt duration is progressively prolonged with increasing age/size (Fig. 2.1A). 2. Batch fecundity increases with age/size but beyond the mid body age/size, it decreases (Fig. 2.1B) (see Pandian, 1994, p 99-101). 3. With increasing temperature, the duration of an intermolt period is decreased (Fig. 2.1C). A unique publication by Beladjal et al. (2003b) demonstrates the role of increasing temperature progressively decreasing Generation Time (GT) in the Morrocan anostraca *Tanymastigites perrieri*; its GT decreases from 52, 21, 13, 14 and to 9% of its reproductive Life Span (LS) at 10, 20, 25, 30 and 35°C, respectively. Consequently, tropical crustaceans grow faster and are more fecund than their temperate and Arctic counterparts. 4. Lipid content of yolk determines clutch size, as it serves as a major energy source for embryonic metabolism. Of yolk constituents used as energy source for embryonic metabolism, lipids contribute 67 to 87% in many crustaceans (Pandian, 1967, 1970a, b, 1972, Pandian and Schumann, 1967). Hence, (i) accumulation of visible lipid droplets in eggs is positively correlated with growth and fecundity but negatively with longevity (Fig. 2.1D). Experimenting with a single clone of *Daphnia pulex*, Lynch and Ennis (1983) also found that life span steadily increased with decreasing food availability. (ii) Fed on alga *Scenedesmus*, *Bosmina freyi* accumulated a higher concentration of ω^3, ω^6 and polyunsaturated fatty acids (PUFA) and was more fecund than that fed on seston containing more organic carbon and low C : N and C : P ratios (Acharya et al., 2005). Briefly, accumulation of lipids in crustacean body diminishes life span but increases fecundity.

Life span (LS) of crustaceans ranges widely from 5-6 day (d) in the cladocera *Moina mucrorura* (see Sarma et al., 2005) to 13 year (y) in polar lysianassoid amphipod *Eurythenes gryllus* (Jonhson et al., 2001) and to 100 y or more in *Homarus americanus* (Martin and Davis, 2001). Known for their short LS (see Sarma et al., 2005) brooding and parthenogenesis, cladocerans are readily amenable to experimentation. Hence, considerable information on their LS and generation time (GT) is accumulated. Mean LS and GT i.e. hatching to puberty or juvenile period) of many tropical cladocerans (data for six species summarized by Pandian [1994, p 86]) and those reported for *B. tripurae* at 20°C by Biswas et al. [2014] and *Macrothrix spinosa* at 29°C, Kumar and Altaff [2000]) average to 21.1 d and 11.7 d, respectively. The GT as a fraction of LS in these tropical cladocerans averages to 47% only. But these LS and GT estimates for many temperate cladocerans (e.g. eight species at 10°C, Bottrell, 1975) are 76.4 d and 49.6 d, indicating that the GT is as much as 65% of their mean LS. Arctic cladocerans invest on GT over 90% of their LS. Briefly, the time investment from hatching to puberty as fraction of the respective LS averages to 47, 65 and 90% for tropical, temperate and Arctic cladocerans, respectively. Presumably, temperature seems to have a profound effect on juvenile hormone, which limits or extends the duration of the juvenile stage (see Chapter, 7.2.1).

TABLE 2.2

Estimated neonate production during the life span of temperate and tropical cladocerans (from Pandian, 1994, permission by Oxford and IBH Publishers)

Parameter	Temperate	Tropical
Adult molt (no.)	5.5	15.0
Berried molt (no.)	4.4	12.0
Neonate produced (no./molt)	3.6	13.4
Absolute neonates produced (no./♀)	15.8	161

As a result of these processes of molting and release of neonates, tropical cladocerans undergo a mean number of 15 adult molts and 80% of these molts are followed by release of neonates, against 5.5 adult molts undertaken by temperate cladocerans (see Pandian, 1994, p 86). This has great implications to progeny production. Table 2.2 shows that the estimated number of neonates produced during the life span of tropical cladocerans is as much as 10 times more than that of temperate cladocerans. This provides a great advantage to aquaculture of fodder crustaceans in tropics.

Like cladocerans, isopods and amphipods do molt as adults and every molt may be followed by release of neonates. Isopods characteristically molt the posterior half of the cuticle first and subsequently the anterior half. Males usually copulate with females immediately following the posterior molt but prior to the anterior molt (Brook et al., 1994, Johnson et al., 2001). GT required to grow and attain sexual maturity decreases with increasing temperature. For example, it decreasss in *Gammarus pulex* from 52 weeks (w) at 5°C to 22, 17 and 3 w at 10, 15 band 20°C, respectively (Sutcliffe, 2010). In another temperate amphipod *Jassa falcata*, GT decreased from 56% at 10°C to 45% at 20°C of the respective LS (Nair and Anger, 1979b). Expectedly, tropical amphipods may invest on GT a smaller fraction of their LS than their temperate and arctic counterparts. Mean LS and GT of 17 temperate isopod species averages to 23 month (m) and 10.6 m, respectively. These values for 13 temperate amphipod species are 15 m and 8.1 m (Pandian, 1994, p 88). Hence, these temperate isopods and amphipods invest 46 and 54% of their respective LS on GT (Table 2.3). A few publications available for a tropical amphipod indicate that the investment on GT ranges from 15% in *Melita zeylanica* (Krishnan and John, 1974), *Eriopisa chilkensis* (Aravind et al., 2007) to 25.6% in *Quadrivisio bengalensis* reared at 17-20°C of their respective LS (K. K. C. Nair and K. V. Jayalakshmy, pers. comm.). Polar and boreal amphipods have a single large well timed brood (Johnson et al., 2001). Notably, the GT values of these sexually reproducing and brooding diecdysic peracarids are less than half the GT of the respective parthenogenic cladocerans.

Life cycle of other crustaceans involves one or more larval stages (see Table 2.11). For example, the most prolific penaeids pass through six sub stages each during the non-feeding nauplius and metanauplius and subsequently three sub stages each during the feeding protozoea and mysis to attain the penultimate post larval stage (e.g. *Penaeus indicus*, CMFRI, 2013). Similarly, the copepods also pass through six sub naupliar and five sub copepodid stages prior to metamorphosis (e.g. *Oithona rigida* Santhanam and Perumal, 2013). Hence, relatively less information is available on their LS and GT. However, the short LS and amenability to culture of branchiopods and copepods have facilitated the collection of relatively more information on these parameters. Among them too, copepods as anecdysic do not molt as adult while branchiopods do. In spinicaudatan branchiopod, the clam shrimp *Eulimnadia texana* reared at 28-32°C, New Mexico and Ohio (USA) invests on GT 17-18% (5 d) (17% by males, 18% by hermaphrodites) of its LS (Zucker et al., 2001). Notably, its LS in natural population was < 14 d (Weeks et al., 2014b). In the bisexual branchiopods too, the investment on GT is shorter at 25°C than at 10°C (Table 2.3).

TABLE 2.3

Investment on generation time as a percentage of life span in some crustaceans

Taxa	Examples	Remarks	Tropical	Temperate	Arctic
Branchiopoda	*Branchipus shaefferi*, 25°C	Bisexual	28[1]		
	Streptocephalus torvicornis, 25°C	Bisexual	31[1]		
	Tanymastigites perrieri, 10°C	Bisexual		48[2]	
Anostraca	*Artemia parthenogenetica*	Bisexual		46[3]†	
	Ukraine strain at 28°C	Parthenogenic		55[3]††	
	Indian strain, tropical	Parthenogenic	22[4]		
	A. franciscana	Bisexual		47[4]	
	A. franciscana	Bisexual		35[4]	
Spinicaudata	*Eulimnadia texana*	Hermaphroditic	17-18[5]		
Cladocera		Parthenogenic	46[7]	65[7]	90[7]
Copepoda	*Onychocamphis bengalensis*	Anecdysic brooder	36[6]		
	Mesocyclops leukarti	Diecdysic brooder	31[7]	41[7]	
	Eucyclops serrulatus	Diecdysic brooder		40-43[8]	75
	Limnocalanus johanseni	Anecdysic shedder			90[7]
	Elaphoidella grandidieri	Parthenogenic	23[9]		
Peracarida				52[7], 54[7]	
Amphipoda	*Melita zeylanica*	Diecdysic brooder	15[10]		
	Eriopisa chilkensis	Diecdysic brooder	15♂, 18♀[11]		
	Quadrivisio bengalensis, 20°C	Diecdysic brooder	26[12]	56[13]	
	Jassa falcata, 10°C	Diecdysic brooder			
Isopoda				46[7]	
Decapoda	*Procambarus fallax*, 20°C	Parthenogenic		48[14]	
Palaemonidae	*Macrobrachium rosenbergii*	Diecdysic brooder	35[15]		
Penaeidae	*Penaeus indicus*	Diecdysic shedder	26[16]		
	P. monodon	Diecdysic shedder	19[16]		

1. Beladjal et al. (2003a), 2. Beladjal et al. (2003b), 3. Golubev et al. (2001), 4. Browne (1980, Browne et al. 1984, 1988), 5. Zucker et al. (2001), 6. Saboor and Altaff (2012), 7. Pandian (1994, p 88), 8. Nandini and Sarma (2007), 9. Nandini et al. (2011), 10. Krishnan and John (1974), 11. Aravind et al. (2007), 12. K. K. C. Nair and K. V. Jayalakshmy, pers. comm., 13. Nair and Anger (1979b), 14. Vogt (2010), 15. Ventura et al. (2011b), 16. CMFRI (2013)

†Oviparous, ††viviparous

A rare combination of oviparity and viviparity in *Artemia* spp provides an opportunity to assess the cost of embryogenesis and brood carriage on GT. Providing rare data, Golubev et al. (2001) reported that within the Ukraine parthenogenic and bisexual strains of *A. parthenogenetica*, the viviparous parthenogenic invests on GT a larger fraction (55%, 41 d) of its LS than the oviparous sexual (46%, 30? d). Interestingly, the difference between 'oviparous' and 'viviparous' *Artemia* is 8 d, which is close to the brooding period in 'oviparous' strain. Similarly, another rare combination of sexual and parthenogenic strain in geographical parthenogenics also provides an opportunity to assess the effect of parthenogenesis on their GT. For example, Nandini et al. (2011) have provided valuable data on the geographic parthenogenic copepod. Expectedly, the parthenogenic strain of *Elaphoidella grandidieri* invests on GT 23% of its life span (25 of 48 d). Unfortunately, data are not available for bisexual strain of geographic parthenogenics (cf Table 3.4).

The short span of summer has enforced monocyclic breeding and univoltnism with all resting eggs in Arctic calanoid copepods *Limnocalanus johanseni* and *Diaptomus sicilis*. These copepods invest on GT 75–90% of their LS (Table 2.3). But the cyclopoid copepod *Mesocyclops leukarti* (33°N) produces two clutches of 27 and 93 eggs at 15°C. It invests on GT 41 and 31% of the respective LS (61 d and 128 d) at 15°C and 27°C, respectively (Pandian, 1994, p 87). Fed on two different diets at 23°C, *Eucyclops serrulatus* also invests on GT 40 and 43% of its LS (Nandini and Sarma, 2007). In Chennai, India, the harpacticoid copepod *Onychocamphis bengalensis* invests on GT 36% of its LS (Saboor and Altaff, 2012).

There seems to be a dearth for information on LS and GT of long living crustaceans, which involve four-five larval stages and many molts within each larval stage. Ventura et al. (2011b) have indicated that in diecdysic brooder *Macrobrachium rosenbergii*, the durations from fertilization to hatching, hatching to completion of larval (PL_{20-120}) and juvenile stages last for 60, 100 and 120 d, respectively, i.e. a GT of 220 d. Assuming the adult LS as 400 d, *M. rosenbergii* invest on GT 35% of its LS (Table 2.3). Scattered information reported by CMFRI (2013) permits that egg shedders like *Penaeus monodon* and *P. indicus* invest on GT 19 and 26% of their respective LS on GT. Briefly, investment on GT across Crustacea is (ia) shorter in bisexuals than parthenogenics, (ib) anecdysics than diecdysics, (ic) oviparity than viviparity and (id) egg shedders than egg brooders and (ii) the investment on GT remains ~30% (<50%), ~50% (<65%) and > 85 (75 to 90%) of respective LS in tropical, temperate and Arctic crustaceans. Interestingly but unexpectedly, the highly predation-proned smaller branchiopods, copepods and peracarids invest on GT a longer fraction of their LS than the larger decapods (see p 42).

Incidentally, there are interesting sex-dependent differences in LS and GT. For example, LS of the male ornamental crab *Porcellana sayana* is 1.5 y but is 1.3 y in the female (Baeza et al., 2013). For the anomuran crab *Aegla paulensis* it is 3.35 y for the female but 2.82 y only for the male (Cohen et al., 2011). The pedunculated balanoid male *Scalpellum scalpellum* invests on GT just 10 d after settlement, while the female requires 1 y for it (Spremberg et al., 2012).

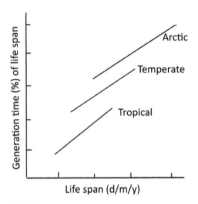

FIGURE 2.2
Generation time as percentage of respective life span in arctic, temperate and tropical cladocerans/crustaceans.

Information described thus far as well as bits and pieces of information from collected from several sources have led to propose a model for GT as a fraction of LS of crustaceans inhabiting Arctic, temperate and tropical waters (Fig. 2.2). For the construction of the LS-GT model, the following three facts were considered: 1. The LS averages to a few days in branchiopods and copepods, a few months in isopods and amphipods and a few years in decapods. Their respective averages were taken as points of interest to draw the horizontal lines. 2. Within each of these three groups, both the LS and GT are shorter in tropical crustaceans than those in their temperate and Arctic counterparts (Table 2.2). Accordingly, the horizontal lines are progressively shifted towards the right. 3. A slanting line connecting the starting points of three horizontal lines fits with trend (Fig. 1 obtained by Sarma et al., 2005) for temperate and tropical cladocerans together.

2.3 Ovigerous Molt and Fecundity

Besides temperature and related to latitudes, factors like (i) life history characteristics, (ii) food availability and quality, (iii) predation and (iv) parasitism may considerably alter the number of ovigerous molts and fecundity. These factors may ultimately determine the life time fecundity by altering one or more of the following: (i) food/predation/parasite induced changes in the fraction of ovigerous females in a population, (ii) alter effective reproductive life span through postponed age/size at sexual maturity and/

or extended interspawning interval and/or (iii) alter the fecundity/clutch and/or egg size. Some examples are described hereunder:

Life History Characteristics In diecdysic crustaceans, every adult molt may not be ovigerous. Many factors alter the number of ovigerous molts. The persisting brooded embryos postponed (e.g. *Orconectes propinques*) or inhibited (e.g. *Palaemon serratus*) ensuing molt and thereby reduced the number of ovigerous molts. Even injection of ecdysterone, the molt hormone, into ovigerous *Palaemonetes kadiakensis* female failed to induce molting due to the elevated titre of Molt Inhibiting Hormone (MIH) (see Hubschman and Broad, 1974). Likewise, no spawning occurred, when embryos were brooded in the egg sacs of copepods (e.g. *Amphiascoides* sp) and in the mantle cavity of cirripedes (e.g. *Elminius modestus*, see Pandian, 1994, p 93). Hence, ablation of all batches of incubated eggs was considered to save the energy cost of brooding and channel it to egg production. Indeed, the relieved *Macrobrachium nobilii* did it. Table 2.4 shows that fecundity in the female, which was repeatedly relieved from brooding her eggs following every adult molt, was 1.5 times more than that of normal female that incubated berried eggs on her appendages (Pandian and Balasundaram, 1982).

TABLE 2.4

Fecundity of normal and relieved (from brooding) females of *Macrobrachium nobilii* (from Balasundaram, 1980).

Parameter	Normal ♀	Relieved ♀
Duration of molt cycle (d)	19.3	17.7
Molt (no./life span)	18.9	20.6
Berried molt (%)	61.0	81.0
Berried molt (no./life span)	11.5	16.7
Fecundity (no./clutch)	2161	2194
Absolute fecundity (no./♀)	24852	36640

In many crustaceans (e.g. astocoid decapods, *O. immunis*, tanaidacean *Leptochelia dubia*, brachyuran *Macropodius rostrata*), two distinct mating morphs are reported. In *Onychocamphis immunis*, copulatory morphs appear during spring and autumn; they molt into non-copulatory morphs in summer and winter, respectively. The first pleopods of the copulatory morph are corneous, hard and sculptured, and transmit the spermatophores. The soft pleopods of the non-copulatory morph are uniramous and unsculptured and incapable of transmitting spermatophore. Consequently, molts during summer and winter are not ovigerous (see Pandian, 1994, p 76-77).

Food Availability and Quality Instead of alternating morphs, the temperate crustaceans and copepods produce larger or smaller broods during different seasons. Encountering the limitation of space within a pouch or sac, the temperate planktonic cladocerans and copepods produce smaller broods with large number of eggs during spring and summer, when food supply is abundant. This adaptation provides a reproductive advantage to the female, and less but adequate yolk reserve for the larvae to successfully pass through the initial stages. Conversely, large eggs in a smaller brood are produced during autumn and winter, when food resources are limited. Yolk reserves in larger eggs insure offspring survival during the first few crucial instars. For example, *Bosmina* sp in Canadian waters produced large (>4 eggs/clutch) brood with smaller (<150 μm length) eggs during the period from May to August but small brood (<2 eggs/clutch) with larger (>350 μm) eggs during autumn and winter (Kerfoot, 1974). In the European Lake Morskie Oko too, *Daphnia* produced larger broods (2 eggs/clutch) from May to September but smaller one (1 egg/clutch) from December to April (Gliwicz et al., 2001). In the cladocera *Simocephalus vetulus* too, egg volume decreased from 10^6/μl at 4°C to 6^6/μl at 20°C (see Pandian, 1994, p 105). In the amphipod, *Gammarus duebeni*, smaller eggs (~6.2 mm³/100 eggs) were produced during summer (18°C) but larger ones (~7.5 mm³/100 eggs) during winter (5°C). Moreover, the number of broods in *G. locusta* was 3.8/generation during winter but 7.8/generation during summer in *G. mucronatus* (Sutcliffe, 2010).

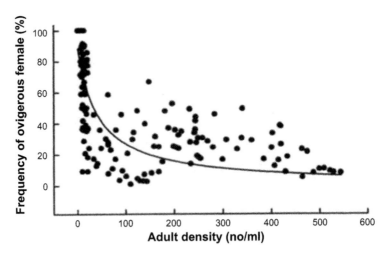

FIGURE 2.3
Frequency of ovigerous female as function of adult density of *Elaphoidella grandidieri* (from Nandini et al., 2011, permission by Dr. S. Nandini).

Population density and consequent competition for food and food quality may have a decisive role on the number of ovigerous molts. Due to intense competition for food in geographic parthenogenic copepod *Elaphoidella grandidieri*, the fraction of ovigerous females decreased from 85% at the density of 2 females/ml to < 20% beyond the density of ~3 females/ml (Fig. 2.3). Adult life span of *Acanthocylops viridis* fed on algae was 30 d with a single of brood containing 19 eggs. When fed *Daphia obtusa*, however, it produced 888 eggs following 12 adult ovigerous molts during its life span of 94 d (see Pandian, 1994, p 103). Incidentally, resveratol, a polyphenolic phytoalexin produced by many plants as an antimicrobial defense compound, did neither extend the life span nor altered fecundity of *D. pulex* (Kim et al., 2014).

TABLE 2.5

Effect of quality and quantity of food on fecundity in selected crustaceans

Food: Name	Berried ♀ (%)	Spawning Pulse/Clutch (No.)	Egg (No./ Batch)	Absolute Fecundity (No.)
		Shedders		
	Calanus finmarchicus (Marshall and Orr, 1952, condensed)			
Starved	51	5.0	16.3	83
Chlorella stigmatophora	47	4.6	15.3	71
Hemiselmis rufescens	77	10	14.3	145
Coscinodiscus centralis	87	20	22.6	463
Skeletonema costotatum	93	40	19.5	783
Syracosphaera carterae	100	35	28.9	1063
Lauderia borealis	100	53	33.4	1779
		Brooders		
	Eurytemora affinis (Heinle et al., 1977, modified)			
Starved		0.9	12	15
Alga (0.375×10^4 cells/ml)		3.7	11	42
Autoclaved detritus		0.8	5	5
Unautoclaved detritus		0.9	7	10
Unautoclaved detritus + alga		4.7	10	50
Protozoa + bacteria		5.4	13	73
Protozoa		5.5	18	103

	Macrobrachium nobilii (Kumari and Pandian, 1991)			
Ration (%)	Egg Size (j/egg)	Clutch (No.)	Egg (No./ Batch)	Total fecundity (No.)
Starved				
20				
45				
60	414	677	3	2033
70	421	1022	3	3066
80	426	1245	3	3755
90	437	1121	3	3362
100	431	1160	3	3481

Table 2.5 shows the effect of selected algae on the number of spawning pulses in an anecdysic egg shedder *Calanus finmarchicus* and diecdysic egg brooders *Eurytemora affinis* and *Macrobrachium nobilii*. The following may be noted: 1. The quality of algal food played a decisive role on the number of ovigerous molts. *Chlorella stigmatophora* as a diet did not promote ovigerous molt. But the other algae *Skeletonema costotatum* and *Syracosphaera carterae* promoted the molts. With reference to egg shedders and brooders, egg shedding *C. finmarchicus* spawned 14 to 33 eggs/pulse with a cumulative fecundity ranging from 71 to 1,779 eggs; however, egg brooding *E. affinis* spawned one to 5.5 eggs/clutch and produced five to 103 eggs for life time, suggesting that the energy cost of egg brooding is indeed very high. 2. Even when starved, the anecdysic (~ 51%) *C. finmarchicus* and diecdysic *E. affinis* continued to produce eggs. In tropical copepod *Sinodiaptomus indicus* too, following starvation, 7% females were ovigerous (Preetha and Altaff, 1996, see also Niehoff, 2004). 3. However, diecdysic *M. nobilii* did not spawn but undertook a single non-berried molt, 56 d after the commencement of starvation, in comparison to 24 d in a female fed *ad libitum*. Hence, mobilization of nutrient reserves from larval stages was adequate to release a single clutch of eggs by starved *E. affinis*, against five clutches in protozoae-fed *E. affinis,* and five pulses of eggs by starved *C. finmarchicus*, in comparison to 53 pulses of eggs, when fed on *Lauderia borealis*. As copepods invest on GT a greater fraction of their LS, they seem to have adequate larval resource to meet the costs of molting and spawning (see p 37) 4. With decreasing ration to *M. nobilii*, (a) the number of adult molts was decreased from 5 to 1, (b) both batch (from 1,160 to 744 eggs) and cumulative fecundities decreased significantly and (c) energy content of an egg (size) was progressively decreased. Clearly, molting, spawning and brooding are energy demanding processes, each of them is met by its reserve mobilization channel. Of the three channels, the molt channel seems to have priority. As time and energy costs of brooding are so high some adult molts may not be ovigerous. When the mobilization channel for oogenesis is limited by food quantity, the prawn reduced the number of ovigerous molts, batch and absolute fecundities and size of an individual egg (Table 2.5).

Predation   To grow and reproduce, the cladocerans have to filter and feed suspended phytoplankton and/or detritus. The threshold levels of food density to initiate growth and first wave of reproduction decrease with increasing body size. Predation is another major factor that advances body size at first reproduction and clutch size. To survive heavy predation in Lake Morskie Oko, *Daphnia* commences breeding with smaller brood at a size of 0.75 mm. But, it begins to produce a large clutch at 1.5 mm body size in Lake Czarny, where fish predation is absent (see Gliwicz, 2003 p 195). Thus, food density and predation are most important factors in deciding the mother's size at first breeding and brood size. With regard to selective predation, the number of eggs in a brood reflects the food availability in the habitat at the time of the new clutch of eggs is deposited in the egg sac or brood chamber.

Occurrence frequency of *D. cucullata* in Lake Mikolajskie, Poland and gut contents of *Osmerus eperlanus* smelt indicates that the smelt selectively predated larger *D. cucullata* (Fig. 2.4a). Similar frequency analysis of clutch size of *Cyclops abyssorum* in Lake Gossenkollersee, Austria and gut contents of *Salmo trutta* clearly showed that the brown trout selectively predated *C. abyssorum*, carrying a larger sac containing more eggs (Fig. 2.4b).

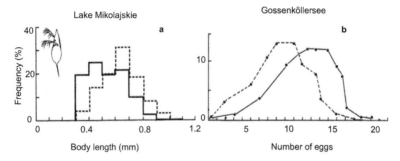

FIGURE 2.4

A. Frequency of egg size (solid line) as a function of body size in *Daphnia cucullata* and gut contents (dashed line) of smelt *Osmerus eperlanus* in Lake Mikolajskie (from Jawinski, 2002, permission by Dr. Z. M. Gliwicz). B. Free hand drawing of frequency of egg size (solid line) as a function of fecundity of *Cyclops abyssorum* and gut contents (dashed line) of *Salmo trutta* in Gossenkollersee (redrawn from Dawidowicz and Gliwicz, 1983).

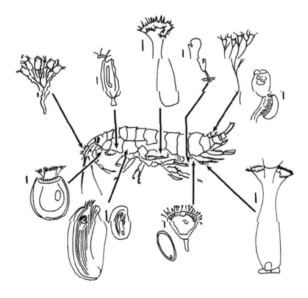

FIGURE 2.5

Sessile and motile epibionts of ciliates on the body surface of *Gammarus locusta*. Scale: bar = 10 μm (from Fenchel, 1965, permission by Science Publishers/CRC Press).

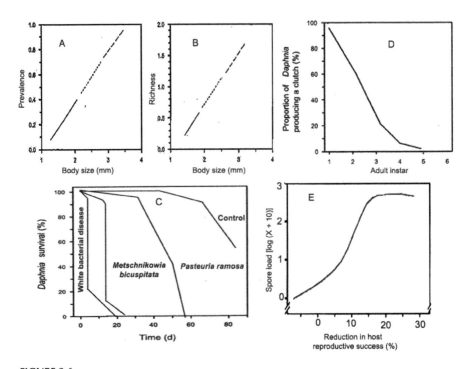

FIGURE 2.6

With increasing body size, both A. parasitic prevalence and B. parasitic species richness increase from smaller *Daphnia longiceps* (solid line) through *D. pulex* (dotted line) to large size *D. magna* (dashed line) (redrawn from Stirnadel and Ebert, 1997) C. Survival trends of infected *D. magna* showing decreases in survival in the following order: white bacterial disease < *Metschnikowia bicuspitata* < *Pasteuria ramosa* (redrawn from Ebert et al., 2000). D. Survival trends of the *D. magna* as a function of adult instar (redrawn from Carius et al., 2001). E. Reproductive success of *D. magna* with increasing parasitic spore load (redrawn from Ebert, 1994).

Parasites   A third important factor is the parasitic reduction of reproductive life span and fecundity. In aquatic animals, crustaceans alone do not secrete mucus to chemically clean themselves. Consequently, their body surface affords one of the best substrata for the epibionts. Considering ciliates alone, a dozen of them inhabit the body surface of *Gammarus locusta* (Fig. 2.5). Not surprisingly, the crustaceans are infected by a variety of parasites that reduce reproductive life span and/or fecundity. For description of parasitic microbes and animals, Brock and Lightner (1990) and Ebert (2005) may be consulted. As an example, the ensuing description is limited to microbial parasites that affect cladocerans alone. The most important microparasites are (i) Microsporidia, (ii) Ichthyosporea, (iii) fungi and (iv) bacteria (Table 2.6). The first two protozoans are intracellular parasites and the others are extracellular. Ovary, fat body, hemocoelom and hemolymph are the sites of these parasites. 1. Both prevalence and multiple parasitic species increase with increasing body size of within a species as well as among smaller and larger cladocerans (Fig. 2.6A, B). 2. Survival trends of infected *Daphnia magna* decreases in the following order: 'white bacterial disease' < *Metschnikowia bicuspidata* < *Pasteuria ramosa* < *Glugoides intestinalis* (Fig. 2.6C). 3. When infected by *P. ramosa*, the proportion of *D. pulex* producing clutches decreases with increasing (adult instar) body size (Fig. 2.6D). 4. Reproductive success of *D. magna* is reduced by increasing parasitic spore load of *G. intestinalis* (Fig. 2.6E).

TABLE 2.6

Microparasites of daphnids, site of infection and prevalence (compiled from Lampert, 2011)

Parasite	Host	Infection Site	Prevalence	
			Mean	Maximum
Microsporidia				
Octosphorea bayeri	*D. magna*	Ovary*		100
Flabelliforma magnivora		Ovary*		
Larssonia sp	–	Fat body*		
Glugoides intestinalis	*D. magna*	Gut wall*	55-99	
	D. pulex	Gut wall*	17-25	
	D. longispina			
Ichthyosporea				
Caullerya mesnili	*D. hyalina*	Gut wall*	14-32	
	D. galeata	Gut wall*	27-50	
Fungi				
Metschnikowia bicuspidata	*D. pulex*	Body cavity†	2-10	
	D. longispina	Body cavity†	0-17	
	D. dentifera	Body cavity†	32	
Ploycaryum laeve	*D. pulicaria*	Body cavity†		80
Bacteria				
Spirobacillus cienkowski	*D. dentifera*	Hemolymph†		12
Pasteuria ramosa	*D. magna*	Hemolymph†	17-35	
	D. magna	Hemolymph†	6	100
	D. pulex	Hemolymph†	12-14	
	D. longispina	Hemolymph†	8-16	

*intracellular, †extracellular

With regard to the construction of the life span vs fecundity model for crustaceans, the following were considered: 1. During the life time, tropical cladocerans produce 10 times more neonates than their counterparts in temperate zone. 2. Fecundity of egg shedders is 2-20 times more than that for egg brooders (see p 77). 3. Parthenogenics produce nearly two times more progenies than gonochorics. Without considering these facts, Blueweiss et al. (1978), who have estimated highly significant relationships between life span (~body size) vs fecundity in many vertebrate groups, reported a non-significant relationship between body size and fecundity in crustaceans. Hence, there are three horizontal lines for each of the tropical, temperate and Arctic crustaceans (Fig. 2.7). Briefly, tropical crustaceans are more fecund than their Arctic and temperate counterparts.

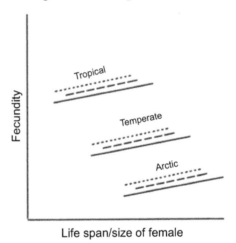

FIGURE 2.7
Fecundity as a function of increasing body size of tropical, temperate and arctic crustacean females (dotted line = parthenogenic brooders and dashed line = anecdysic egg shedders and solid line = diecdysic brooders.

2.4 Energy Allocation

Maternal investment on egg production over a biologically meaningful period of time has been estimated for some crustaceans. It is expressed as a fraction of assimilated energy (see Pandian, 1987, p 86, e.g. Kumari and Pandian, 1987). Using energy (e.g. *Macrobrachium nobilii*, Pandian and Balasundaram, 1982), carbon (e.g. *Euphausia pacifica*, Lasker et al., 1970), nitrogen (e.g. *Tigriopus brevicornis*, Harris, 1973) or dry weight (e.g. *Calanus finmarchicus*, Gaudy, 1974) as a unit of measurement, data on energy allocation for egg production have been reported (Table 2.7). Understandably, smaller crustaceans, which are heavily predated, invest a large fraction of their assimilated energy on egg production. Remarkably, parthenogenic crustaceans invest as much as 28% (*Artemia salina*) and 62% (after deducting 7% on exuvia) of their assimilated energy on egg production. Despite their diel vertical migration to 60 m depth (Lampert, 2011), the fraction of energy expended by them on metabolism remains low at around 25%; the low value reported by Richman (1958) has been confirmed by Buikema (1975). As they pass through limited number of larval (six nauplius + six copepodid) stages and do not molt as adults, copepods also invest 16% of their assimilated energy on egg production. In fact, values for the daily reproductive effort of copepods range from 7% of their body weight to 100% (Ambler, 1985). Not surprisingly, the sessile cirripede *Balanus glandula* invests on egg production 13% of its assimilated energy, as its larvae undertake a high risk to reach suitable substratum. In capture fisheries, the reproductive output of prawn has been assessed, considering the weight of absolute fecundity as a function of body weight. These values range from 12% in shrimp to mostly 20% in others (see Li et al., 2011).

TABLE 2.7

Assimilated energy allocated (%) for egg production in crustaceans (from Pandian, 1994, with permission by Oxford and IBH Publishers)

Species	Egg Production	Exuvia	Growth	Metabolism	Basis
		Anostraca			
Artemia salina	28	4	4	54	Energy
Branchinecta gigas	8	11	–	–	Energy
		Cladocera			
Daphnia pulex	69	–	6	25	Energy
		Copepoda			
Pareuchaeta norvegica	19	1	27	53	Energy
Tigriopus brevicornis	23	0.4	5	73	Nitrogen
Calanus finmarchicus	12	0.9	25	62	Dry weight
		Cirripedia			
Balanus glandula	13	2.5	11	73	Energy
		Mysidacea			
Metamysidopsis elongatus	13	7	19	55	Energy
		Isopoda			
Cirrolana harfordi	8	–	32	59	Energy
		Amphipoda			
Calliopius laeviusculus	7	5.3	19	35	Carbon
		Euphausidacea			
Euphausia pacifica	1	7	6	86	Carbon
		Decapoda			
Macrobrachium nobilii	14	7	7	–	Energy

2.5 Gametogenesis and Fertilization

For detailed description of oogenesis in peracarids and penaeids, Johnson et al. (2001) and Carbonell et al. (2006) may be consulted. Notably, both endogenous (ovarian) and extra-ovarian (hepatopancreas) vitellogeneses occur prior to and after folliculogenesis in crustaceans (e.g. *Aristeus antennatus*, Carbonell et al., 2006). Following the regular oogenesis, many crustacean females produce haploid oocytes (Fig. 2.8). In apomictic (Simon et al., 2003) parthenogenic crustaceans, the synapsis of homologous chromosome does not occur, resulting in absence of crossing over. Consequently, diploid oocytes are produced. Contrastingly, the unisexual fishes have a different pathway of producing diploid oocytes. In them, oogenesis commences with premeiotic endomitosis. Hence, even with regular meiosis I and II, they produce diploid oocytes. In the red shrimp *A. antennatus*, the spermatophore placement during the breeding season increases the ovarian differentiation and vitellogenesis from 20% in March to 92% in June; hence, the placement of spermatophores provides a signal to accelerate the ovarian development (Carbonell et al., 2006).

For a detailed account on spermatogenesis in peracarids, Johnson et al. (2001) may be referred. Figure 2.8 briefly summarizes the cytological events that produce haploid spermatids. Subsequently, they may pass through either spermiogenesis in ~ 50% of crustaceans or spermatophorogenesis in others. As a result of spermiogenesis, spermatozoa are produced, each of which consists of a head and flagellated tail (Fig. 2.9A). In ostracods, as many as seven different spermatozoa are generated. Ostracods are also known to generate long spermatozoa. For example, *Propontocypris monstrosa* of 0.6 mm body length produce (10 times longer) sperm, which measures 6 mm in length (Cohen and Morin, 1990). Some shrimps like *Penaeus occidentalis* with 50 million sperm (300 mg sperm in 46 g shrimp) are categorized as high count species, while others like *P. stylirostris* are low count species with ~ 10 million sperm (120 mg in 39 g shrimp) (Alfaro-Montoya and Vega, 2011). In other crustaceans including one in ostracods, non-flagellated, non-motile sperm called 'spikes' are produced. It may be rewarding for molecular biologists to study the genetic mechanism that switches on either the spermatophorogenic pathway or spermiogenic pathway in crustaceans. At the anterior portions of the vas deferens (Hinsch and Walker, 1974), the non-flagellated sperm (Fig. 2.9B) are then encapsulated in a spermatophore, which may be pedunculated or non-pedunculated. A spermatophore is an elongated, flask-shaped structure containing a few spikes (Fig. 2.9C). Spermatophores are held together in a row on a strip of membrane by a peduncle and are structurally species specific (e.g. some macrurans, Subramoniam, 1981). Being an integral part of spermatogenesis in vertebrates, apoptosis eliminates abnormal germ cells. Demonstrating the presence of a disintegrin and metallophase (ADAM) in the Chinese mitten crab *Eriocheir sinensis*, Li et al. (2015) have reported the role of Es-ADAM 10 and Es-ADAM 17 in regulation of sperm quality.

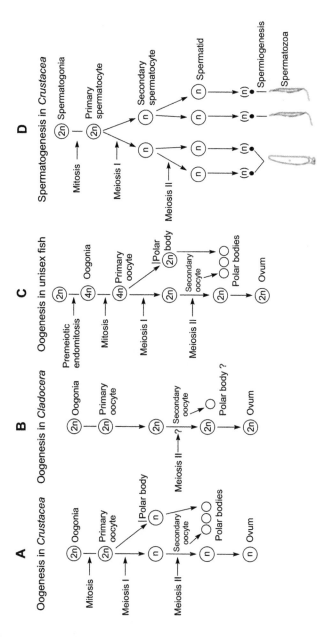

FIGURE 2.8

Gametogenesis in Crustacea. A: Oogenesis in Crustacea. B. Oogenesis in Cladocera. C. Oogenesis in parthenogenic Cladocera. Note the absence of meiosis I but meiosis II occurs to produce diploid eggs (e.g. *Artemia*, see Dumont and Negrea, 2002, p 131). C. For comparison, meiosis in unisexual fishes is shown. Note that the initial premeiotic endomitosis and regular oogenesis produce diploid eggs. D. Spermatogenesis in Crustacea. Note that spermatids of many crustaceans undergo spermiogenesis to produce a flagellated spermatozoa but those of others go through spermatophorogenesis to produce spermatophore containing non-flagellated 'spikes'.

The very first ultrastructure studies on testicular structure were undertaken in notostracans and spinicaudatans by Scanabissi et al. (2005, 2006). In *Eulimnadia texana*, as in branchiopods, the testis is a double structure, located in the hemocoel through the entire body length on each side of the midgut (diffused, as in planarians). Hailing from the testicular wall, sperm mature centripetally towards the lumen. Within the lumen, 37-92% degenerated sperm cells at different stages were mixed with normal sperm cells (Fig. 2.9D, E). These degenerated sperm cells could be involved in the primitive spermatophore-like structure (Scanabissi et al., 2006). Mature sperm cells of notostracans (e.g. *Triops concriformis*, Scanabissi et al., 2005) and spinicaudates (e.g. *E. texana*, Scanabissi et al., 2006) are amoeboid, with a few pseudopodia and contain a dense cytoplasm and few organelles (Fig. 2.9B). Acrosome reaction of the non-motile spermatozoa is regarded as an indicator of fertilization (Li et al., 2010).

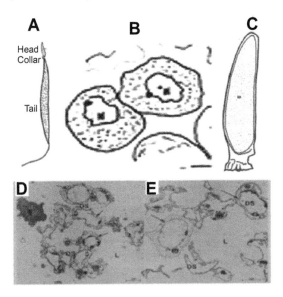

FIGURE 2.9
A. Free hand drawing of a flagellated sperm of an amphipod, B. Non-flagellated sperm of a spinicaudate (redrawn from Weeks et al., 2014a), C. Free hand drawing of a pedunculated spermatophore of a crab, D and E. Cross sections through male gonad of *Eulimnadia texana* E. Partially degenerated and completely degenerated (marked by DS) sperm cells. Bar = 2 μm (from Scanabissi et al., 2006) (for Fig. D and E permission by S.C. Weeks).

In many macrurans and brachyurans with seminal receptacle, fertilization is internal. The non-pedunculated spermatophores are transmitted in a fluid medium. Their sperm are activated by a secretion of the unfertilized eggs or that released by the female during ovulation (see Pandian, 1994, p 44). In

Solenocera agassizii, the modified and united endopodites of the first and second pairs of pleopods and appendixes masculina transmit spermatophore to thelycum; for a comparative account on the structure and function, Villalobos-Rojas and Whertmann (2014) may be consulted. Non-flagellated, non-motile spermatozoa packed in pedunculated firm spermatophore that are ejaculated, transmitted by one or other appendage of a male and placed in the vicinity of external openings of the oviduct, resulting in external or rather epizoic fertilization or on the sternal pouch-like seminal receptacle (e.g. Anaspidacea, Stomatopoda, Decapoda, Table 2.8). Dupré and Barros (2011) may be referred to know the course of events followed by a spike to fertilize an egg.

TABLE 2.8

Sperm transfer strategy in crustaceans (from Pandian, 1994 updated and modified, permission by Oxford and IBH Publishers)

Taxa	Nature of Sperm	Transmitting Structure	Fertilization	Sperm Storage
1. Flagellated Sperm Transfer Strategy				
Ostracoda	Long, flagellated, motile	Sclerotized penis near caudal furca	Internal	Nil
Cirripedia	Flagellated, not highly motile	Long protrusible penis	Internal in mantle cavity	Activated only at the proximity of ova
Mysidacea*	Flagellated, but not motile	Penis + fourth pair of pleopod	Marsupium	?
Cumacea*	Flagellated, but not motile	Penis originating from coxa	internal	–
Isopoda*	Flegellated, motile	Genital apophyses/ appendix masculina	Internal	Sperm fertilize egges in oviduct
Amphipoda*	Flagellated, not highly motile	Penis papillae	Marsupium	Activated only at the proximity of ova; storable for 4 d
2. Non-flagellated Sperm Transfer Strategy				
Anostraca	?	Eversible copulatory processes	Internal in brood pouch	Not storable
Spinicaudata	Non-flagellated, non-motile		Epizoic	
Cladocera	Non-flagellated, non-motile	Modified post-abdomen	?	?
Copepoda	Non-flagellated, non-motile	Fifth thoracic legs	Epizoic	Transmitted as firm storable spermatophore
Branchiura	?	?	?	Transmitted as spermatophore

Table 2.8 Contd.

Taxa	Nature of Sperm	Transmitting Structure	Fertilization	Sperm Storage
Stomatopoda	Non-flagellated, non-motile	Penis attached to the last thoracic legs	Epizoic/ internal	Transmitted as sperm into sternal pouch-like seminal receptacle
Anaspidacea	?	First two pairs of pleopods	Epizoic	Transmitted into sternal pouch-like seminal receptacle
Euphausiacea	Non-flagellated, non-motile	Petasma	Epizoic	Transmitted as firm storable spermatophore
Decapoda	Non-flagellated, non-motile, sperm in pedunculated spermatopore	Third pereiopods e.g. carideans	Epizoic	Into pleopods
	Non-flagellated, non-motile, sperm in non-pedunculated spermatopore in fluid	Cylinder-like first pleopod with a groove e.g. macrurans	Internal	Into sterna pocket-like receplacle or seminal receptacle
	Non-flagellated, non-motile sperm in non-pedunculated spermatopore in fluid	Cylinder-like first pleopod with a groove into which piston like second pleopod pumps e.g. Brachyura, crayfish and lobsters	Internal	Into sterna receptacle as sperm plug which is storable

*transferred as spermatophore

2.6 Morphology and Anatomy

With the advent of molecular biology, conventional zoological studies on morphology and anatomy have been unfortunately neglected. With regard to crustacean reproduction, they are important from points of sperm transmission and regeneration (Chapter 4). Unlike their terrestrial taxonomic counterparts possessing a bilaterally symmetrical pair of ovaries and testes, many crustaceans have a single median ovotestis (e.g. Notostraca: García-Velazco et al., 2009). However, many crustaceans like the anostracan *Streptocephalus dichotomus* possess bilaterally symmetrical pair of ovaries (Munuswamy and Subramoniam, 1985b). A detailed account on this is provided elsewhere.

The flagellated motile sperm are transmitted by the penis in ostracods, cirripedes, isopods, amphipods, mysids and cumaceans (Table 2.8).

Interesting information is provided for the cirripede *Semibalanus balanoides* (Walley, 1965). The long erected penis searches and locates the receptive female in its vicinity and is capable of massive and multiple inseminations. For the first time, a penis of a dwarf male permanently attached to a female has been described; it is three-four times longer than the size of the dwarf male (700 μm) of *Verum brachiumcancari* collected from 500 m depth (Buhl-Mortensen and Hoeg, 2013). In many ostracods, the male reproductive system reaches a third of body volume and a size of 35% of their body length (Cohen and Morin, 1990).

TABLE 2.9

Reported field and laboratory-made observations on spermatophore placement efficiency of selected copepods, euphausiid and penaeids (compiled from Pandian, 1994, p 44)

Species/Reference	Remarks	Placement Efficiency (%)
Pareuchaeta norvigica	Sampled 10,500 females	79
Centropages typicus	300 mating pairs in laboratory	80
Calanus helgolandicus	Prodigious egg shedder. Equal in laboratory and field	81
Euterpina acutifrons	Fertility remained unaffected in successive egg sacs	83
Acartia tonsa	Maintained for 7 y and extended for 32 filial generations	84
Thysanoessa raschi	Fertility of the euphausiid was comparable to copepods	85
Penaeus merguiensis	Sampled 10,684 females	78
P. japonicus	Observed from ocean-spermatophored females	68

The appropriate placement of spermatophore has been estimated for the egg shedding calanoids, euphausiids and penaeids as well as egg brooding copepods from large samples collected from laboratory and/or field observations. These values range from 68 to 85% (Table 2.9). Hopkins and Machin (1977) estimated that the placement of a single, double and treble spermatophores in *Pareuchaeta norvegica* was 72, 20 and 7%, respectively. Gonado Somatic Index (GSI) values reported for male crustaceans range from 0.2 to 2.5 (see Pandian, 1994, p 46). With specific reference to egg shedding copepod *Calanus helgolandicus*, a maximum of 100 eggs are fertilized by sperm from two spermatophores of 1.8% body weight. Contrastingly, the semen transferred by *Balanus balanus* and *S. balanoides* amounts to 50% of its body weight to successfully fertilize 300 eggs. Hence, the spermatophore strategy of sperm transfer is more precise and economic than that by sperm transfer in the form of semen by the penis. Not surprisingly, many crustaceans have chosen the pathway of spermatophore formation, instead of regular spermiogenic pathway. Understandably, not only many crustaceans but several aquatic invertebrates belonging to many major and minor phyla have adopted spermatophore strategy. For example, sperm transfer from male to female in the form of spermatophore is ubiquitous in the majority of annelids, crustaceans, onychophores, chelicerates, rotifers, gastrotrichs, kinorhynchs, pogonophores, phoronids and by spermatozeugma in many molluscs (Adiyodi and Adiyodi, 1989, 1990).

In many crustaceans, separate matings are obligatorily required for each spawning. It costs time and energy but offers scope to select genetically better mates and increases genetic diversity. Sperm maintenance in Types 2 and 3 readily ensures fertilization but limits the choice of female and reduces genetic diversity. Some points highlighting the events in these types are listed: Type 1. Non-storable sperm/spermatophore: (i) In the absence of storage facility, each clutch of egg has to be fertilized by the same or other male e.g. Anostraca. Type 2. Storable sperm/spermatophore at the cost of male: (i) Seminal fluid within the spermatophore keeps the sperm alive for a few days in copepods, prawns, lobsters. For example, a single impregnation of spermatophores is adequate to release fertilized eggs in seven-13 pulses at intervals of 6 d/pulse in *Cyclops*. But a second impregnation is required after the release of 4th pulse of eggs in *Mesocyclops thermocyclopoids*, which releases as many as 15 clutches (see Muthupriya and Altaff, 2004). Type 3. Storable sperm/spermatophore at the cost of female: (i) With the presence of highly secretary spermatheca, viability of sperm is retained for a long time. In anecdysic crabs, the female mates only once; the sperm remain viable in it for > 1 y and are used repeatedly in spawning e.g. *Callinectes sapidus*. (ii) Transmolt retention of sperm within spermatheca facilitates their use to fertilize successive spawnings without requiring further impregnation after the normal molt, e.g. *Portunus sanquinolentus, Paratelphusa hydrodromus*. (iii) In terrestrial isopods, the sperm remain viable for > 2 y, i.e. a single impregnation is adequate for a series of spawning e.g. *Trioniscus pusillus, Oniscus asselus*. (iv) Sperm of two males received at different ecdyses can be stored for > 2 y e.g. *Armadillium* spp (see Pandian, 1994, p 46-47).

2.7 Brooders and Shedders

Majority (>96%) of crustaceans brood fertilized eggs in their body. In fact, the sub-terranean lepidomysid *Spelaeomysis bottazzii* broods both embryos and nauplioid larvae (Ariani and Wittmann, 2010). However, calanoid copepods (other than Notodelphyoida), branchiurans (parasitic), syncarids, some euphausiids (see Pandian, 1994, p 94) and penaeids do not brood but simply shed their eggs. They constitute less than 4% of crustaceans. Hence, egg brooding is a hallmark strategy of crustaceans. During the checkered history of evolution of egg brooding crustaceans seem to have explored different sites for egg carriage without disturbing the posture and motility very much (Fig. 2.10). Mantle cavity (e.g. *Balanus*), projected sac (e.g. copepod), dorsal pouch (e.g. *Daphnia*), marsupium (by opposing thoracic appendages, isopods and amphipods, however see Thomson, 2014), thoracic (e.g. *Nematoscelis difficilis*) and abdominal appendages (e.g. *Procambarus*) are the selected sites for the egg carriage (Table 2.10). As the increasing volume of developing embryos compresses the female's internal organs including the gut, the brooding females of isopods and tenaids may not feed at all. Hyperliid amphipods have special internal brood chambers or ventral pockets not made up of oostegites. At least a dozen sphaeromatids lack oostigites and brood their embryos in too large ventral pockets (see Johnson et al., 2001). The parthenogenic cladocerans carry the heaviest brood, i.e. 25% of its body weight. Not surprisingly, its posture is visibly tilted.

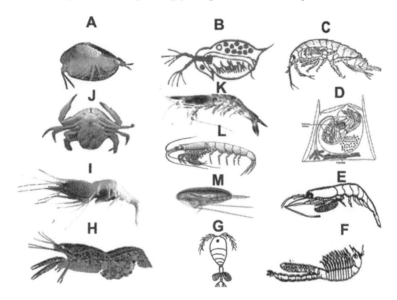

FIGURE 2.10

Brooding sites in Crustacea. A. *Eulimnadia texana* showing dorsal egg pouch (courtesy S.C. Weeks). B. *Daphnia* (courtesy W. Lampert, International Ecology Institute) showing dorsal pouch. C. *Gammarus locusta* showing marsupium (courtesy T. Fencel, International Ecology Institute) D. A section through a barnacle to show embryo being incubated in the mantle cavity (free hand drawing), E. *Nematoscelis difficilis* (showing thoracic appendages carrying berried embryos (redrawn from Nemoto et al. 1972), F. Ventral egg sacs in *Streptocephalus dichotomus* (courtesy, N. Munuswamy), G. A cyclopoid copepod showing lateral pouches (personal observation), H. The crayfish *Procambarus* carrying embryos on the abdominal appendages (courtesy, G. Vogt), I. *Plesionika spinipes* showing the incubation of eggs on pleopods (courtesy CMFRI), J. *Portunus pelagicus* showing embryos carried on abdominal appendages (courtesy: M. Ikhwamuddin), K. *Metapenaeus barbata*, an egg shedding penaeid (courtesy CMFRI), L. *Meganyctiphanes norvegica*, an egg shedding euphausid (courtesy, Bulletin of the Scripps Institution of Oceanography), M. *Calanus finmarchicus, an egg shedding copepod* (courtesy, Kurt Tande, Northeast Fisheries Science Center News).

TABLE 2.10

Sites of egg brooding, clutch size and brood mortality in selected crustaceans (from Pandian, 1994, modified and updated, permission by Oxford and IBH Publishers)

Taxonomic Group	Clutch (No.)	Clutch as % of Body Weight	Brood Loss (%)
Egg Brooders			
In mantle cavity			
Conchostraca	–	–	–
Ostracoda	3-85	–	–
Cirriipedia	300	9	–
In attached pounch			
Cladocera	21	25	19
Notodelphyoida	–	–	–
In projected sac			
Cephalocarida	–	–	–
Anostraca	110	9	–
Notostraca	–	–	–
Copepoda	78	9	9
In marsupium			
Mysidacea	80	–	10
Cumacea	162	–	–
Teniidacea	–	–	–
Isopoda	230	–	22
Amphipoda	250	–	26
In thoracic appendages			
Leptostraca	–	–	–
Stomatopoda	–	–	–
Euphausiacea	95	18	–
In abdominal appendages			
Prawns, lobsters	81,000	13	41
Crabs	72,000	15	54
Egg Shedders			
Podocopa†		–	
Calanoidea	2000	–	
Branchiura	–	–	
Syncarida	–	–	
Euphausiacea	296	–	
Penaeidea	725,000	–	

†Egg depositors, –Information is required

Brooders afford protection, ventilation and grooming of the brooded eggs. Eggs are simply held in the mantle cavity (well ventilated but not exposed), dorsal pouch or marsupium (relatively more ventilated but semi-exposed) but are firmly fastened to the setal hairs of abdominal appendages (well ventilated but completely exposed). Eggs are also held in ventral abdominal (e.g. *Branchinecta*), thoracic (e.g. *N. difficilis*) or projected sac of copepods (Fig. 2.10). In decapods, fanning frequency to ventilate attached eggs is accelerated with advancing development in *Macrobrachium idae* (Pandian and Katre, 1972). The frequency is increased from 2,676 time/hour (h) to 8,760 time/h during the initial and terminal periods of incubation in *M. nobilii* (Pandian and Balasundaram, 1980).

Brood pouch, sac and marsupium are relatively safer anlagen but limit brood size and ventilation. Their brooding capacity ranges from ~20 eggs in cladocerans to ~80 in copepods or to ~250 in isopods and amphipods or at the maximum to ~400 in euphausiids (Table 2.10). Ventilation is greatly improved, when eggs are brooded on thoracic (e.g. Stomatopoda) or abdominal (e.g. Decapoda) appendages. These appendages can carry up to a million eggs. Brooded eggs serve as an excellent substratum for fouling epibionts, namely fungi (Fig. 2.11) and others. Aquatic organisms clean their body chemically by continuously secreting mucus. However, crustaceans do not secrete mucus. Hence, they have to clean their body including their brooded eggs by physically scratching, and removing fungi and other epibionts (see Bauer, 1979). For example, *Homarus americanus* broods eggs for 9 m. Preening of eggs by the brooding females is effective in reducing infection by epibionts (see Brock and Lightner, 1990).

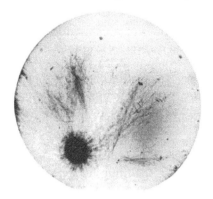

FIGURE 2.11
A developing embryo of *Macrobrachium nobilli* infested by *Lagenidium*. Note the hyphae (from Balasundaram, 1980).

Egg loss during the brooding period ranges from 9% in the sac of copepods to 22-26% in marsupium of isopods and amphipods, and to 41 to 54% on appendages of decapods (Table 2.10). With regard to brood mortality, the following have to be noted: 1. Egg predation: Cyclopoid copepods enter the brood cavity of daphnids to eat eggs and sap liquids from ovaries. The larger daphnids are more vulnerable (Gliwicz and Stibor, 1993). Hence, it is difficult to decide whether the loss is for eggs or embryos (Gliwicz, 2003). The two valves of the carapace of daphnids have to be kept open at an optimum level to permit exchange of water but to prevent the loss of brooded eggs in the pouch. There are other predators, which eat the brooded eggs. The nemertean *Carcinonemertes mitsukurii* (Shields and Wood, 1993) and nicothoid copepod *Choniosphaera indica* (Gnanamuthu, 1954) are predators of brooded eggs of *Portunus pelagicus*. The nemertean *C. errans* predated the brooded eggs of *Cancer magister* at the rate of ~1 egg/d and thereby reduced the brooded embryos to <30% on the day of hatching (Wickham, 1980). 2. A second reason for the loss of brooded eggs was that the crustacean eggs imbibe water and swell in volume, especially during the terminal phase of brooding. The volumetric increase of an egg is in the range of 1.6 time (0.8 mg water/egg) in *Pagurus benrhardus* (Pandian and Schumann, 1967), 2.5 time (0.8 mg/egg) in *Crangon crangon* (Pandian, 1967), and 2.7 time (2.3 mg/egg) in *Homarus gammarus* and (1.8 mg/egg) *H. americanus* (Pandian, 1970 a, b). In the signal crayfish *Pacifastacus leniusculus*, egg volume increased from 9 to 12.5 mm^3 and egg case strength decreased from 108 to 38 g (Pawlos et al., 2010). These lead to breaking of the egg membrane and release of the larvae in batches. 3. A third cause for the mortality of brooded eggs is the infection by epibionts. Many species of fungi belonging to *Lagenidium, Haliphthoros, Atkinsiella* and *Fusarium* are important pathogenic epibionts on the eggs of shrimps, lobsters and crabs (see Brock and Lightner, 1990). In the marsupium of isopods, the entire brood can be destroyed by fungus (*Plectospira*) or parasitic isopods (*Clypeoniscus* and *Ancyananicus*) (see Johnson et al., 2001).

Strikingly, the egg shedding crustaceans produce and broadcast three to 20 times more eggs than their respective taxonomic counterparts. For example, egg shedding euphausiids broadcast three times (~300 eggs) more eggs, in comparison to brooding (~100 eggs) euphasiids (Table 2.10). Remarkably, the calanoid copepods shed 10 times more eggs than the egg brooding copepods. These data provide an idea of the costs of time and energy of adult molts and egg brooding. Carrying ~10 to 20% of its body weight, continuous cleaning of surface of eggs and incessant ventilation cost time and energy, which remain to be estimated. Irrespective of being less speciose (~400 species, cf 470 species Atyidae), the egg shedding penaeids are global in distribution and occupy different trophic levels of the food chain at various depths of ocean (5,000 m bsl) (see CMFRI, 2013). Saving energy by shedding eggs and adopting anecdysic molt cycle, the calanoid copepods are prolific breeders and constitute a major source of food for fishes in nature and aquaculture.

There are reports on injury and death of pelagic spawning fishes (e.g. *Gadus morhua*, see Pandian, 2014 section 2.4), in which oocytes are hydrated within the females; however, no report is yet available for pelagic spawning penaeid females. Apparently, hydration in penaeid eggs occurs after spawning. Hence, imbibition of water or hydration of eggs as a mechanism of hatching, the following may be noted. Regarding the timing of hydrated in pelagic eggs, there is a subtle difference between fishes and crustaceans. The oocytes are hydrated prior to spawning within the mother fish but the penaeid eggs are hydrated after spawning (Fig. 2.12). Both in demersal fishes and egg brooding crustaceans, that produce pelagic larvae, the eggs are hydrated during embryogenesis. Devoid of pelagic larval stage, the lobsters and the semi-terrestrial *Ligia oceanica* release their demersal young ones at an advanced stage and as neonates, respectively. Hence, the level of hydration during embryogenesis is low, as in lobsters or almost nil, as in *Ligia*. However, hydration of all brooding crustaceans during embryogenesis may serve as a mechanism of egg hatching. Conversely, the pelagic eggs of calanoid copepods, penaeid shrimps and euphausiids may have different mechanism of hatching, as their eggs remain hydrated since spawning.

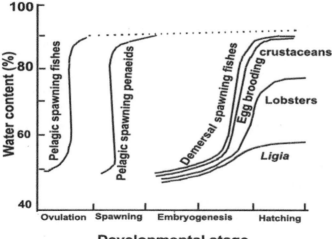

FIGURE 2.12
Hydration of oocytes/eggs of pelagic spawning fishes (e.g. *Solea solea*, Pandian and Flüchter, 1968) and penaeids, as compared to demersal spawning fishes (see Pandian and Flüchter, 1968) as well as egg brooding crustaceans (e.g. *Crangon crangon*, Pandian, 1967), which produce pelagic larvae. Devoid of pelagic larval stage, the larvae of lobsters (Pandian, 1970 a, b) and *Ligia* neonates (Pandian, 1972) contain less water.

2.8 Larval Development

In about 60% crustaceans, the development is direct (e.g. cladocerans) with a 'manca' stage (peracarids) but it is indirect in others. The occurrence or absence of specialized ontogenetic digressions labels it as metamorphic or anamorphic (Table 2.11). The indirect development involves the following: 1. A single larval stage, the nauplius or metanauplis. 2. Two stages (i) nauplius followed by copepodid, (ii) mysis → zoea, (iii) zoea → glaucothoe/grimothea, (iv) zoea → megalopa or larva → praniza. 3. Three stages namely (i) nauplius → cypris → kentrogen, (ii) protozoea → zoea → post larva, (iii) phyllosoma → puerulus → pseudibaccus. 4. Four stages namely (i) nauplius → calyptosis → furcilia → cryptopia, (ii) nauplius → protozoea → mysis → mastigopus, (iii) epicardium → microniscus → cryptoniscus → bopyridium. Nauplius is the most common first larval stage in many crustaceans (cf p 32). Notable are the shifts in the larval stage from nauplius to metanauplius in anamorphics, from nauplius to mysis in mildly metamorphics and from protozoea/pseudozoea to zoea in metamorphics. Secondly, four larval stages are involved in diecdysic egg-shedding marine penaeids and parasitic epicarideans. Thirdly, nauplius is followed by cypris stage in sessile balanoids. Notably, *Metanephros* lobsters hatch at an advanced stage with restricted swimming ability and lack a zoeal stage; within a short natatory period, they settle within hours or days after hatching (Robey and Groeneveld, 2014).

Density of crustacean eggs is higher than seawater. Unless brooded, these eggs shall sink slowly in seawater but rapidly in freshwater. Contrastingly, many crustacean larvae are not denser than sea water. Hence, these larvae drift readily as plankton. Low density of freshwater reduces the scope for drifting and makes it energetically expensive. Hence, freshwater crustaceans reduce the number of planktonic larval stages. In seawater, the planktonic larval stages, for penaeids last for nine to 30 d involving three to 11 molts (Pandian, 1994, p 68). The triglycerides have a density of 0.92 mg/l (Lewis, 1970). Hence lipid levels in the eggs determine the density of eggs rather than the egg size. Accumulation of lipids in eggs as an energy source for embryonic metabolism and its use during embryogenesis along with imbibition of water maintain the density of egg fairly constant during incubation.

Egg size offers valuable information on reproductive strategy of a species. Many decapods show a clear correlation between egg size and type of larval development ranging from normal (through quashi-) direct abbreviated development (Hernáez et al., 2010). The pantropical decapod genus *Macrobrachium* comprises of ~240 species with its diverse spread of >100 species in East and Southeast Asian countries. Their size ranges from 6 to 32 cm in body length. An estimated 200,000 t of *M. dacqueti* worth a billion US $ are produced and marketed. There are three groups of *Macrobrachium*: 1. Those which complete the entire life cycle within sea. 2. Amphidromus species adopting r-strategy by producing small but large number of eggs, which are released in the brackish water/sea; after completion of the Prolonged Larval Development (PLD) of 11+ larval stages, they return to freshwater. 3. Those adopting k-strategy by producing large but small number of eggs: their larvae pass through Abbreviated (2-4) Larval Development (ALD) stages (Table 2.12). Using multilocus phylogenetic analysis, Wowor et al. (2009) have traced the origin and evolution and suggested that the Asian *Macrobrachium* have invaded freshwater two-three times and caves at least twice. In his review, Anger (2013) has traced the origin and evolution of Neotropical *Macrobrachium*.

TABLE 2.11

Larvae and larval development in crustaceans (from Pandian, 1994, with permission by Oxford and IBH Publishers)

Taxanomic Group	Larval Stages
	Anamorphic
Branchiura	Nauplius
Euphauciacea	Nauplius → Calyptopis (Protozoea) → Furcilia (Zoea) → Cryptopia (post larva)
Cephalocarida, Branchiopoda, Mystacarida	Metanauplius
	Mildly Metamorphic
Copepoda	Nauplius → Copepodid (post larva)
Penaeidae	Nauplius → Protozoea → Mysis (Zoea) → post larva
Nephropasidae	Mysis → post larva
	Metamorphic
Thoracica	Nauplius → Cypris
Sergestodae	Nauplius → Elaphocaris (Protozoea) → Acanthosoma (Zoea) Mastigopus → (post larva)
Caridae, Stenopodidea	Protozoea → Zoea → post larva
Scyllaridae	Phyllosoma → Puerulus → Pseudibaccus
Anomura	Zoea → Glaucothoe → post larva
Brachyura	Zoea → Megalopa
Somatopoda	Pseudozoea → Alima → Synzoea → Somatopodid (post larva)
Gnathiidae	Larva → Praniza (post larva)
	Strongly Metamorphic in ♀
Notodelphoida, Caligoida, Lernaepodida	Nauplius → copepod (post larva)
Epicaridae	Epicardium → Microniscus → Cryptoniscus → Bopyridium
	Strongly Metamorphic
Monstrilloida	Nauplius → Parasitic → Larva
Acrothoracica, Ascothoracica	Nauplius → Cypris
Rhizocephala	Nauplius → Cypris → Kentogon
	Epimorphic
Cladocera, Anaspidacea, Spelaegriphacea, Mysidacea	In *Leptodora* alone metanauplius
	Epimorphic with Manca*
Neabliacea, Thermasbaenacea, Cumacea, Tanaidacea, Amphipoda, Isopoda, Ostracoda	

*Lacks the last thoracic appendages

TABLE 2.12

Prolonged (PLD) and Abbreviated (ALD) Larval Development in *Macrobrachium* (compiled from Wowor et al., 2009)

Species	Habitat	Larval Stage (No.)	Fecundity (No./Clutch)
M. shokitai	FW	2	37
M. asperuhim	FW	2	42
M. totonocum††	FW	3	18
M. sintangense	FW	4	63
M. lanchesteri	FW	2*	226
Palaemon debilis	SW	7	392
M. latidactyloides	BW	8	1054
M. nippone	BW	9	4100
M. equidens	BW	10	3500
M. dacqueti	BW	11	1,50,000*
M. lar	SW	11†	40,000
M. latimanus	BW	11	2906
M. idae	BW	12	2905
M. semimelinki	SW	12	532

*maximum, average will be low
†Mexican sp, Mejía-Ortíz et al. (2010)

The extreme environmental changes in temperature, salinity and oxygen levels in high-shore rockpools demand highly challenging adaptation, especially during reproduction and development. To assess the effects of the extreme fluctuations, McAllen and Brennan (2009) have made a rare but interesting study to estimate the time duration required for the appearance of ovary, egg sac, nauplius release and nauplius-copepodid 1. Following mating, the harpactoid *Tigriopus brevicornis* produced 12-14 broods, each up to 150 eggs. Its development was the quickest at 23°C but produced the lowest number of copepodid 1 (Table 2.13). At salinity 35 g/l too, it was fastest and ensured the highest survival. Surprisingly, ovary and egg sac did appear but no nauplius was released at anoxic O_2 levels. Incidentally, it is not clear, why *T. brevicornis* did not produce cysts at these extreme conditions.

TABLE 2.13

Effect of temperature, salinity and oxygen levels on time (h) duration required to complete the selected stages of *Tigriopus brevicornis* in a high-shore rock pool (compiled from McAllen and Brennan, 2009)

Reproductive Stage	Temperature (°C)			Salinity (g/l)			O_2 concentration (ml/l)		
	5	12	23	5	35	70	9.2	1.0	0.0
Ovary appearance	204	120	112	156	112	248	120	216	196
Egg sac appearance	416	332	224	318	256	0	253	414	396
Nauplius release	627	460	308	494	392	0	403	624	636
Nauplius-Copepodid1	887	604	392	0	530	0	541	0	0
Survival (no.)	54	72	50	0	64	0	50	0	0

2.9 *In vitro* Incubation and Translocation

Increasing volume of developing eggs due to imbibition of water at terminal stage encounters space problem. To reduce it, eggs are hatched in three-22 batches in some decapods (see Pandian, 1994, p 64). In *Macrobrachium nobilii*, 24, 18, 27, 23, 7 and 2% of eggs were hatched on 12, 13, 14, 15, 16 and 17th day after berrying (Balasundaram and Pandian, 1981). The rhythm of hatching profile is under maternal control. A hatching substance characterized by relative stability and readily diffusibility is released by the balanoid mother (Crisp and Spencer, 1958) to stimulate embryonic movements and thereby hatching is initiated by the mother. The maternal substance released by crabs is a peptide (Christy, 2011).

Any attempt to synchronize hatching, say *in vitro* culture of embryos affords many benefits: 1. Egg brooding is an expensive activity and relieving the mother from the task of brooding increases the reproductive output by 1.5 times (see p 39) 2. The larvae hatched on the last batch contain less reserve yolk energy than those that hatch in the first batch (e.g. 8% *Macrobrachium idea*, Pandian and Katre, 1972, 30% *Homarus americanus* Pandian, 1970b). Developing embryos are attached to the ovigerous setae of their mother and/or each other by funiculus and egg coat. Using delicate tweezers, Giovagnoli et al. (2014) removed 10-30 embryos from a mother of *Palaemonetes argentinus* without harming the mother or injuring the embryos. More than 98% of these embryos, after *in vitro* incubation, successfully completed the development and hatched, in fact one day earlier than those hatched by the mother. Similarly, simple *in vitro* culture of embryos of some decapods has also been successful (e.g. Grapsid crab, Bas and Spivak, 2000, *M. rosenbergii*, Damrongphol et al. 1990). However, it is a laborious procedure to remove hundreds and thousands of embryos using the delicate tweezers. Simple plucking of embryos from mother harms the mother and injures the embryos.

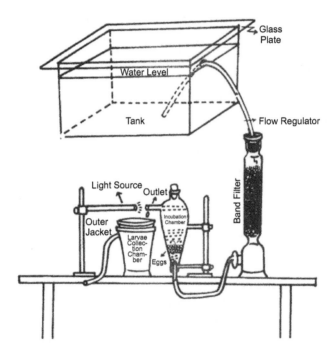

FIGURE 2.13

An incubator for successful incubation and synchronized hatching of 3-d and 12-d old *Macrobrachium nobilii* embryos (from Balasundaram, 1980)

TABLE 2.14

Effects of egg mass removal from the mother *Macrobrachium nobilii* and egg development (from Balasundaram, 1980)

Age of Egg (h)	Developing Eggs	Mother
Fertilized eggs before berrying	100% mortality within 4 h of incubation	Normal
0	100% mortality within 6 h of incubation	Died immediately or later following the egg removal
3	100% mortality within 6 h of incubation	Died immediately or later following the egg removal
12	Normal development; 10% abnormality in shape	Too sensitive; when eggs were force-relieved, some died
24	Normal development; no abnormality in shape	Too sensitive; often abandoned egg mass
48	Normal development; no abnormality in shape	Too sensitive; often abandoned egg mass
72	Normal development; no abnormality in shape	Successfully withstood the stress of egg mass removal

To minimize the loss of yolk reserve in hatchlings, efforts have been to develop an incubator and to synchronize hatching. Sandoz and Rogers (1944) attempted to simultaneous hatch all the 50 incubated eggs of *Cancer magister*, providing a surface area of 0.5 cm^2/egg. Costlow and Bookhout (1960) increased the density (to 110 eggs), providing surface area of 0.09 cm^2/egg, affording running water and agitating water in a shaker. Table 2.14 shows that *M. nobilii* eggs can be removed from female *M. nobilii* only after 3 d after brooding without harming the eggs or 12 d after brooding without harming the mother. The incubator developed by Balasundaram and Pandian (1981) consists of an overhead tank containing aerated and antibiotic-added (amphicillin, 5 ppm to ward off bacteria) water (Fig. 2.13). From the tank, water is filtered through sterilized sand and allowed to enter (at the regulated speed by a clip screw) at 3.5 l/h into an incubator flask, in which disinfected 3-d or 12-d old eggs (200-300) are incubated over a perforated screen firmly fixed at a lower level of the flask. With turbulence generated by inflowing water into the flask, the eggs rotate in vertical circles, providing a surface area of 1 cm^2/egg. The hatched larvae were attracted to a light source focused against the opening of the outgoing water from the flask. On completion of their incubation period, 70% of 3-d old eggs were hatched simultaneously on the 12th day. Simultaneous hatching increased from 83, 90 to 98% with increasing egg age from 6 to 9 and 12 d, respectively (see Balasundaram and Pandian, 1981).

It is also possible to translocate developing embryos from marsupium of a female to another to know whether or not the mother recognizes the presence of alien embryos among its own embryos. In fact, intra-specific and inter-specific translocations have been made. The presence of mantle cavity in barnacles and clam shrimps as well as marsupium in amphipods and isopods provide excellent opportunity to undertake simple but meaningful experiments. Ooviviparity occurs in some crustaceans and may have evolved independently in many groups like Anostraca, Cladocera and Peracaridae. Cladocerans like *Daphnia* and *Chydorus* undergo normal development under *in vitro* incubation. But *Moina* and *Polyphemus* do not, as they receive maternal nutritive fluid (see Pandian, 1994, p 61).

Many gammarids can discriminate conspecific embryos from those of alien species (e.g. *Gammarus mucronatus*, Borowsky, 1983). *Corophium bonnellii* can distinguish conspecific embryos from similar-sized embryos of *C. volutator* and *Lembos webskeri*, and replace conspecific embryos into its marsupium (Shillaker and Moore, 1987). In the semi-terrestrial *Orchestia gammarellus*, hatching is under the maternal control. The osmotic concentration of marsupial fluid improved embryonic survival up to 96%, in comparison to 86% in those incubated *in vitro* in seawater (Morritt and Spicer, 1996). Hence, amphipods may neither receive maternal nutrition nor 'adopt' the translocated embryos of alien species. In *Tanais dulongi*, mothers inject a large 'yolk' dose through gonophores, upon which the embryos feed prior to their departure from the marsupium (see Johnson et al., 2001). *Leptomysis lingvura* was induced to successfully 'adopt' the translocated equal-sized embryos of G. *mucronatus* (Wittmann, 1978). To behavioral ecologists and functional morphologists, the study of adoption of translocated conspecific and interspecific embryos of cladoderans and peracardians may be rewarding.

In the almost terrestrial *Ligia oceanica*, the mother did not recognize the presence of alien embryos and continued to incubate, when embryos of the same age but from another female were translocated. However, the alien embryos were rejected, when the translocation involved embryos of different developmental stage (pers. obser.). The crustacean mothers secrete a hatching substance to stimulate embryonic movements and to initiate the hatching process (Crisp and Spencer, 1958). It is likely that the embryos at an early stage of development were rejected by *L. oceanica*, in which the translocated embryos were at the stage of awaiting hatching. *Ligia* is known to receive maternal nutrition (see Pandian, 1994, p 62). This observation is also confirmed by the fact that the loss of yolk on embryonic metabolism is 62% in *L. oceanica* (Pandian, 1972), in comparison to 72% in other decapods (Pandian, 1967, 1970a, b). Pandian (1972) considered that the 10% of maternal nutrients, though small in fraction, may be an obligate necessity.

3

Sexual Reproduction

3.1 Sexuality and Sex

Sexual reproduction is characterized by the presence of sex, gametogenesis, meiosis and fertilization. In parthenogenics, females are always present but males sporadically. Ameiotic oogenesis usually occurs but meiotic gametogenesis occurs also sporadically. Asexual reproduction involves budding, fragmentation and fission (see Chapter 1). Unlike many authors, who consider parthenogenesis as asexual reproduction, this presentation includes parthenogenesis as a sexual mode of reproduction. In all facultative parthenogenic crustaceans, occasional males and females are produced; in these parthenogenic males and females, meiotic gametogenesis and fertilization occur.

Known for their sexual diversity and plasticity, crustaceans are a fascinating group of invertebrates. Barring three species of rhizocephalan cirripedes (Shukalyuk et al., 2007), they reproduce sexually. In them, sexuality includes gonochorism, parthenogenesis and hermaphroditism (Fig. 3.1). Within parthenogenics, there are obligate and facultative parthenogenics. Facultative parthenogenics include geographic and cyclic parthenogenics. Within cyclic parthenogenics, there are monocyclics, dicyclics and polycyclics. However, facultative parthenogenesis is invariably interspersed within sexual ontogenetic pathway. Hermaphroditism also includes simultaneous and sequential, which again includes protandrics and protogynics. The size advantage model (Ghiselin, 1969) proposes that sequential hermaphroditism occurs, when a small or young individual reproduces most efficiently as a member of one sex but changes sex, when it is older and larger. The model predicts protogyny, where there is sexual selection for larger male and protrandry, where selection favors larger females with high fecundity. Within each of these sequentials, there are diandrics with primary and secondary males and digynics with primary and secondary females. Besides the known mating systems, a rare androdioecy (mixtures of hermaphrodites and males) is reported from ~ 33 species within simultaneous hermaphrodites (Weeks et al., 2006c).

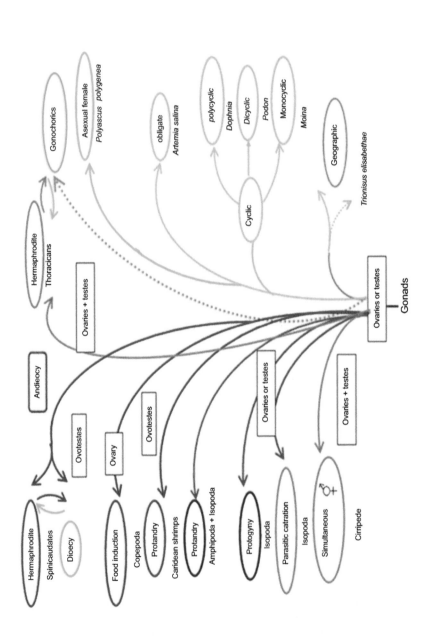

FIGURE 3.1

Sexuality, modes of reproduction and gonadal morphology of crustaceans. Note the presence of ovaries and/or testes as well as ovotestes.

"Sex change is a unique but an intriguing phenomenon, especially from the point of genetic basis of sex determination. Instantaneously, this unique phenomenon poses a series of challenging questions regarding the presence of sex chromosome systems and sex determining genes" (Pandian, 2011, p 160). In crustaceans, protandrics may become simultaneous hermaphrodites in a few carideans (e.g. *Lysmata wurdemanni*, Bauer, 2000). Hence, sex is differentiated at larval stage in gonochorics but up to adults in sequential hermaphrodites. In copepods, abundance of food may induce sex change from male to female (Gusamao and McKinnon, 2012). Protandric and protogynic sex change occurs in parasitic bopyrids. Incidentally, parasitic castration is limited to female function alone in simultaneous hermaphrodites (e.g. *L. amboinensis*, Calado et al., 2006). Cryptoniscus larva of *Ione thoracica* that settles first on a definitive host metamorphoses into a female but all the subsequent ones into males (Pandian, 1994, p 82).

3.2 Gonads and Ovotestes

In many crustaceans, the gonads consist of either discrete organs (e.g. amphipods) (Fig. 3.3A) or tubular structures (e.g. balanoids, Fyhn and Costlow, 1977, Fig. 3.2B) of ovaries or testes. The others have either tubular (e.g. tadpole shrimp) or discrete (e.g. heterologous in sex changing carideans or bilobbed in spinicaudates) ovotestes, which may be delimited or undelimited type (cf Sadovy and Shapiro, 1987). In the former, exemplified by sex changing carideans, the ovotestis consists of distinct heterologous ovarian and testicular lobes (Fig. 3.3C). In the latter, they remain intermingled (Fig. 3.2D) but effectively separated by ovarian or testicular wall (Fig. 3.2C). The delimited type of ovotestis may be amenable to ovariectomy or castration. However, the location of the caridean gonads, just above the heart and below the dermal layer in cephalothorax, inhibits surgical ovariectomy or castration. But, it may still be possible to achieve castration in crustacean males by injection of busulfan or supplementation of it in the diet (see Pandian, 2011, p 149) or by artificial infection by castrating micro-parasites (see later). Chemical means of achieving ovariectomy or castration shall open a new avenue to undertake innovative research and to identify organs (other than gonads) harboring Primordial Germ Cells or their derivatives Oogonial Stem Cells/Spermatogonial Stem Cells (see Pandian, 2013, p 174). With increasing importance of experimental studies, the need for description of gonadal morphology is obvious. Incidentally, visualization of anatomy of reproductive system now appears possible with the introduction of confocal laser scanning microscopy and two-photon microscopy (Fitzer et al., 2012).

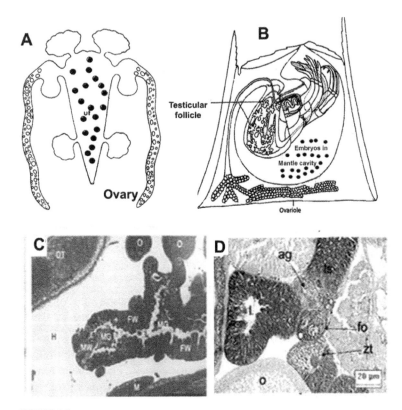

FIGURE 3.2
A. Pair of ventro-laterally placed tubular ovaries, which are connected to dorso-medially located uterus/ovisac in *Streptocephalus dichotomus* (permission by N. Munuswamy). B. Free hand drawing of a barnacle to show the different locations of ovariolar and testicular follicles. Note the presence of embryos being brooded in the mantle cavity. C. Longitudinal section through posterior region of the hermaphroditic gonads of *Eulimnadia michaeli*. The gonads include a female wall (FW) that is interrupted at the posterior-most tip by a male wall (MW). From the MW, sperm (MG) are released into the gonadal lumen (L) (from Brantner et al., 2013b, permission by S.C. Weeks). D. Histological section through ovotestis of *Triops* species showing the presence of Ag = germinal area, o = oocyte, fo = follicular duct and zt = testicular zone (from Garcia-Velazco et al., 2009, per com).

In many lower crustaceans, the ovaries and testes or ovotestes are tubular in structure. Hermaphrodites with or without males are known to have originated from sexual ancestors independently many times (e.g. four times in spinicaudates, Weeks et al., 2014a). As a result, a pair of tubular ovaries and testes is present in thoracic balanoids but bilobed delimited pair of ovotestes (cf Sadovy and Shapiro, 1987) in spinicaudates. In the anostracan, *Streptocephalus dicotomus*, the gonad is compartmentalized into a dorso-median uterus/ovisac, to which a pair of ventro-laterally placed tubular (vitellarium) ovaries are connected (Fig. 3.2 A). The germinal epithelium gives rise to both oogonial and nurse cells, which are subsequently differentiated (Munuswamy and Subramoniam, 1985a, b). With a sessile mode of life, the symmetry is lost in balanoids. The paired ovaries consisting of branched tubules (ovarioles) are located between the mantle cavity and basal membrane (Fig. 3.2B) but the testes, rather testicular follicles in the main body. A symmetric pair of bilobed (i.e. heterologous) ovotestes, which are positioned adjacent to the digestive tract within the hemocoel, is present in *Eulimnadia azisi, E. gibba, E. gunturensis* and *E. michaeli* (Fig. 3.2C) (Brantner et al., 2013b). Scanabissi et al. (2006) have described the presence of symmetrically paired tubular testes in *E. texana* male. However, the ovary and testis seem not distinctly separated into heterologous zones in notostracans (Fig. 3.2 D).

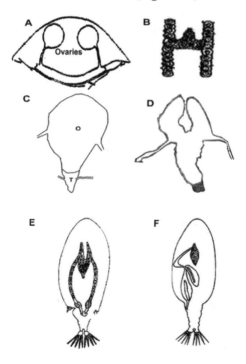

FIGURE 3.3
A. Free hand drawing of a section to show the locations of a pair of ovaries in an amphipod. B. the triangular mid-piece is the generarium of the ovaries in the mysid *Mesodopsis orientalis* (redrawn from Nair, 1939). C. A representative of the anterior ovary and posterior testis in ovotestis of a protrandric simultaneous hermaphroditic shrimp. D. A picture to show a Y-shaped ovary in marbled crayfish (redrawn from Vogt et al., 2004). E. Female reproductive system showing a single mid-ovary in *Heliodiaptomus viduus* (from Altaff and Chandran, 1994, permission by K. Altaff). F. Male reproductive system showing a single mid-testis in *Allodiaptomus raoi* (from Bama and Altaff, 2004, permission by K. Altaff).

In amphipods, a pair of discrete ovaries in female or testes in male is placed in symmetric anlagen (Fig. 3.3A). In sex changing protandric and protogynic isopods, a pair of ovaries or testes are also symmetrically placed (Brook et al., 1994). The distantly placed ovaries are bridged together to form a triangular mid-piece in the mysid *Mesodopsis orientalis* (Fig. 3.3B), *Mysis oculata* and *Neomysis japonica*. Caridean sequentials may have also originated from their respective gonochoric ancestors independently many times (for fishes see Frisch, 2004). In sex changing carideans, a spindle shaped ovotestis with a pair of oviducts is present (e.g. *Lysmata californica*, Bauer, 2000). In the protandric carideans, oocytes are generated from the posterior end of the ovary and are moved anteriorly, as vitellogenesis proceeds (Fig. 3.3C). In the parthenogenic *Procambarus fallax*, the ovarian component is a Y-like structure (Fig. 3.2D). During the male phase of protandric carideans, the function of testes is elaborated and the ovaries remain almost vestigial. But the ovary alone functions during the female phase. In protandric simultaneous hermaphroditic carideans, the functions of both ovaries and testes are simultaneously developed in the ovotestes. Notably, single median ovotestes is limited to a single median ovary in female (*Heliodiaptomus viduus*, Altaff and Chandran, 1994, Fig. 3.2E) or testis in the male (e.g. *Allodiaptomus raoi*, Bama and Altaff, 2004) copepods (Fig. 3.2F). In many hermaphroditic notostracans, the presence of a single median ovotestis is also reported (e.g. *Triops* sp, Garcia-Velazco et al., 2009).

3.3 Gonochorism

Most crustaceans are dioecious and dimorphic. In them, sexual dimorphism ranges from microandry (e.g. dwarf males in balanoid *Scalpellum scalpellum*, Spremberg et al., 2012, parasitic isopod *Parabopyrella* sp, Calado et al., 2006) to macroandry (e.g. Blue color chelate *Macrobrachium rosenbergii*). There are many hypotheses to explain sexual dimorphism. The standard explanation for the macrogyny is that large females produce large broods/eggs and gamete production in smaller males is not constrained by size. The size advantage model (Ghiselin, 1969) has successfully explained the adaptive significance of larger female size. Because males invest a little on their offspring relative to females, the parental investment theory predicts that males grow small (Trivers, 1972). In species, in which males compete for mates, territories and other ecological resources to attract females, the contest and/or competition favors large males (Andersson, 1994). Large aggressively dominant males may have greater access to many females. However, these hypotheses are inadequate by themselves, simply because the selective process is more general than patterns and fecundity increases with female body size in many species, in which males are large/larger than females (Harvey, 1990).

In crustaceans, the sex may be expressed as female, male, hermaphrodite or intersexual. Female and male sexes are readily distinguishable, for example, by (i) size, (ii) antennules, (iii) asymmetrical fifth pair of pleopods, (iv) number of urosomes, (iv) caudal furca, (vi) genital complex and (vii) tegumental glands (e.g. *Sinodiaptomus indicus*, Dharani and Altaff, 2002) in adult copepods. The presence of Primordial Germ Cells (PGCs), responsible for manifestation of sex, has been detected on the second day of embryonic development in *M. rosenbergii* (Damrongphol and Jaroensastaraks, 2001) and as Primary Gonial Cells in *Pandalus platyceros* (Hoffman, 1972). Detection and description of PGCs in other crustacean groups are needed urgently. The expression of PGCs or their derivative Oogonial Stem Cells (OSCs) alone, and PGCs or their derivative Spermatogonial Stem Cells (SSCs) alone in the gonads results in the manifestation of ovary or testis and thereby female or male sex. However, the simultaneous expression of PGCs/OSCs and PGCs/SSCs in ovotestis/gonad manifests hermaphrodite (see Pandian, 2012, p 109).

3.3.1 Sex Ratio

That a single male can inseminate many females has pushed female ratio up to such an extent that male ratio is < 0.07 in deep sea copepods like *Stenhelia* (Pandian, 1994, p 78). Other than spatial and temporal variations, the following life history features that alter sex ratio are: (i) parthenogenesis, (ii) sequential hermaphroditism, (iii) parasitic castration and (iv) androdioecy. In protandrics, male ratio progressively decreases with corresponding increase in female ratio, as size/age increases. The reverse is true in protogynics. For example, the ratio is 0.67 ♀ : 0.33 ♂ in protogynic tanaid *Kalliapseudes schubartii* (Pennafirme and Soares-Gomes, 2009). In facultative parthenogenics, males are generated only, when environmental conditions become unfavorable. In them, the male ratio is as low as 0.01 (e.g. *Pleuroxus hamulatus*, Pandian, 1994, p 78). The ratio is zero in obligate parthenogenics *Daphnia middendorffiana* and *D. tenebrosa*, which are endemic to the Arctic. The ratio can be high in a few cirripedes, whose female holds a 'harem' of one-13 'dwarf' males. Parasitic castration may significantly alter sex ratio. Interestingly, reproduction is restricted to a single female 'the queen' in many colonial species of the Atlantic alpheid *Synalpheus* displaying a complex social organization (Duffy, 1996, 2007).

In view of their commercial importance, sex ratio of a few shrimps from capture fisheries are reported (Table 3.1). Female ratio ranges from 0.32 in *Potimirim brasiliana* to 0.99 in *Argis lar*. Sex ratio is skewed in favor of female in *Processa bermudensis* and *Exopalaemon carinicauda*, but in favor of males in *P. brasiliana* and *Latreutus fucorum*. Notably, these ratios are closer to 0.66 ♀ : 0.34 ♂ or 0.33 ♀ : 0.67 ♂, which may have an implication in sex determination. Male ratio is high in *P. brasiliana*, despite the fact that there were no males beyond the size of 17 mm CL; it is not clear whether it is due to small sample size. No male was found beyond a particular size in shrimps, whose sex ratio is skewed in favor of females. There were only 40 specimens of male out of 4,172 *Argis lar*. It is likely that the males die soon after fertilizing the female. In meiobenthic copepod *Asellopsis intermedia* too, males die in good numbers immediately following copulation (Pandian, 1994, p 88-89).

TABLE 3.1

Sex ratio of shrimps from capture fisheries

Species/Reference	Specimens collected (No.)	Sex Ratio ♀ : ♂	Observations
Potimirim brasiliana Rocha et al. (2013)	821	0.32 : 0.68	No ♂ beyond 17 mm (CL) but ♀ up to 24 mm
Latreutus fucorum Martinez-Mayen and Roman-Contreras (2011)	1092	0.39 : 0.61	–
Processa bermudensis Martinez-Mayen and Roman- Contreras (2013)	2720	0.60 : 0.40	–
Exopalaemon carinicauda Oh and Kim (2008)	~ 4000	0.68 : 0.32	No ♂ beyond 18 mm (CL) but ♀ up to 21 mm
Argis lar Seo et al. (2012)	4172	0.99 : 0.01	No ♂ beyond 24 mm (CL) but ♀ up to 36.5 mm

3.3.2   Spawning and Fecundity

Spawning causes an enormous drain of energy and nutrient resources from females. As a result, interspawing intervals range from a few hours (24 h) in *Moina micrura*, brooding ~ 5 eggs (Murugan, 1975a, b) through a few days (19 d) in *Macrobrachium nobilii* carrying a few hundred eggs (Pandian and Balasundaram, 1982) to a few months (6-12 m) in *Panulirus longiceps cygnus*, producing 500,000 eggs (Chittleborough, 1976). In general, fecundity of many crustaceans is related to body size. In view of diversity of their morphology, a host of parameters has been used to represent mother's size: (i) body length: Amphipoda (Sheader, 1978), (ii) prosoma length: Copepoda (Hopkins, 1977), (iii) carapace width (CW): crabs, e.g. *Callinectes sapidus* (Graham et al., 2012), (iv) carapace length (CL): shrimps: *Potimirim brasiliana*, (Rocha et al., 2013), (v) weight: Cirripedia (Barnes and Barnes, 1968) and (vi) volume: Amphipoda: (Sutcliffe, 2010), Decapoda: *M. lamarrei* (Katre, 1977), *Palaemon longirostris* (Beguer et al., 2010).

The total number of oocytes contributing to fecundity is assured by waves of oogonial proliferation and subsequent oocyte recruitment (see Pandian, 2013, p 24). With regard to fecundity, the terms used in fishery biology have to be introduced to crustacean biology: 1. Batch Fecundity (BF) or a clutch is the number of eggs produced per spawning. It is a function of body size (L). Because BF is related to the volume of space available in hemocoelomic cavity to accommodate ripe bilateral ovaries, ovotestes or median ovary ($F = aL^b$), geometry suggests that the length exponent would be 3.0. Unlike freely broadcasting oviparous fishes, > 96% of crustaceans brood their eggs on one or the other site of their body. Further, spawning is preceded by a molt in diecdysic crustaceans. Though these features complicate BF relation to body size, BF increases with increasing body size in several crustaceans (Fig. 3.4A, B).

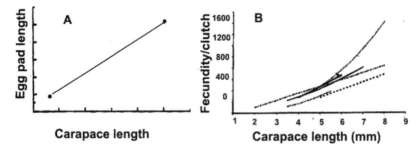

FIGURE 3.4
A. The relationship between egg pad length and carapace length in a representative decapod e.g. *Crangon uritai* (redrawn from Li et al., 2011). B. A representative figure to show the shifted trends of fecundity-body length relation in different populations of a decapod e.g. *Potimirim brasilina* (redrawn from Rocha et al., 2013).

Secondly, Potential Fecundity (PF) is the maximum number of oocytes commencing to differentiate and develop. However, due to one or other environmental factor like food supply, a fraction of these developing oocytes may be resorbed through atresia. Munuswamy and Subramoniam (1985a) noted that unfertilized oocytes of *Streptocephalus dichotomus* are resorbed (see also Murugan et al., 1996). Describing the ovarian development of the red shrimp *Aristeus antennatus*, Carbonell et al. (2006) reported that 10% of oocytes in a batch of oogenesis remain undeveloped at the resting stage; hence, the realized fecundity is 90% only. The realized fecundity decreased from the 'PF' of 3,581 eggs in *M. nobilii* fed *ad libitum* to 2,033 eggs in the prawn receiving 60% ration (Kumari and Pandian, 1991). Data reported by CMFRI (2013) indicate that the oocyte atresia decreased from 10% at sexual maturity to 1.3% in large females of *Metapenaeus dobsoni*. Incidentally, Robey and Groeneveld (2014) have used the terms 'potential' and 'effective' fecundities to estimate the loss of brooded embryos. Egg brooding and loss of brooded egg are unique features of crustaceans. Tropea et al. (2012) have used the term 'actual fecundity' to denote the realized fecundity of the crayfish *Cherax quadricarinatus*. Relative Fecundity (RF) is the number of oocytes, eggs or neonates/unit body weight of a crustacean mother. Unlike BF, the RF provides scope for comparative analyses of reproductive potential/reproductive effort at intra- and inter-species levels. Fecundity of a species or population may vary temporally and spatially (Fig. 3.4 B).

Thirdly, determinate fecundity means the presence of synchronized development of a fixed number of oocytes. Egg brooding crustaceans may be characterized by determinate fecundity. Indeterminate fecundity means the presence of unsynchronized, unlimited number of developing oocytes. The protogynic tanaid *Kalliapseudes schubartii* simultaneously broods from different embryonic stages, indicating that it has indeterminate fecundity (Pennafirme and Soares-Gomes, 2009). It is likely that all egg shedding crustaceans possess indeterminate fecundity. However, it has to be verified. Fourthly, hydration of fully developed oocytes after ovulation but prior to spawning is an important event in pelagic spawning fishes (Pandian, 2014, p 32-33). In egg shedding crustaceans, the number of spawning pulses and interspawning intervals are known (see Table 2.5). However, hydration of egg shedding crustaceans occurs after spawning (Fig. 2.12). Incidentally, the term hydration should not be confused with wetting of the cyst in natural habitats (e.g. Weeks et al., 2014a).

Fecundity is decisively an important factor in recruitment. Considering decapods, its relation to body size or a related parameter is described. It is primarily determined by the available space within hemocoelom to accommodate the ripening oocytes in the ovaries and secondarily by the space available on the abdominal appendages to brood the embryos. Unlike pelagic spawning fishes, hydration is likely to occur after spawning in penaeid eggs (cf Fig. 2.12, 6.3). Expectedly, a perfect linear relationship is found between egg pad length and CL in *Crangon uritai* (Fig. 3.4A). Hence, size at sexual maturity may serve as an index of fecundity. For example, fecundity increases from 34,500 to 75,000 and to 2,48,000 eggs in *Metapenaeus dobsoni*, *Penaeus semisulcatus* and *P. monodon*, which attain sexual maturity at the size of 6.4, 13.0 and 19.6 cm Body Length (BL), respectively (Table 3.2). Within a species, it increases from ~ 200 eggs/clutch to 780 eggs/clutch with increasing CL from 3.82 to 6.0 mm in *P. brasiliana* (Rocha et al., 2013). In fact, a similar linear relationship is reported for the protogynic isopod *Gnorimosphearoma luteum* (Brook et al., 1994) and protandric simultaneous hermaphrodite *Lysmata boggessi* (Baeza et al. 2014), and hermaphroditic cirripede *Scalpellum scalpellum*, in which fecundity increases from 100 to 250 eggs with increasing body size (Buhl-Mortensen and Hoeg, 2013). In a rare anomuran crab *Pachycheles monilifer*, a symbiont on bryozoa or polychaete, the linear relationships are also apparent (Leone and Maltelatto, 2015). However, the first factor that alters the level and angle of the linear relationship within a species is different populations. Figure 3.4 B represents an example of it. The second factor that may alter the level of this linear relationship is related to egg shedders and anecdysics (Fig. 3.5A, B). This is also true of crabs. For example, the diecdysic grapsids spawn a few hundred eggs/clutch (see Hartnoll, 2009). But fecundity increases from 58,600 to 556,000 eggs/clutch with increasing size from 77 to 140 mm (CW) in an anecdysic crab *Portunus sanguinolentus* (Yang et al., 2014).

TABLE 3.2

Estimated life span, reproductive size range and fecundity of Indian penaeid shrimps of commercial value (compiled from CMFRI, 2013)

Species	Life Span (m)	Reproductive Size (cm, BL)	Fecundity (Eggs/Spawning)
M. dobsoni	36	6.4-12.5	34,500-1,59,600
M. monoceros	>18	6.4-19.5	52,000-5,48,000
M. affinis	36	19.0	36,000-1,59,000
P. stylifera	12-24	7.5-13.5	40,000-2,36,000
P. semisulcatus	>18	13.0-26.3	75,000-2,83,000
P. indicus	>24	13.0-22.3	23,400-2,268,000
P. monodon	18-36	19.6-31.5	2,48,000-8,11,000

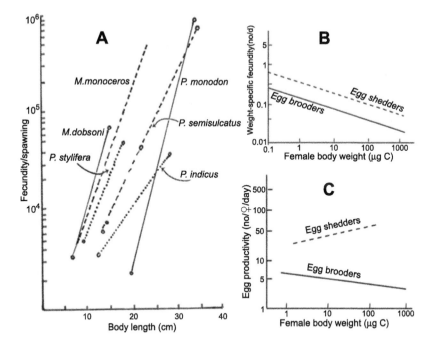

FIGURE 3.5
A. Fecundity as function of body length in selected Indian penaeid prawns (drawn using data of CMFRI, 2013), B. Decreasing trends of fecundity (no/d) as a function of body weight of egg shedding and brooding copepods. C. Egg productivity (no/♀/d)- body weight relation reported for egg shedding and brooding copepods (from Kiorboe and Sabatini, 1995, permission by International Ecology Institute).

Though it may be significant, the linear relationship between body size and fecundity does not explain all the variables (see also Warburg, 2013). For example, the variables explained by this relationship ranges from 24% in *C. sapidus* (Graham et al., 2012) to 60% in *Metanephrops mozambicus* (Robey and Groeneveld, 2014). In fact, fecundity varied so widely within each size class (e.g. Fig. 3.8D) that a positive relation could be found, only when mean fecundity of each size class was considered in *Processa bermudensis* and *Latreustes fucorum* (Martinez-Mayen and Roman-Contreras, 2011, 2013). In the anostracans *Streptocephalus torvicornis* (Beladjel et al., 2003a) and *Tanymastigites perrieri* (Beladjal et al., 2003b), no correlation between body length and clutch size was found. Fecundity (in number) was better correlated with body weight of *P. longirostris* (Beguer et al., 2010). Further, the linear relationship became flat, when fecundity in dry weight was related to CL in *Exopalaemon carinicauda* (Oh and Kim, 2008) and *Argis lar* (Seo et al., 2012). A small but subtle difference between relative fecundity and weight-specific fecundity/reproductive output is to be noted. The former refers to the number of eggs/clutch but the latter to weight of eggs/clutch in relation to size (e.g. Oh and Kim, 2008) or weight of female (Kiorboe and Sabatini, 1995). In protandric simultaneous hermaphrodite *L. boggessi*, fecundity increased significantly with body length and body weight but decreased with its body volume (Baeza et al., 2014); this was also true in the snapping shrimp *Alpheus estuariensis* (Costa-Souza et al., 2014). Considered on log scale, the weight specific fecundity, i.e. fecundity weight per unit weight of female per day linearly decreased as a function of the female's body weight in many egg brooding but increased in egg shedding anecdysic copepods (Fig. 3.5C).

All these differences in fecundity-body size relationship can be traced to one or another of the following: 1. The difference between space available in hemocoelom and the brooding site. CMFRI (2013, p 326) is perhaps the only literature source that provides adequate information for this analysis in the egg shedding penaeid *Metapenaeus monoceros*. Accordingly, the increase in number was 4,959, 9,713 and 96,941 ova for every mm increase in body length (BL), g body weight and g ovary weight, respectively. Incidentally, the ovary weight increased fairly at the constant rate of 9.7 mg/g ovary in 10.2 g prawns to 10.3 mg/g ovary in 45.56 g prawns. Notably, the correlations obtained for all three trends namely (i) ova number vs body length, (ii) ova number vs body weight and (iii) ova number vs ovary weight, were all linear and significant. This implies that all the differences in the fecundity-body size relations are dependent on the egg brooding site only. Apparently, Oogonial Stem Cells (OSCs) in the egg shedding penaeids and possibly calanoids of all ages/sizes continue to undergo mitotic division more or less at the same rate to produce an increasing number of ova. Unfortunately, data like those available for penaeids are not available for egg brooding crustaceans. Hence, it is not clear whether the reported fecundity in older/larger brooder is due to the diminishing production of oogonia with increase in age and size and/ or due to egg loss due to incubation period. Secondly, contrasting trends emerge, when fecundity was considered by weight and not by number. The publication by Ikhwanuddin et al. (2012) seems to provide a partial explanation namely the embryonic stage at which the relation is estimated. As expected of the anecdysic crab, an unrecognizable correlation between fecundity-body size was noted in *Portunus pelagicus* at the commencement of embryogenesis. But the relation shifted to almost horizontal, when fecundity (in number) was correlated with body weight at embryonic stage 1. However, it finally became almost linear at the terminal end of brooding (Fig. 3.6). Understandably, the authors, who have estimated the relationship at different stages of embryonic development, have obtained either a linear or flat relationship. A vast majority of authors have chosen to estimate fecundity in number of eggs, which is an important measure required for recruitment analysis. A few have chosen to estimate fecundity in weight in an attempt to assess reproductive output. Only a very few copepodologists have gone for weight-specific fecundity estimates. To facilitate inter-specific and intra-specific comparisons, the estimate on relative fecundity in number of eggs/ body weight of mother is recommended. Estimates with these parameters provide information on the number of eggs for recruitment analysis and the RF (considering weight of the mother) facilitates both intra-specific and inter-specific comparisons.

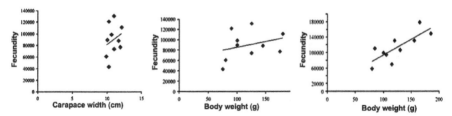

FIGURE 3.6
Shifts from an unrecognizable correlation from (left panel) fecundity-carapace relationship at embryonic stage 1 to that between fecundity and body weight at stage 2 and to the same at stage 3 in *Portunus pelagicus* (from Ikhwanuddin et al., 2012, permission by M. Ikhwanuddin).

Among diecdysic shrimps, egg carrying palaemonid shrimps may produce a few hundred to thousand eggs/clutch. But the egg shedding penaeids shed thousands to millions of eggs/spawning (Fig. 3.5A). Compiling relevant information for the anecdysic 16 species of egg brooders and 29 species of egg shedders among copepods, Kiorboe and Sabatini (1995) found that egg productivity was 5.3 and 40.0 eggs/female/d in egg brooders and egg shedders, respectively. In other words, there are 7.5-fold differences in the number of eggs produced per day and 2.5-fold differences in weight specific fecundity between egg shedders and egg brooders. This implies that an egg brooder produced ~ 3 times bigger eggs than an egg shedder. Hence, there is a clear tendency for increase in egg productivity with increasing body size, among egg shedders, while there is a decreasing one among egg brooders (Fig. 3.5C). Clearly, the egg brooders are trapped in reproductive senescence, while the egg shedders escaped from it.

Further, the fecundity-size relation may also be influenced by egg loss, egg size, food availability, and so on. The egg loss ranges from 5.6% in *C. uritai* (Li et al., 2011) to 54% (see Table 2.10). An important factor is the egg size. In general, a small brood consists of larger eggs and large ones of smaller broods. The relation between body size and egg size ranges from non-linear correlation in the lobster *Metanephrops mozambicus* (Robey and Groeneveld, 2014), estuarine crab, *Chasmagnathus granulata* (Gimenz and Anger, 2001), caridean shrimp *P. longirostris* (Beguer et al., 2010) and *L. boggesii* (Baeza et al., 2014) to the one, in which egg diameter decreases with increasing size in *C. sapidus* (Graham et al., 2012). Gimenz and Anger (2001) reported significant variability in egg size among the broods of *C. granulata* reared under identical conditions. Mothers do not control the egg quality and zoea size (*C. sapidus*, Koopman and Siders, 2013) and zoea size in *Cherax quadricarinatus* (Tropea et al., 2012). In fact, no recognizable correlation was apparent between mother size on one hand and energy content as well as fatty acid contents of zoea, on the other hand.

3.3.3 Reproductive Senescence

Uniquely, crustaceans have to share the available assimilated energy between (i) growth including molting, (ii) oogenesis including mate searching and (iii) brooding embryos (Fig. 3.7). Understandably, the intense competition for resources between these components during the checkered history of evolution has eliminated one or more of these components. The energy cost of mate searching, courting and so on is one more reason (besides motility, see Chapter 1) that may have driven some crustaceans to opt for parthenogenesis or hermaphroditism. The diecdysic penaeids and some euphausiids have eliminated the expensive brooding component. On attaining sexual maturity, the anecdysic chydorids (males alone, see Dodson and Frey, 2009), copepods and most brachyuran crabs have opted to forego somatic growth bypassing through a non-molting adult phase. As a climax, the calanoid copepods have eliminated both the growth and brooding components to have all the available resource solely for egg production (Fig. 3.7). Of the estimated 11,500 copepod species, belonging to 200 families and 1,650 genera, the calanoids comprise ~ 1,800 species that are wide spread in coastal waters and constitute one of the most important taxa among primary consumers and serve as food for large crustaceans and fish larvae (Kusk and Wollenberger, 2007).

The intense competition by the three components for resource has led some crustaceans to opt to parthenogenesis and some others to hermaphroditism. But gonochorism seems to have been trapped by senescence. It must be noted that parthenogenogenesis reduces genetic diversity and selfing in hermaphroditics results in inbreeding depressions (e.g. spinicaudates, Weeks et al., 2006a). In this analysis, reproductive senescence is considered using the following indices: 1. Post-reproductive sterile females and males and 2. Decreasing fraction of ovigerous females in populations. Whereas these indices may only indicate the possible presence of reproductive senescence, it is the number of Oogonial Stem Cells (OSCs) per unit of ovary, which is the correct pointer of the senescence.

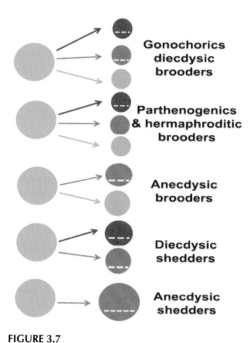

FIGURE 3.7

Resource ● partition between somatic growth inclusive of molting ●, gametogenesis inclusive of mate searching ● and egg brooding ● in crustaceans. Note the elimination on the costs of (i) mate searching in parthenogenics and hermaphrodites ●, (ii) growth and molt in anecdysics ● like copepods and crabs and (iii) egg brooding in diecdysic egg shedders ● like penaeids and (iv) molting as well as egg brooding ● ● in calanoid copepods.

(i) Post-reproductive Period Beladjal et al. (2003b) were perhaps the first to estimate the senescence period in an anostracan *Tamymastigites perrieri*. However, many authors, who have reported fairly useful information on post-reproductive period in crustaceans, have not glanced at their data from the point of reproductive senescence. Available evidence indicates the almost ubiquitous existence of a shorter or longer period of reproductive senescence across crustaceans (Table 3.3). The duration ranges from 9% in diecdysic *Corophium insidiosum* reared at 15°C to 24% of adult life span in *Jassa falcata* at 20°C. Hence, temperature plays an important role in limiting or extending the duration of reproductive senescence. With the presence of spermatheca in these amphipods, the need for insemination following every molt may not be obligatory. With no facility to store sperm in the anostracan *Streptocephalus torvicornis*, the post-reproductive males were sustained over 63% of adult life span, which is difficult to comprehend. In anecdysic copepod, *Euterpina acutifrons*, another factor namely food availability plays a role in limiting or extending the duration of post-reproductive period. Algal density of 6.10^3/ml seems to be the optimum, when the period is 28% of adult life span; at the two extremes of starvation and maximal food density of 6.10^5/ml, the

Reproduction and Development in Crustacea

periods are extended to > 50% of adult life span. Notable is the presence of (~ 5%) post-reproductive senescence in Marmokrebs (see also Table 3.8), which is claimed as a model that does not suffer senescence (Vogt, 2010).

TABLE 3.3

Estimated post-reproductive period in crustaceans

Species/References		Estimated Post Reproductive Period as (%) Adult Life Span	Remarks
From Cultured Crustaceans			
Anostraca *Streptocephalus torvicornis* Beladjal et al. (2003a)		63	Post-reproductive period for males lasted maximally for 206 d
Branchinecta orientalis, Atashbar et al. (2012)	12°C 27°C	2 7	
Copepod *Euterpina acutifrons* Zurlini et al. (1978)	20°C	28-57	Post reproductive period accounted for 54, 28, 57 and 48% of its adult life span when fed at algal density (no./ml) of 6.10^2, 6.10^3, 6.10^4 and 6.10^5, respectively
Eucyclops serrulatus Nandini and Sarma (2007)		40-60	Post reproductive life span accounted for 40 and 60% in *Scenedemus* and *Scenedemus + Brachionus*-fed copepods, respectively
Amphipoda *Corophium insidiosum* Nair and Anger (1979a)	10°C 15°C 20°C	14 9 16	For female
Jassa falcata Nair and Anger (1979b)	10°C 15°C 20°C	10 7 24	For female
Decapoda *Procambarus fallax*		5	Parthenogenic
From Field Observations			
Copepoda *Asellopsis intermedia* Lasker et al. (1970)			Following copulation, males died and became scarce, whereas brooded females survived even 7 m after insemination
Decapoda *Exopalaemon carinicata* Oh and Kim (2008)			Males were absent beyond 18 mm (CL) size but females continued to appear up to 22 mm size
Argis lar Seo et al. (2013)			Males were absent beyond 25 mm size, but females survived up to 35.5 mm size
Potimirim brasiliana Rocha et al. (2013)			Ovigerous females appeared up to 23.5 mm size but non-ovigerous females and males were absent beyond 21 and 17 mm sizes, respectively.

Field collected data of a few shrimps and a copepod indicate the disappearance of males immediately after copulation (e.g. *Asellopsis intermedia, Argis lar,* Table 3.3). A single spermatophore suffices the fertilization of all the eggs spawned in three pulses of *Cherax quatricarinatus* (Peixoto et al., 2011). Notably, the shrimps breed once in a season. However, there are others like the penaeids, which are multiple spawners in a season. More research is required to assess the post-reproductive period of shrimps in the context of the number of inseminations required in single or multiple spawners within a breeding season.

(ii) Ovigerous Female In crustaceans, the fraction of ovigerous females varies very widely from 26% *Sessarma ricordi* to 78% in *Grapsus adscensionis* (Hartnoll, 2009). Irrespective their respective peaks, the fraction of ovigerous females decreased from 13% at 8.5 mm (CL) to 1% at 12.5 mm in *Alpheus estuariensis* (Costa-Souza et al., 2014), from 20% at 16 mm (CL) to 2% at 20 mm in *Potimirim brasiliana* (Roche et al., 2013), 24% at 2.4 mm to 2% at 3.8 mm in *Latreutes fucorum* (Martinez-Mayen and Roman-Contreras, 2011), 42% at 3.8 mm to 2% at 8.4 mm in *Processa bermudensis* (Martinez-Mayen and Roman-Contreras, 2013) and 12% at 21.5 cm to 2% at 35.5 mm in *A. lar* (Seo et al., 2012). This may indicate that the large ovigerous females are either predated or unable to produce eggs.

(iii) Fecundity With increasing body size and/or advancing age, not only does the fraction of ovigerous females decrease but also the fecundity decreases. In many branchiopods, fecundity decreases beyond the mid adult age (Fig. 3.8A) The same holds true for tropical, temperate and Arctic cladocerans (Fig. 2.1B, 3.14A) and cyclopoid copepod *Eucyclops serrulatus* fed on *Scenedesmus* or *Scenedesmus + Brachionus* (Nandini and Sarma, 2007). It is not clear whether senescence is induced in those reared in laboratory. In the cirripede *Bendria purpurea* too, the reproductive senescence is clearly apparent (Fig. 3.8F). It also holds true of the diecdysic decapods *Macrobrachium totonacum* (Mejia-Ortiz et al., 2010). Reared at two different densities, the egg shedding *Parvocalanus crassirostris* also shows the trends indicating that the increasing density imposed the onset and progress of reproductive senescence (Fig. 3.8E). The decrease in fecundity per female in the high density culture indicated that nine times greater accumulation of metabolic wastes and excretory products (as water was changed only once in both the densities) reduced fecundity by reducing fecundity per clutch and/or extending interspawning interval. Walker's (1979) data indicate that water quality especially from aged conditioned water extends the interspawning intervals. The interval in *Pseudocalanus minutus* was extended from 1.3 d in fresh medium to 3.4 d in conditioned medium. It is not clear whether the accumulated wastes decreased fecundity also. Briefly, the stress due to brooding and/or excretory waste imposes reproductive senescence. Hence, the trends obtained for the egg brooding copepods (Fig. 3.5E, Fig. 8.2A), euphausiid (Fig. 3.8D, Fig. 8.2B) and balanoid (Fig. 3.8F) and palaemonid (Fig. 8.2C) clearly show

that the egg brooding crustaceans collected from the field do undergo reproductive senescence with advancing age or increasing body size.

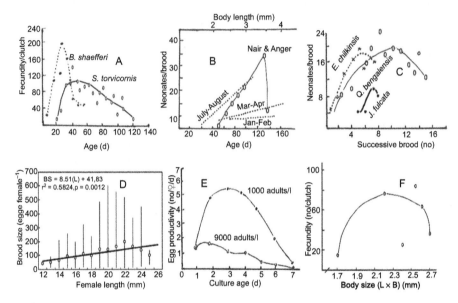

FIGURE 3.8
A. Fecundity as a function of age in *Branchipus shaefferi* and *Streptcephalus torvicornis* (drawn from data of Beladjal et al., 2003a), B. Fecundity as a function of season in *Corophium insidiosum* (redrawn from Shaeder, 1978), over which the data reported for fecundity as a function of age of *C. insidiosum* reared at 15°C by Nair and Anger (1979a) are superimposed, C. Clutch size as a function of successive broods of F_1 and F_9 in *Quadrivisio bengalensis* reared at 20°C (K. C. C. Nair and K. V. Jayalakshmy, pers com) and *Eriopisa chilkensis* reared at ~ 28°C (Aravind et al., 2007), *Jassa falcata* reared at 15°C (Nair and Anger, 1979b), D. Brood size as function of body length in *Euphausia pacifica* (from Gomez-Gutierrez et al., 2006, permission by International Ecology Institute) E. Egg productivity as a function of culture age of *Parvocalanus crassirostris* (redrawn from Alajmi and Zeng, 2014). F. Fecundity as a function of body size in sessile cirripede *Bendria purpurea* (drawn from data reported by Utinomi, 1961).

In the amphipod *Corophium insidiosum*, linear (increases in elevation and slope) relationships between fecundity and body size have been reported by Shaeder (1978, see also for the same in other amphipods, Sutcliffe, 2010). However, data reported by Nair and Anger (1979a) for *C. insidiosum* reared at 15°C show a trend like those obtained for the other crustaceans (Fig. 3.8B). In fact, data of K. K. C. Nair and J. V. Jayalakshmy (pers. comm.), who have reared the tropical amphipod *Quadrivisio bengalensis* at 20°C up to 14 generations, also display similar trends confirming the reproductive senescence (Fig. 3.8B). It holds true in another tropical amphipod *Eriopisa chilkensis* (Aravind et al., 2007) (Fig. 3.8C) as well as in another temperate amphipod *Jassa falcata* (Nair and Anger, 1979b).

Notably, Wilber (1989) reported rare data for the sperm count relation to body size in the stone crab *Menippe* sp. Following an initial increase up to mid-body size, the sperm count remained at the same level (Fig. 3.9). Apparently, male crustaceans may not enter into reproductive senescence but may rather prefer death, as in *Asellopsis intermedia*.

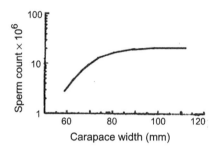

FIGURE 3.9
Relationship between sperm count and body size in the stone crab *Menippe* sp (redrawn from Wilber, 1989, permission by International Ecology Institute).

The described analysis confirms the occurrence of reproductive senescence and entry of older/larger females into menopause. The decreasing egg production in older/larger crustaceans suggests the minimization or cessation of Oogonial Stem Cells (OSCs) production in older crustaceans, as much as it occurs in fishes (Fig. 3.10). These findings may have a couple of implications: 1. Broodstock maintenance in aquaculture and 2. Maintenance of OSCs: In the context of maintenance of OSCs, a classical observation by Kaczmarczyk and Koop (2011) in *Drosophila melanogaster* has relevance. In this fruitfly, there are early reproducing (S) and late reproducing (L) strains. Egg laying begins in S strain early in the first w itself but ceases also early during the 7th w. On the other hand, it commences late by the 7th w but continues beyond 9th w in the L strain. On the whole, the life time fecundity of the S strain is less than a tenth of the L strain. Both of them commence with 2.5 OSCs/ovariole. But by the 9th w, the S strain has only 0.1 OSC/ovariole, in comparison to 1.94 OSCs/ovariole in L strain. Kaczmarczyk and Koop have traced the causes for the efficient maintenance of OSCs in the L strain to a few genes. In the signaling pathway of *Dpp*, a member of TGF-β morphogens, plays a central role in maintenance of OSCs. The mutant OSCs, that lack *Dpp* receptor or transcriptional effector, undergo precocious differentiation (Xie and Spralding, 2000). A second TGF-β homolog , *Gbb* also contributes to the maintenance of OSCs. Contrastingly, the cell-autonomous *bag-of-marbles* (*bam*) gene acts on OSCs to promote their differentiation. *Dpp* signaling acts to prevent *bam* expression on the OSCs. The activity of *Dpp* and *Gbb* signals is known to decline with age (Zhao et al., 2008). Apparently, mutation or non-expression of one or other of these genes may also cause early fertility and reduce the ability to maintain the OSCs, as age advances or body size increases (see Pandian, 2012, p 87).

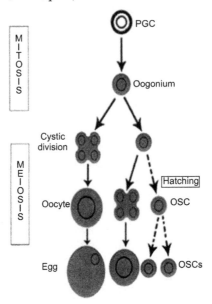

FIGURE 3.10
Oogenesis in animals. Note the generation of oogonial stem cells during adult stage (from Pandian, 2012, permission by CRC Press).

A similar pattern of egg laying occurs in some crustaceans. For example, high fatty acid containing alga-fed and low fatty acid containing alga-fed *Bosmina freyi* are similar to S and L strains of *D. melanogaster* (see Fig. 2.1D). Similarly, *Scenedesmus*-fed and *Scenedesmus* + *Brachionus*-fed *E. serrulatus* (see Fig. 3.14B) may also be equal to S and L strains of *D. melanogaster.* With their amenability to culture and short life span, the parthenogenic cladocerans and gonochoric copepods may serve as ideal crustaceans to locate and describe the role of *Dpp* and *Gbb* or their equivalent in maintenance of OSCs in crustaceans and to possibly postpone the onset of reproductive senescence.

3.3.4 Special Groups

Commercially important prawns and crabs as well as culture of 'fodder' copepods are described here. The penaeid shrimps comprise of 400 species. Being globally distributed from shallow waters to abyssal zones (up to 5,000 m depth), this group comprises of high economic value species and accounts for a third of crustacean fisheries. Incidentally, those, who are interested in management of capture prawn fisheries, may consult the excellent handbook by CMFRI (2013). India is an important country not only for capture but also culture of prawns. Table 3.2 summarizes the available information on the estimated life span, reproductive size range and fecundity of selected Indian prawns. Many of them live for 2-3 y, grow quickly to attain sexual maturity and are multiple-spawners during the peak breeding season. During recent years, Mexico has made many advances, especially in genetics of multiple spawners. The available information is summarized below: 1. Multiple spawning penaeids generate a large number of offspring, which become spawners themselves (Wyban and Sweeney, 1991). 2. Using quantitative genetic approach, the multiple spawning capacity is found to be an inheritable trait in *Penaeus vannamei* (Ibarra et al., 2005). 3. Heritability of many traits have been determined in *P. vannamei*. They are the (i) number of spawnings (0.54), (ii) egg vitellin (0.47), (iii) egg acylglycerides (0.34) in adults and (iv) egg protein (0.18), (v) fecundity (0.17), (vi) egg diameter (0.07), (vii) oocyte diameter (0.57) and (viii) oocyte maturity (0.71) in sub-adults (Arcos et al., 2003, 2004). 4. The phenotypic 'predictor trait' are defined by the levels of vitellin (VIT) and acylglycerides (Ibarra et al., 2005). 5. Partial or complete transcribed sequences (cDNA) of vitellogenin (*Vg*) gene have been isolated and characterized in *P. vannamei*, *P. japonicus*, *P. merguiensis*, *P. monodon* and *P. semisulcatus* through reverse transcription of the expressed (mRNA) *Vg* gene (s). The length of the complete cDNA in these penaeids is consistent between 7,898 and 8,012 bp (see Ibarra et al., 2007). 6. The analysis of these sequences has revealed the presence of one *Vg* gene in both ovary and hepatopancreas. An increasing VTG-mRNA during the gonadal development in ovarian and hepatopancreatic vitellogenin has been observed in *P. japonicus* and *P. semisulcatus*. These organs are involved in transcription of *Vg* gene and synthesis of VTG during penaeid maturation (see Ibarra et al., 2007). 7. Differential expression of *VTG*-mRNA in ovary and hepatopancreas suggests that they are the primary and secondary sites of vitellogenin synthesis and are required to complete vitellogenesis. Hence, VTG protein level in hemolymph may serve as an indicator of the capacity for multiple spawning.

Apart from these, research in Lebanon recommends a feed containing 42% protein and 16 mg protein/kj for the cray fish *Cherax quadricarinatus* (Ghanawi and Saoud, 2012). The Brazilians have found that a single spermatophore suffices to fertilize eggs of three successive spawns in *C. quadricarinatus*. The spermatophore can be extruded manually by gently pressing around the coxae of the fifth pair of pereiopods (Peixoto et al., 2011) or electrically using an electric pulse (9 v) applied on the same location (Nakayama et al., 2008).

In the xanthid crabs, sex ratio is biased towards male in *Menippe* sp. (Wilber, 1989). Following a fixed number of molts (e.g. ~ 25 in *Callinectes sapidus* (Graham et al., 2012), many anecdysic brachyuran crabs undertake terminal molt and become sexually mature. In them, growth is determinate.

A factor like temperature or food availability that accelerates molting frequency decreases intermolt duration and growth, as well. Hence, sexually mature crabs may be of different sizes as measured by Carapace Width (CW) or weight. Consequently, the fraction of ovigerous females in a population increases from 57% in 'small' (20-35 g) to 73% in 'large' (50-70 g) females. At the soft stage immediately following terminal molt, a female may be inseminated by more than one male and the sperm stored in the seminal receptacles are used to fertilize eggs up to 18 spawnings, six each/spawning seasons during its life time; in all *C. sapidus* may have three spawning seasons (Graham et al., 2012, Koopman and Siders, 2013). The bill box crab *Limnopilos nayanetri* stores sperm received during the first copulation in the proximate receptacle but those of subsequent copulations in distal dorsal bursa (Klaus et al., 2014). Contrastingly, each spawning is preceded by a molt in the tropical marine diecdysic grapsid crabs (e.g. 24 adult molts in *Grapsus adscensionis*, Hartnoll, 2009), except in *Metopograpsus messor* (Sudha and Anilkumar, 1996). Exceptionally, with a long intermolt period lasting for ~ 300 d (anecdysic?), *M. messor* has 14 to 18 clutches/molt, in comparison to other grapsids with a single clutch/molt.

Culture of 'Fodder' Copepods A central part of the oocyte maturation is the vitellogenesis. In crustaceans, vitellogenesis occurs in two phases: Vitellogenesis 1 is a slow process and is dependent on the ovarian yolk source. Vitellogenesis 2 is characterized by rapid accumulation of the yolk, which is provided by an extra-ovarian source. In egg shedding calanoids, a batch of oocytes matures synchronously in ventral layer of the ovary and is all released during one spawning event. The speed of oocyte maturation and release of eggs is dependent on temperature and food supply.

In the tropical diecdysic cyclopoid copepod *Oithana rigida*, the number of clutch increased from 12 to 21 and absolute fecundity from 442 to 605 eggs with increasing body size (BL) from 0.75 to 0.84 mm (Santhanam and Perumal, 2012). However, a recalculation of the data revealed that batch fecundity decreased from ~ 37 eggs in smaller size classes to ~ 29 eggs in the largest size class, indicating the onset of reproductive senescence. When offered increasing mixed algal density from 1,000 to 20,000 cells/ml to *Nannocalanus minor*, egg productivity increased from three to 32 eggs/♀/d (Santhanam et al., 2015). Similarly, the productivity of *Paracalanus parvus* increased from eight to 32 eggs/♀/d with increasing temperature from 18°C to 30°C (Santhanam et al., 2013).

Using carbon as unit of feed, Niehoff (2004) elucidated the effect of food quality and quantity on maturation of oocytes and clutch size in the temperate copepod *Calanus finmarchicus*. On doubling the density of dinoflagellate *Heterocapsa* sp from 150 to 300 µg C/l, the fraction of ovigerous females increased from 50 to 80% , the clutch size from 25 to 57 eggs and egg productivity from 12-17 to 51 eggs/♀/d. These values were 40% for ovigerous females, 18 eggs/clutch and 9 eggs/♀/d, when diatom *Thalassiosira weissflogii* served as diet at the density of 150 μg C/l. Fed on *Heterocapsa* sp, the calanoid females were at the GS4 developmental stages of the ovary, in which young (OS1 and OS2) and maturing (OS3 and OS4) oocytes were also numerous. But the ovary had scarcely a few OS1 and OS2 and was packed loosely with OS3 and OS4 oocytes in those fed on *T. weissflogii*. The minimum structural weights of 8 μg C and 76 μg N were required to initiate reproduction.

For egg production, copepods require (n = 3) polyunsaturated essential fatty acids, which cannot presumably be synthesized by them (Fraser et al., 1989). Hence, chemical composition of food is a key factor in reproduction of decapods (Harrison, 1990). In an interesting publication, Lacoste et al. (2001) showed that the egg productivity of *Calanus helgolandicus* (20 eggs/♀/d) and hatching success (90%) were normal, when the calanoid was fed on 'efficient' pyremnesiophyte *Isochrysis galbana* (ISO) or dinoflagellate *Prorocentrum minimum* (PRO). However, with 'deficient' diatom *Phaeodactylum tricornutum* (PHA) or chlorophyte *Dunaliella tertiolecta* (DUN) served as diet, both egg productivity (< 2-1 egg/♀/d) and hatching success (0-20%) were decreased. The decrease in the productivity was reversible, when DUN was replaced by PRO. Apparently, gonadal development was blocked temporarily at GS3 stage but it became irreversible, when OS2 and OS3 oocytes reached a point of no return due to prolonged feeding with the 'deficient' DUN diet. Ornithine content of the eggs was abnormally high when ornithine metabolism and polyamine pathway were affected beyond recovery. These were followed by atresia and necrosis of oocytes.

To study the key biological parameters that influence egg productivity of *Parvocalanus crassirostris*, Alajmi and Zeng (2014) selected five densities at 1,000, 3,000, 5,000, 7,000 and 9,000 adults/l. Water (26°C) was changed once a d and the copepods were fed on *Isochrysis* and diatom *Chaetoceros mulleri* at 1 : 1 ratio. Egg productivity followed a steep increase to ~ 5 eggs/♀/d on the 3-4 d, but began to decline on the 5th d and reached 2 eggs/♀/d on the 7th d of culture at lower densities of 1,000 to 3,000-5,000 adults/l (Fig. 3.8E). Conversely, the productivity remained ~ 2 eggs/♀/d until the 4th d and subsequently dropped to almost zero on the 7th d at higher densities of 7,000 and 9,000 adults/l. Total egg production also increased from 5 eggs/♀ at higher densities to 20 eggs/♀ at lower densities. Hatching success also followed more or less similar trends.

Regarding the composition, the ratios were ~ 0.44, 0.26, 0.26, 0.04 for nauplii: copepodids: females: males at lower densities. But on the 15th d of culture, the ratio became 0.33 for nauplii, 0.16 for copepodids, 0.28 for females and as much as 0.23 for males. These ratios imply that the male ratio trebled from 0.15 at lower densities to 0.45 at higher densities. Alajmi and Zeng recommended stocking seven adults/ml. Of course, there are others, who have recommended stocking density of up to 12/adults/ml for *P. crassirostris*, which, being a calanoid, has all the assimilated energy for reproduction alone (see Fig. 3.7). These values may be compared with two adults/ml for the parthenogenic copepod *Elaphoidella grandidiera* (Nandini et al., 2011), which, being a diecdysic brooder, has to share the available energy among molting, growth and embryo brooding.

3.4 Parthenogenesis

Parthenogenic mode of reproduction is limited to about 0.1% of all species (Bell, 1982). Cyclic parthenogenesis arose within Branchiopoda during the Permian, when Cladocera evolved as a taxon (see Decaestecker et al., 2009). Reporting the discovery of four clones of *Procambarus clarkii*, a close relative of parthenogenic *P. fallax*, Yue et al. (2008) suggested its recent origin. Hence, it is likely that parthenogenesis in branchiopods is more ancient than in decapods and has originated independently in these crustaceans. Parthenogenic reproduction is considered as a ticket to a swift extinction (see Normark et al., 2003). However, *Darwinula stervensoni*, a freshwater ostracod with nearly world-wide distribution reproduces parthenogenically for at least 25 million years (m y). Despite accumulation of mutations, it has undergone a slow molecular evolution at half as fast as other sexually reproducing invertebrates (Schon et al., 1998).

3.4.1 Taxonomic Survey

Table 3.4 shows that parthenogenesis occurs predominantly in Anostraca, Cladocera and Ostracoda, sporadically in Copepoda and Isopoda and rarely in Decapoda. In the absence of males in limited samples (e.g. *Paraprotella saltatrix*), or from female biased sex ratio (e.g. *Corophium volutator*, Schneider et al., 1994), some amphipods were suspected to be parthenogenic. Similarly, the hermaphrodite *Triops cancriformis* was thought to be parthenogenic in the absence of males. In general, the taxons like Notostraca, Spinicaudata and Cirripedia, which employ hermaphroditism, may not be parthenogenic.

Geographic parthenogens occur in Anostraca, Cladocera, Ostracoda, Isopoda and perhaps Copepoda. It is characterized by the existence of parthenogenic and bisexual morphs within a single species and is generally accompanied by polyploidy. The two morphs have different geographic distributions. The parthenogenic triploid females (3n = 24) of *Trioniscus elisabethae coelebs* are found only in northern Europe. They produce unreduced clonal eggs and reproduce employing ameiotic parthenogenesis. Males do appear sporadically and may even mate with 3n females but no fertilization occurs. The second morph is bisexual and diploid (2n = 16) and is found in southern Europe. Where the two morphs are sympatric, as in France, there is absolute reproductive isolation between them.

TABLE 3.4

Occurrence of parthenogenesis in Crustacea

Taxon	Observations
Anostraca	Geographic: *Artemia parthenogenetica*
Cladocera	Obligate and facultative including geographic and cyclic
Ostracoda	Almost all Darwinuloids and a few Cytheroids and Cypridoids (Cohen amd Morin, 1990), *Eucypris virens*
Copepoda	Geographic (?): *Elaphoidella grandidieri*
Isopoda	Geographic: *Trioniscus elisabethae coelebs*
Amphipoda	*Paraprotella saltatrix*: males are absent but sample is limited to 70 specimens (Takeuchi and Guerra-Garcia, 2002), *Corophium volutator*: female-biased sex ratio
Decapoda	Obligate: *Procambarus fallax* Facultative: *Orconectes limosus*

Among anostracans, *Artemia parthenogenetica* is a renowned example for displaying parthenogenesis. Campos-Ramos et al. (2009) recorded the mixture of sexual and parthenogenic cysts in a commercial lot from the Great Salt Lake, Utah (USA). On rearing the nauplii, they found that about 65% was skewed to *A. parthenogenetica* at temperatures from 10°C to 32°C but above the threshold of 33°C, the hatching proportion of *A. parthenogenetica* decreased leaving mostly the sexual *A. franciscana* at temperatures between 33°C and 36°C (Fig. 3.11). Rearing six *Artemia* species each in six different Salinity-Temperature (S-T) combinations, Browne and Wanigasekera (2000) found that the gonochorics *A. persimilis* reproduce in all six S-T combinations, *A. salina* and *A. franciscana* in four, *A. sinica* in three but the parthenogenic *A. parthenogenetica* only in two combinations. Both the African field survey and experimental observations show that the parthenogenic species and populations of *Artemia* are characterized by low phenotypic plasticity. However, molecular evidence suggests that parthenogenesis in *Artemia* is relatively ancient with a single parthenogenic lineage branching from an Old World sexual ancestor of about 5 m y ago. Automictic recombination, that can occur in diploid but not in polyploids, appears to have played a key role in the long term maintenance of the parthenogenic lineage (Browne, 1992). The occurrence of triploid (India, Turkey, Browne et al., 1984) and tetraploid (Spain, Italy) and pentaploid (China) (Zhang and King, 1993) *A. parthenogenetica* is reported. The triploids from India and Turkey have larger broods and shorter generation time than the two diploid populations of Spain and France (Browne et al., 1984).

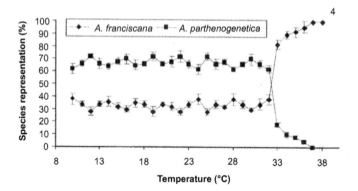

FIGURE 3.11
Hatching of nauplii of *Artemia franciscana* and *A. parthenogentica* at different temperatures but from the same commercial sample of Great Salt Lake (from Campos-Ramos et al., 2009) (pers. comm.)

Thanks to their calcified bivalved carapace, a rich and almost continuous fossil record of ostracods, since their origin 400 m y ago is available for investigations (Yamaguchi and Endo, 2003). A minimum of 17 of 600 freshwater ostracod species include sexual and parthenogenic populations (Chaplin et al., 1994). As a non-marine representative of the large group ostracods, *Eucypris virens* has attracted much attention (e.g. Adolfsson et al., 2009; Bode et al., 2010; Rossi and Menozzi, 2012). From these investigations, the following may be summarized: 1. Parthenogenic forms are considered to have originated from sexual ancestors without reversing, as it requires more evolutionary steps to re-acquire sexual reproduction than it does to lose it (Domes et al., 2007). The oldest parthenogenic lineages may have originated 10 m y ago. 2. Parthenogenic lineages are effective colonizers in short-lived erratic and/or predictably unstable habitats, whereas sexual lineages are better competitors in more permanent habitats with geological stability. 3. Parthenogenic forms are relatively more common, often more successful and more widely spread than the sexual forms. Reproductive performance of parthenogens is two-three times higher than their sexual relatives (see also Table 2.3). 4. Life span of males is shorter than females, as it undergoes the terminal molt sooner or die earlier than females.

5. Parthenogenesis, hybridization and polyploidy are an intertwined phenomenon that may contribute separately and together to the success of clonal taxa in new habitats. 6. Unlike in the isopods, hybridization between diploid parthenogenic and sexual gamete occurs more widely among the ostracods (Turgeon and Herbert, 1994, 1995). 7. Parthenogenic reproduction is more often coupled with elevation in ploidy. Both diploid and triploid parthenogenic lineages have originated multiple times from several sexual

lineages. For example, of 12 parthenogenic clusters of *E. virens* of Europe, 10 are apparently derived from independent origins of parthenogens. Only two of parthenogenic clusters do not have a triploid representative, indicating 10 independent transitions to polyploidy. Triploid, but not diploid parthenogenic clones dominate the habitats in northern Europe. The wide geographic distribution of triploids is more due to the elevated ploidy than to parthenogenesis. Ploidy elevation promotes range expansion of these lineages more effectively than parthenogenesis per se. 8. In *E. virens*, the mean number of alleles per locus does not differ between parthenogenic and sexual populations but on average heterozygosity is higher in the parthenogens, 9. Resting eggs have fundamental roles in processes of extinction and recolonization. Both parthenogenic and sexual females produce resting eggs. Investments on resting eggs do not seem to cost more in terms of fecundity for females with different reproductive modes (however cf p 36) and 10. In ostracod parthenogenics, reproduction is believed to be apomictic.

In copepods, three harpacticoids *Canthocamptus staphylinus* (Sarvala, 1979), *Epactophanes richardi* and *Elaphoidella grandidieri* (Nandini et al., 2011) are reported to be parthenogenics. The effect of food density on life span, ovigerous female and life time fecundity in *E. grandidieri* is described by Nandini et al. (2011).

The slough crayfish, also called Marmokrebs and marbled crayfish, *Procambarus fallax* is the first and perhaps only one of more than 10,000 decapods, reported to employ parthenogenesis. It attains sexual maturity at the age of 6-7 m and size of 4-8 mm BL excluding chelae and 1.5 to 15 g body weight (Vogt, 2011). During its reproductive adult life span from 7 m to 3 y, it spawns seven times (Vogt, 2011), each time bringing forth ~ 400 eggs (Vogt, 2010). Embryogenesis lasts for 17-28 d. To his credit, G. Vogt has made a standard procedure to confirm that *P. fallax* is indeed an obligate parthenogenic. 1. Testicular tissues, ovotestis or male gonoducts, gonopores or gonopods were never found either in juvenile or large adult specimens. 2. *Wolbachia* -like feminizing microorganisms were not found in the ovaries (Vogt et al., 2004). 3. Sequences of six microsatellite loci, three of them being heterozygous, revealed that all the 400 batch mates were genetically identical with each other and the mother (Vogt, 2011).

Despite the isogenicity among the batch mates (Martin et al., 2007), an analysis of nuclear microsatellite loci revealed a broad range of variations in body color, growth, life span, ages at successive spawning, behavior and number of sense organs, even when reared at identical conditions. As an example of these phenotypic variations, ages at successive spawnings by eight females are listed in Table 3.5. Thus Vogt et al. (2008) demonstrated that individual genotype can map to different phenotypes and thereby generate variability among the batch mates in *P. fallax*. Yet, it was not shown whether these phenotypic variations are inheritable.

TABLE 3.5

Life span and ages at successive spawnings in parthenogenic *Procambarus fallax* (compiled from Vogt et al. 2008)

Spawning (No. and Day)	B5	B1	B4	B3	B2	B5	B6	B7
Life span (d)	910	437	571	610	626	568	626	626
1st on d	315	157	183	168	160	328	531	531
2nd on d	394	267	390	392	369	507	–	–
3rd on d	507	375	516	502	497	–	–	–
4th on d	643	–	–	–	–	–	–	–
5th on d	850	–	–	–	–	–	–	–
Post-productive life span (%)	7	14	10	18	21	11	15	15

Notably, there has been a tendency for the members of Paracambaridae to switch to parthenogenesis. The spiny-check crayfish *Orconectes limosus*, a relative of *P. fallax* switched to facultative parthenogenesis, when females were physically separated from males over a period of more than 10 m (Buric et al., 2011). In another close relative *P. clarkii*, the occurrence of natural clones has been reported; four clones, heterozygous at most of the five microsatellite loci, each consisting of two-six genetically identical females were found (Yue et al. 2008). Deviations from bisexuality are documented at least in 12 of the 29 genera belonging to the families of Astacidae, Parascidae and Paracambaridae, to which *P. fallax* belongs (see Vogt, 2002). Martin and Scholtz (2012) have reported intersexuality in *P. fallax*.

3.4.2 Cladocera

With 620 species of Cladocera and 150 species rich *Daphnia* that radiated within these order and genus, respectively, the cladocerans are ubiquitous components of inland waters all around the world (Decaestecker et al., 2009). Cyclic parthenogenesis occurs in some lineages of cladoceran *Daphnia* complex and in a branchiopod *Cyclestheria* (Taylor et al., 1999). The life cycle of these crustaceans involves alternation between parthenogenic and sexual reproduction. Males are generated alternating with every one (polycyclic, e.g. *D. magna, Moina brachiata*) or two (dicyclic, the first one during early summer and a second long one in autumn, e.g. *Podon polyphemoides, Sida crystalline, D. galeata, Ceriodaphnia reticulata, Alona quadrangularis*). Lineages of monocyclic *Daphnia pulex*, for example, engage in a bout of sexual reproduction for every ≈ 5-20 parthenogenic generations per year (Haag et al., 2009). *D. cucullata*,

Moina micrura and Pleuroxus denticulatus are other examples for monocyclic species, which have a single autumnal phase of sexual reproduction. Unlike the cyclic parthenogenic female rotifers, the same *Daphnia* female produces sequentially ameiotic and meiotic eggs. Skipping the costly production and maintenance of males provides a short term evolutionary advantage over sexual reproduction for a larger (egg) gamete-producing parthenogenic female. The infrequent sexual reproduction in cyclic parthenogenesis seems to adequately purge the accumulated harmful mutations and decrease the extinction rate. Hence cyclic parthenogenesis confers additional long term advantages that neither the obligate parthenogenesis nor sexuality alone imparts. Hence the cyclic parthenogenesis makes 'the best of both modes of reproduction'. Not surprisingly 28 out of 30 temperate *Daphnia* species reproduce by cyclic parthenogenesis, in comparison to four out of six *Daphnia* species, that inhabit the Arctic, reproduce by obligate parthenogenesis (Colbourne et al., 1998). Notably, another cladoceran, *Holopedium* reproduces by automixis in the Arctic but by cyclic parthenogenesis in temperate zones (Herbert et al., 2007).

With the exception of some obligate parthenogenic lineages *Daphnia* switches to sexual reproduction, when triggered by high density (crowding, Fitzsimmons and Innes, 2006), diminishing food availability (LaMontagne and McCauley, 2001) and/or short day-length (Stross and Hill, 1969). Normally, a parthenogenic female produces subitaneous eggs that develop into parthenogenic daughters in the brood pouch. But the females can also be stimulated to produce two sexual (amphigenic) oocytes that undergo meiosis and must be fertilized to develop further (Lampert, 2011, see also later). The fertilized eggs move into the ephippium and cease development after a few cleavages. Some obligate parthenogenic lineages can also produce ephippia containing parthenogenic eggs alone (Fig. 3.12). In fact, Weider (1987) noted that tetraploids *D pulex* are less fecund with smaller broods than their sympatric diploids, suggesting that greater amount of yolk is accumulated in each 4n clonal egg. The complicated reproductive modes and life cycles of the obligate and cyclic parthenogens as well as sexuals provide ample opportunities for *Daphnia* to survive and flourish in a wide variety of limnetic ecosystems.

Incidentally, it is known that with the expression of meiosis inhibitor gene (see Decaestecker et al., 2009), the sex chromosomes WZ (shown within a box Fig. 3.12) in parthenogenic cladocerans do not undergo regular meiosis I. However, a single polar body is formed corresponding to meiosis II (see Fig. 2.8B). Initiation of sexual reproduction commences parthenogenic reproduction of sexual males and females, however with a short or long time gap (Fig. 3.13). It is likely that during meiosis II, disjunction of W and Z chromosomes occurs followed by endomitosis, which is not uncommon in the process of clonal egg production (see Pandian, 2011, p 90). As a result, the female egg carries WW chromosomes with its polar body harboring Z chromosome alone. Similarly the male egg carries ZZ chromosomes but its polar body holding W chromosome alone. With the combinations of WW and ZZ chromosomes in sexual females and male, respectively, the meiosis inhibitor gene is no more expressed. But following sexual reproduction, WZ female is generated, in which the meiosis inhibitor gene begins to express. While this possibility does exist, cytogenetic evidence is urgently required.

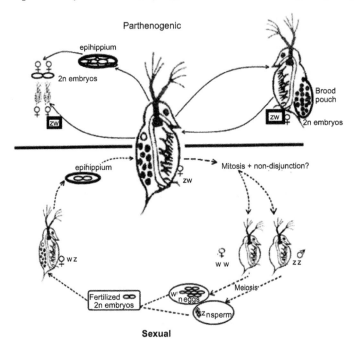

FIGURE 3.12

Parthenogenic (upper) and sexual (lower) modes of reproduction in a cyclic parthenogenic cladoceran. Note the presumed designation of sex chromosomes in parthenogenic and sexual cladocerans. The proposed ZW chromosomes are boxed to mark meiosis suppressor gene expression.

Herbert (1978, 1987) has proposed that the transition from the obligate sexual form to cyclic parthenogenesis may have occurred in the following sequence: production of diapausing and subitaneous eggs → brooding the embryos → environmental sex determination and meiosis suppression in subitaneous eggs. Incidentally, *Cyclestheria* lacks the full complement of the complicated daphnian amphiteric parthenogenesis and may represent a transitional stage; its parthenogenetic females produce both meiotic and ameiotic eggs but the meiotic eggs are either reabsorbed or fail to develop (Roessler, 1995). Endemic to America and unique among parthenogenic animals, cyclic parthenogenesis like that of *Daphnia* lineages is not prevalent. It may be more due to the constraints encountered by the same female to sequentially produce ameiotic and meiotic eggs than the advantages conferred by cyclic parthenogenesis. However, phylogenetic and palaeontological evidences indicate that cyclic parthenogenesis is the ancestral breeding system of *Daphnia* genus with an evolutionary history of more than 100 m y (Taylor et al., 1999).

Within the genus *Daphnia*, the obligate parthenogenesis has evolved at least on four independent occasions by employing mostly hybridization and rarely other mechanisms like contagious spread of a meiosis suppressor gene with or without involving polyploidization (Table 3.6). The lineages of *Daphnia* that employ obligate parthenogenesis have totally abandoned male sex. They have also accumulated deleterious mutations and have low genetic diversity (e.g. *D. miranda*). In the laboratory cultures, not a single male appeared up to 180 generations in *M. brachiata* and *D. magna* (Flossner, 1972) as well as up to 666 and 780 generations in *D. pulex* and in *M. macrocopa*, respectively (Banta and Brown, 1929a, b, c). In North America, transitions to obligate parthenogenesis have occurred rarely, e.g. one out of 15 *Daphnia* species complexes but multiple times in *D. pulex* complex of Europe at the altitudes of the Alps through hybridizations (Dufresne et al., 2011). In North America, the transition has arisen through a sex limited dominant mutation that suppressed meiosis in subitaneous eggs (Herbert and Finston, 2001). Hybrids between *D. pulex* and *D. pulicaria* invariably reproduce by obligate parthenogenesis in North America (Herbert and Crease, 1983, Herbert et al., 1993). Interestingly, "*D. pulex* reproduces by obligate parthenogenesis in eastern Canada but switches to cyclic parthenogenesis in the west with an abrupt transition coincident with the forest/prairie ecotone" (Herbert et al., 1993). Conversely, *D. pulicaria* reproduces by cyclic parthenogenesis in eastern Canada but by obligate parthenogenesis on the prairies. Arctic (above 54-61°N) and alpine populations of *D. pulex* and *D. pulicaria* reproduce by obligate parthenogenesis. But their temperate populations reproduce either sexually or cyclic parthenogenically. To link between habitat differences and relative fitness in the form of direct competition, the undertaken field investigations in ponds of southern Finland indicate almost clear geographic isolation between obligate parthenogens and cyclic parthenogens. Of 55 ponds investigated, 39 are inhabited by cyclic parthenogens, 14 obligate parthenogens and only two by both of them. Both of them have almost equal relative fitness in their respective habitats (Lehto and Haag, 2010). Briefly, the obligate parthenogenic populations of *D. pulex* dominate the polar regions of the entire holarctic region. By contrast, the populations in the temperate regions of Europe invariably reproduce by cyclic parthenogenesis.

TABLE 3.6

Mode of origin and ploidy level in parthenogenic cladocerans

Taxon	Origin	Ploidy	Habitat	References
Obligate Parthenogenic Lineages				
Polar D. pulicaria	Hybrid	>3n	Arctic lakes	Weider et al. (1999)
D. 'pulex' X D. pulicaria	Hybrid	2n	Temperate lakes	Herbert and Finston (2001)
D. pulicaria X D. 'pulex'	Hybrid	>3n	Arctic lakes	Dufresne and Herbert (1994)
D. middendorffiana	Hybrid+	>3n	Arctic lakes	Dufresne and Herbert (1997)
South America				
D. 'pulicaria' A, B, C	Hybrid	>3n	Alpine lakes	Mergeay et al. (2008)
D. truncata X D. pursilla††	Hybrid	–	Desert pond	Herbert and Wilson (2000)
Facultative Parthenogenic Lineages				
Panarctic *D. 'pulex'*	Contagious	2n	Temperate ponds	Innes and Herbert (1988)
D. pulicaria	Contagious	2n	Temperate ponds	Innes and Herbert (1988)
D. tenebrosa	Hybrid	>3n	Arctic lakes	Dufresne and Herbert (1994)
D. cephalata+	Spontaneous	2n	Temperate ponds	Colbourne et al. (2006)
*D. thompsoni**	Hybrid	>3n	Alpine lakes	Herbert and Wilson (1994)

+multiple parent, ††*Daphniopsis*, *Ctenodaphnia*, >3n = triploids and polyploids

In oligotrophic ponds, the time and food availability constraints may impose immediate investment on dormant eggs. Hatchlings from dormant eggs of *Daphniopsis ephemeralis* can consist of parthenogenic females as well as sexual females and males (Schwartz and Herbert, 1987). The obligate parthenogenic *D. pulicaria* in Spain produces epihippial dormant eggs in the first clutch itself (Perez-Martinez et al., 2007). These alternative strategies result in low reproductive output (2 eggs/epihippium) and may have similar effects. But sexual species suffer from the two-fold cost of males, albeit ensuring genetic diversity.

To encounter and sustain life, *D. pulex* seems to adopt the strategy of *D. ephemeralis*, i.e. early investment in sexual reproduction. Innes (1997) undertook an interesting investigation on sexual reproduction on *D. pulex* in Canadian temporary ponds. His observation is briefly summarized in Table 3.7. In two temporary ponds, the frequency of sexual males commenced as early as in May and progressively increased from 25% on May 3 to 43% on May 18 in males and from 5 to 38% in females. Similarly the male broods also increased from 44% on May 3 to 56% on May 10. Thus sexual males, females and male broods appeared immediately after the population was reestablished from dormant cysts in early spring. Of course, only females hatched from dormant cyst. The early presence of sexual males and females suggests that they were produced during the first few broods of epihippial females. Consequently, this *D. pulex* lineage reproduced sexually alone and the Spanish lineage *D. pulicaria* parthenogenically alone.

TABLE 3.7

Sexual reproduction in *Daphnia pulex* in a temporary pond (compiled from Innes, 1997) *approximate values

Date	Ovigerous ♀ (%)	Fecundity (No./Clutch)	♂ (%)*	♀ (%)*	♂ Broods (%)*
3rd May	56	13	25	5	44
6th May	66	24	22	22	46
10th May	43	26	28	18	56
14th May	75	29	42	4	7
18th May	23	26	43	38	–
21st May	28	0	28	26	22

Marine Cladocerans Of 620 species of Cladocera (Forro et al., 2008), only eight are truly marine. They are arranged in two orders: Ctenopoda and Onychopoda and five genera, *Penilia, Evadne, Pseudoevadne, Podon* and *Pleopus* (Onbe, 1999). A very interesting type of parthenogenic paedogenesis occurs in a few cladocerans. In the podonid *Evadna normanni*, parthenogenic embryos within the ephippium of the mother, carry their own eggs in their brood pouches. By then, embryos have passed the blastula stage in their pouches (see Pandian, 1994, p 121). *Penilia avirostris* is the most abundant (> 2,000 indivi/m^3) and widely distributed species. In it, the gamogenic females are larger than parthenogenics, whereas males are smaller than females. Males (0.0-12.2 indivi/m^3) and gamogenic females (0.0-5.5 indivi/m^3) occur almost throughout the year on the southeast coast of Brazil at low densities of 0.0-1.3% of the population during the three consecutive years (Miyashita et al., 2010). Parthenogenic females carry one-12 embryos (mean 3.6)/♀ but gamogenic females one (78% of the females) or two resting eggs; the latter has been observed, when population density peaks (> 8,000 ind./m^3). Miyashita et al. have found a positive correlation (r = 0.31) between brood size and body length in *P. avirostris*. However, Marazzo and Valentin (2003) have found it only up to females carrying six embryos/clutch and Atienza et al. (2008) have reported a negative correlation. The biokinetic range for the dicyclic *P. polyphemoides* falls between 0.2 and 27°C, and 2.5 and 31.5 g/l salinity. However, sexual morphs and resting eggs appear in Chesapeake Bay at temperatures between 11 and 17°C (see Pandian, 1994, p 123).

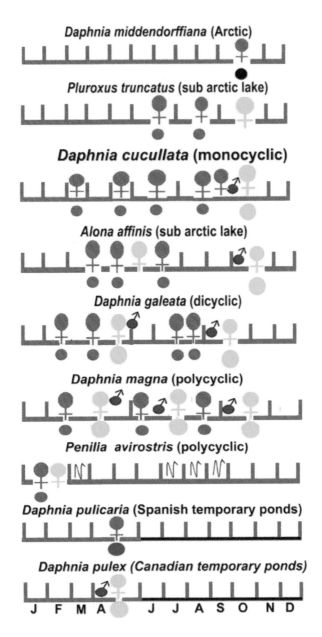

FIGURE 3.13
Occurrence of sexual females and males in selected cladocerans.
♀ = parthenogenic female, ♀ = sexual female, ♂ = sexual male
● = parthenogenic ephippium, ● = sexually produced epihippium,
Ɲ = months during which no collection was made. During all
other months, parthenogenic and sexual females and males were
simultaneously present in *Penilia avirostris*. Horizontal black line
indicates the period during which no water was available in the pond.

Chydorid Cladocerans The life cycle of most chydorid cladocerans is autumnal monocyclic. The chydorid males become mature after the third instar and stop molting, while the female mature after two molts and have constant clutch size of 10 embryos following every adult molt (Dodsen and Frey, 2009). The observations reported by Nevalainen and Sarmaja-Korjonen (2008) from their investigation on gamogenic periods of 26 chydorid cladocerans from nine lakes in Finland (60°N- 70°N) may briefly be summarized: 1. In *Alona costata, A. exigua, A. nana, Alonella excise, Anolopsis elongata* and *Ancistropus emarginatus* sexual reproduction was synchronized to a single autumnal period in all the lakes. 2. A combination of ~11.5°C and ~ 8 h photoperiod induced sexual reproduction. 3. The duration of sexual life cycle lasted from 1 w in *A. costata* to 10 w in *Alona affinis* during autumn. 4a. During autumn, sexual females alone occur in *A. guttata* (Lakes Hauklampi, Iso Lehmalampi, Iso Majaslampi, Tuhkuri), *A. affinis* (in Lake Kalatoin), *Pleuroxus truncatus* (in Lake Jourjarv) but (4a) during summer in *A. rectangulata* (Lake Hamptrask). 4c. Conversely, males alone appear during autumn in Lake Kangaslampi. It is difficult to comprehend how these temporally isolated sexual females and males may produce sexual ephippia.

Most available observations on the appearance of sexual males and females in different taxons of Cladocera are summarized in Fig. 3.13, from which, the following may be noted: 1. In *Daphnia pulex*, sexual females and males appear during autumn in relatively more permanent lakes but during spring in temporary ephemeral ponds. 2. Sexual females alone appear during February in ctenopodan *Penilia avirostris*. From these observations, the following may be inferred: Sexual males and females are not simultaneously generated. Notably, the males appear first in many cases but the female during summer in chydorids and February in ctenopods. The constellation of sex chromosomes and cytogenetic mechanism of producing sexual males remains to be known. 3. The appearance of sexual males and females in *D. pulex* and *D. ephemeralis* during spring but sexual females during summer in chydorids like *A. affinis* indicates that the role played by environmental factors like decreasing temperature, photoperiod, food availability and crowding may vary from habitat to habitat. The inbuilt genetic mechanism to generate parthenogenic females and males may also respond differently to different environmental cues (see Lampert, 2011). 3. It is difficult to comprehend the appearance of sexual females alone during summer and even during autumn, as in chydorid species.

Expectedly, attempts have been made to elucidate the genetic mechanism that transforms parthenogenic to sexual reproduction. For example, Xu et al. (2009) generated and characterized an expressed sequence tag (EST) data set from *D. carinata*. A set of 1,495 clusters were generated by sequencing 3,072 randomly chosen clones from cDNA library of parthenogenic juvenile. More research is in the pipeline to unravel the mystery of the genetic mechanism that transforms parthenogenics to the sexual mode of reproduction.

3.4.3 Aging and Senescence

By producing clonal eggs, parthenogenics may progressively accumulate increasing quantum of deleterious mutations. Hence, they may suffer reproductive senescence more than sexually reproducing crustaceans. In both parthenogenic diecdysic egg brooding crayfish and cladocerans, fecundity increases up to mid age and subsequently decreases, as their respective age is advanced (Fig. 3.14A, B). Irrespective of change in temperature (Fig. 2.1B) or food density (Fig. 3.14A), the bow-like trends obtained for fecundity-age relationship clearly indicates that even the optimal temperature or food density may not eliminate the onset and progression of reproductive senescence in these parthenogenics.

However, Vogt (2010, 2011) claimed that the obligate parthenogenic Marmokrebs *Procambarus fallax* does not undergo aging-related reproductive senescence. Hence, it can be used as an ideal model for analysis in various biological research. In support of this claim, he produced many evidences; of them, three are relevant: (i) The presence of stem (or satellite, Vogt, 2010) cells in heart, hepatopancreas and neurons, (ii) The global DNA methylation in selected tissues and (iii) Equal (~ 7% body weight) reproductive output in 6 m and 2-3 y old females. With due credit to his 20 years of dedicated contributions, the following from Vogt's own data may be noted:

1. That two of eight phenotypes spawned only once and none of them more than five times (Table 3.5), against the expected seven spawnings (Vogt, 2011) may indicate that the Marmokrebs were reared under identical but stressful conditions. Regarding aging and senescence, the maximal life span of its con-generic *P. erythrops* is > 16 y, whereas that of *P. fallax* (1,610 d) is < 4.5 y. Of the maximal life span of 1,610 d, spawning lasts up to 1,530 d, leaving 5% of its adult life span spent as post-reproductive menopause period. This period ranges from 7 to 21% of the respective adult life span in different phenotypes (Table 3.5).

2. Though varied among batch mates, the global DNA methylation level in hepatopancreas decreases from the mean of 2.01% in 188 d-old adult to 1.65% in 626-d old adult. This decrease is ~ 18% , when the age of crayfish is advanced by another 438 d. Similar decreases have also been noted in other tissues (Vogt et al., 2008). In support of the DNA methylation, Vogt (2010), under the impression that the lobsters live for 100 y or more, has resorted to the supporting data from telomerase activity of *Homarus americanus* (Klapper et al., 1998). However, Klapper et al. have used mid-aged two lobsters weighing ~ 400 and 800 g only.

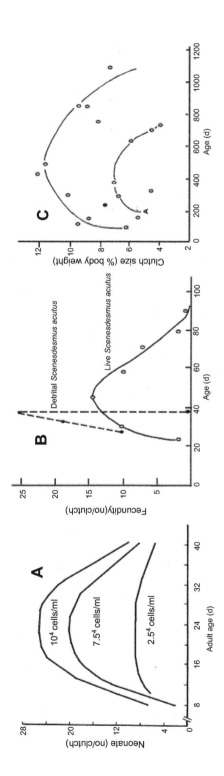

FIGURE 3.14

A. Simplified free hand drawing to show that fecundity decreases beyond mid-adult age in *Ceriodaphnia cornuta* fed at different algal densities (redrawn from Richman, 1958). B. Fecundity as a function of age in the geographic parthenogenic copepod *Elaphoidella grandidieri* fed on live *Scenedesmus acutus* (drawn using data from Nandini et al., 2011). C. Clutch size as function of age in marbled crayfish (drawn from data reported by Vogt, 2010).

3. BrdU-labelled migratory stream and lateral proliferation of neurons of the 'neurogenic niche' in deuterocerebrum is a strong evidence for the presence of stem cells in juvenile Marmokrebs (see Fig. 6F of Vogt, 2010). However, their presence is rather an indication of non-sexualization of brain (Le Page et al., 2010, see also Pandian, 2012 p 2-4, 140-142). The number of Oogonial Stem Cells (OSCs) is the correct index of reproductive senescence. For example, commencing with 2.5 OSCs/ovariole in *Drosophila melanogaster*, OSCs decreased to 0.01/ovariole at the age of 11 w in the senescence prone S strain but it was maintained at 1.94 OSCs/ovariole in the senescence-resistant L strain (Kaczmarczyk and Koop, 2011). Hence, research may have to be undertaken to estimate the number of OSCs per unit ovary and to prove that *P. fallax* does not undergo aging and senescence.

4. With regard to equal reproductive output by a 6 m-young and 2-3 y-old Marmokrebs, the values (collected from two founder lineages [see his Fig. 3] and plotted against body length [in his Fig. 4]) reported by Vogt (2010) on clutch size (as% body weight) are plotted against age in Fig. 3.14C. Apparently, the reproductive output decreased with advancing age in the presumably two founder lineages. Hence, the egg brooding Marmokrebs also undergoes reproductive senescence, as other egg brooding crustaceans, especially parthenogenics do it.

As if to confirm that the parthenogenics do undergo senescence, the full length senescence-associated protein (SAP) gene of *D. pulex* has been cloned. *DpSAP* is specifically expressed in the joints of the second tentacle of parthenogenic and sexual females, suggesting a role in the reproductive transformation of *D. pulex* (Liu et al., 2014).

3.5 Hermaphroditism

There are > 65,000 animal species that employ hermaphroditism (Jarne and Auld, 2006). A majority of them have originated from dioecious ancestors (Eppley and Jesson, 2008, however, see also Iyer and Roughgarden, 2008). In Crustacea, hermaphroditism includes (i) simultaneous and (ii) sequential hermaphrodites. Simultaneous hermaphrodites may change from androdioecy to dioecy (e.g. thoracic barnacles, Spremberg et al., 2012) or from dioecy to hermaphroditism (e.g. spinicaudates, Weeks et al., 2009). However, sequentials undergo natural sex change once in their life time in a single direction from male to female in protandrics and female to male in protogynics.

TABLE 3.8

Distribution of protandric, protandric simultaneous and protogynic hermaphrodites in Crustacea (compiled from Brook et al., 1994, Tsai et al., 1999, Vogt et al., 2004, Baeza et al., 2010)

Protandrics	
Cirripedia	: *Gorgonolaures muzikae, Synagoga sondersi*
Isopoda	: *Anilocra physodes, A. fontinalis, Cuna insular, Emetha audouinii, Ichthyoxenus fushanensis, Hemioniscus balani, Liriopsis pygmaea, Munidion pleuroncodis, Nerocila californica, Philoscia elongata, Rhyscotus ortonedae, R. parallelus*
Amphipoda	: *Acontioistoma marinonis, A. tuberculata, Ocosingo borlus, O. fenwicki, Scolopostoma prionoplax, Stegocephalus inflatus, Stomacontion punga punga*
Decapoda	: *Argis dentata, Athanas indicus, A. kominatoensis, Atya bisulcata, A. serrata, Paratya curvirostris, Calocaris mecandreae, Campylonotus rathbuni, C. semistriatus, C. capensis, C. vagans, Caridina richtersi, Chorismus antarticus, Crangon crangon, C. franciscorum, Hippolyte inermis, Melicertus kerathurus, Pandalus borealis, P. danae, P. gracilis, P. goniurus, P. hypsinotus, P. jordani, P. kessleri, P. montagu, P. montagu tridens, P. nicolsi, P. platyceros, Pandalopsis dispar, Parastacus nicoleti, P. nippenensis, Procssa edulis, Samastascus spinifrons, Emerita analoga, E. asiatica, Solenocara membranacea*
Protandric Simultaneous	
Cirripedia	: *Chelonobia patula*
Decapoda	: *Lysmata amboinensis, L. boggessi, L. californica, L. debelius, L. hochi, L. holthuisi, L. galapagensis, L. grabhami, L. nayaritensis, L. nelita, L. nitida, L. pedersoni, L. rathbunae, L. seticauda, L. wuurdemanni, Exhippolysmata ensirostris, E. oplophoroides, Thor manningi*
Protogynic	
Isopoda	: *Cyathura carinata, C. polito, C. profunda, Paraleptosphaeroma glynni, Gnorimosphaeroma luteum, G. oregonense, G. naktongense*
Tanaidacea	: *Apseudes hermaphroditicus, Hageria rapax, Heterotanais oerstedi, Leptochelia dubia, L. forresti, L. neapolitana, Kalliapseudes schubartii*

In Crustacea, sequentials include protandrics, protogynics and protandric simultaneous hermaphrodites. There are ~ 57 species of protandrics distributed in Cirripedia, Isopoda, Amphipoda and Decapoda, 19 species of protandric simultaneous hermaphrodites, limited to a single species in Cirripedia and 18 species of carideans as well as 14 species of protogynics belonging to Isopoda and Tanaidacea (Table 3.8). The crustacean sequentials include 63% protandrics, 16% protogynics and 21% protandric simultaneous hermaphrodites. These values of 84% (63 + 21) protandry and 16% for protogyny are quite a contrast to 71% protogyny and 19% protandry in teleostean fishes (Pandian, 2012, p 105). Surprisingly, protogyny is also limited to seven species of Isopoda and seven species in Tanaidacea. The number of sequentials could be more, as bopyrid isopods may prove to be sequential hermaphrodite (see section 7.3).

3.5.1 Simultaneous Hermaphroditism

The occurrence of simultaneous hermaphroditism is limited to lower crustaceans namely, Notostraca, Spinicaudata and Cirripedia.

Notostraca In geographic populations of *Triops* sp, the tadpole shrimp is gonochoric in some populations but androdiocious hermaphrodites in others (e.g. *Triops* sp, Garcia-Velazco et al., 2009). Within androdioecious populations, the frequency of males ranges from 0 to 21% in *T. longicaudatus* (Grigarick et al., 1961) and *Triops* sp (Garcia-Velazco et al., 2009), which may be compared to 13% male frequency in androdioecious hermaphrodites of *Eulimnadia texana* (Weeks et al., 2001). In his systematic review, Longhurst (1955a) has recorded the occurrence of androdioecy in *T. cancriformis*, *Lepidurus apus* and *L. arcticus* (cf. Wojtasik and Brylka-Wolk, 2010). In its species specific or geographic population specific form, androdioecy is prevalent among entamostracan crustaceans that inhabit ephemeral ponds. By employing androdioecy, they have avoided inbreeding and escaped from Muller's ratchet. As 'living fossils', the primitive genera *Triops* and *Lepidurus* have survived over 150 m y (Longhurst, 1955b).

Spinicaudata Thanks to the enthusiastic endeavors made by S. C. Weeks and his colleagues during the last few years, very interesting information has been accumulated on sexuality and mating system of spinicaudates. Employing androdioecy, an unstable mating system, the spinicaudates have successfully sustained it and survived for the last 24+ m y. They have a short life span of ~ 25 d and mature by 7-8 d (Weeks et al., 2014b) and produce hundreds to thousands of eggs in several clutches (Weeks et al., 1997). The fertilized eggs, rather cysts are resistant to desiccation and can be dormant for decades. A single male can fertilize from one to 12 hermaphrodites per day (Weeks et al., 2006c). They display a broad range of reproductive strategies from pure dioecy to pure hermaphroditism. Hence, they are excellent organisms to test models of stability of the unique androdioecy. In them, sex is determined by an incipient ZW chromosomes (Weeks et al., 2010). Three sex chromosomal systems are known. Males are homozygous ZZ but the hermaphrodites can be either homozygous WW monogenics or heterozygous ZW amphigenics. These WW and WZ hermaphrodites are phenotypically similar but genotypically different. When selfing, the monogenics breed true, producing 100% monogenic offspring. But when crossed, produces 25% ZW amphigenics: 50% ZZ males and 25% WW monogenics. In males, the first two pairs of thoracic appendages undergo differentiation into claw-like claspers (Fig. 3.15 A), which are used to hold on to carapace of hermaphrodites during mating. In the absence of claspers, the hermaphrodites cannot outcross with other hermaphrodites (Weeks et al., 2006 c).

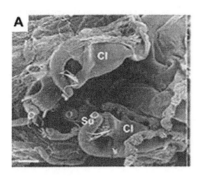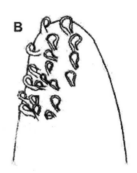

FIGURE 3.15
A. First two thoracic legs modified as claspers (CL) with sucker-like (SU) distal projections in *Eulimnadia texana* male (from Scanabissi et al., 2006, permission by S. C. Weeks). B. Free hand drawing to show the tip of cincinnuli on the endopod of first pleopod during the male phase of *Lysmata californica* (redrawn from Bauer, 2000).

In the genus *Limnadia*, the hermaphrodites outcompeted both males and females to form all- hermaphroditic *L. lenticularis* and *L. yeyetta* (see Weeks et al., 2014a). Two independent origins of hermaphroditism from dioecy were described (Weeks et al., 2009). Recently, a third independent derivation of hermaphroditism from dioecy was documented in *Cyzicus gynecia* (Brantner et al., 2013a). Providing evidence for the fourth all-hermaphroditic genus *Calalimnadia*, Weeks et al. (2014a) showed *C. mahei* collected from Mauritius Island were reared in France for two generations. From a total of 10,038 offspring originated from 23 female lines, no male was found. The phylogenetic reconstruction of 40 species spinicaudates has revealed that hermaphroditism has originated four times from the ancestors (Weeks et al., 2014a).

Hermaphrodites outcompeted females but not males to form androdioecy in 13 androdioecious species. From their observations on *Eulimnadia* over 3 y, Weeks et al. (2014b) reported sex ratio of androdioecious *E. texana* as 0.17 monogenic ♀ : 0.60 amphigenic ♀ : 0.23 ♂. From an another laboratory study, Weeks et al. (2008) recorded the male ratio of 0.21 for 14 *Eulimnadia* species. These ratios are closer to the Mendelian sex ratio in ZW female heterogametic sex determination system (Weeks et al., 2010). However, the male ratio reported by others ranged from 0.00 to 0.31 and averaged to 0.05 in 13 *Eulimnadia* species from the fields of USA (Sassaman, 1995). Similarly, it also ranged from 0.00 to 0.30 with an average of 0.15 for *E. feriensis* and *E. dahli* collected from Australian fields (Weeks et al., 2006b). In the laboratory-reared *E. texana* too, the ratio ranged from 0.00 to 0.33 and averaged to 0.24 (Weeks et al., 2014b). Using male ratio data complemented with genetic assays of 15 *Eulimnadia* species from all over the world, Weeks et al. (2006b, 2008, 2009) suggested that *E. azisi* is all-hermaphroditic but *E. gibba*, *E. michaeli* and *Eulimnadia* sp are androdioecious (see Brantner et al., 2013b). Secondly, these wide ranges in male ratios suggest that within an androdioecious species, there can be hermaphroditic populations. In many populations of India and Thailand, the male ratios of *Eulimnadia* spp confirm that there are hermaphroditic populations within androdiocious *E. michaeli* (Table 3.9). The following have become apparent from the extensive investigation by Brantner et al. (2013b): *E. michaeli* KK4A populations of both India and Thailand and KK6A and KK8A of Thailand as well as *Eulimnadia* species (Karuveta, Kerala, India) are pure hermaphrodites. On the other hand, *E. gibba* population at Veli, Kerala may employ androdioecy. The male ratios of *E. michaeli* KK2A of Thailand (with male ratio of 0.47) and other Indian populations KK2A (with the male ratio of 0.1) of *E. michaeli* have male ratios that are difficult to comprehend.

TABLE 3.9

Sex ratios of F_1 (*) and F_2 († or ††) offspring of *Eulimnadia* populations from different geographical locations in India (* or †) and Thailand (††) (compiled from Brantner et al., 2013b)

Species	Population	Male Ratio	Monogeny (M/ Amphigeny (AM)	Androdioecious (A)/ Hermaphroditic (H)
		Sex ratio in F_1		
E. azisi	Ghat*	0.0		H
Eulimnadia sp	Karuvetta*	0.0		H
E. gibba	PEDI*	21.6	AM	A
Eulimnadia sp	Veli*	14.5		A ?
		Sex ratio in F_2		
E. azisi	Ghat†	0.0	100 M	H
Eulimnadia sp	Karuvetta†	0.0	100 M	H
E. michaeli	KK4A†	0.0	100 M	H
E. michaeli	KK4A††	0.0		H
E. michaeli	KK6A††	0.0		H
E. michaeli	KK8A††	0.0		H
E. michaeli	KK8A†	7.5		
E. michaeli	KK2A†	9.9		
E. michaeli	KK9A††	12.5		
E. michaeli	KK7A††	14.3		
E. michaeli	KK2A††	47.2		

Thecostraca This taxon comprises of three major groups: Cirripedia, Ascothoracidae and Facetotecta (Martin and Davis, 2001). Within suspension-feeding cirripedes, sexuality ranges from gonochorism in Acrothoracica to hermaphroditism in Thoracica. Gonochorism is maintained in ectoparasitic Ascothoracidae and endoparasitic Rhizocephala and Facetotecta (?) (Hoeg et al., 2009). Within thoracican genus *Octolasmis*, *O. angulata* is dioecious but *O. warwiokii* is androdioecious (Yusa et al., 2012). Hence, thoracicans offer an interesting testing model for theories on evolution of reproductive strategies

(Spremberg et al., 2012). For egg production in cirripedes, Barnes (1989) may be consulted.

To reduce the high cost of semen production of up to 50% body weight (see p 53), dwarf males that attain sexual maturity without somatic growth have been evolved not only in thecostracans but also in parasitic bopyrids (see later); for example, the thecostracan dwarf (or complemental/apetural) males grow to half the length and one eighth the size of the female/hermaphrodite. With precocious sexual maturity, they enjoy a higher fertilization success per sperm. The cyprid larva settling on the mantle rim of con-specific hermaphrodite metamorphoses into a dwarf male but that on a substratum becomes a hermaphrodite. When exposed to surplus of hermaphrodites with empty receptacles, not more than 50% cyprid larvae of *Scalpellum scalpellum* develop into dwarf males. Hence, a genetic sex determination is suggested (Spremberg et al., 2012). In *Balanus galaetus* (Gomez, 1975) and sexually dimorphic cyprid of *Ulophysema oresundense* (Melander, 1950), the existence of genetic sex determation mechanism has also been suggested.

In the barnacle *Bathylasma alearum*, 83% hermaphrodites carried dwarf males (Foster, 1983). The number of dwarf males per host ranges from one in *Paralepas klepalae* (Kolbasov and Zovina, 1999) to 13 in *O. warwiokii* (Yusa et al., 2001). A positive relationship between the number of dwarf males and body size of hermaphrodites in *Koleolepas owis* (Yusa et al., 2001) and *Verum brachiumcancri* (Buhl-Mortensen and Hoeg, 2013) has been reported, indicating that the large hermaphrodites receive more sperm. To facilitate cross fertilization, the hermaphrodites settle within distance of 50 mm from one another (Yusa et al., 2001). In *S. scalpellum*, 19% solitary hermaphrodites harbor no males altogether, 81% of them carried 3.5 dwarf males/hermaphrodite. For gregarious hermaphrodites, these values were 38% with no males and 62% carry 3.2 males/hermaphroditie (Spremberg et al., 2012). Mathematic models suggest that the existence of hermaphrotitism in relatively large mating groups, androdioecy in smaller groups and dioecy in still smaller groups (Urano et al., 2009).

3.5.2 Sequential Hermaphrodites

Protogyny   Monandry is reported from a few isopods e.g. *Gnorimosphaeroma oregonense* (Brook et al., 1994) and *G. naktongense* (Abe and Fukuhara, 1996) and diandry in tanaids e.g. *Hargeria rapax* (Modlin and Harris, 1989). Whereas *G. oregonense* and *G. naktongense* produce only a single brood prior to sex change, the diandric *H. rapax* produces three broods. In *H. rapax*, primary and secondary males measure < 2.3 and > 2.3 mm body size, respectively and the female measures 1.6-3.6 mm. Transitional requires 8-11 d to change sex from female to male.

Protandry   While the protandric monogynic decapods are free-living, ascothoracican protandrics are parasitic. In *Gorgonolaureus muzakie*, a gorgonian parasite, eggs and larvae are brooded within the carapace of mature female; the bivalved free swimming male (?), after inseminating one or more sessile females, settles on the gorgonian host and changes sex to female (Grygier, 1982).

From his field and histological investigations, Subramoniam (1981) confirmed the occurrence of protandric digyny in the mole crab *Emerita asiatica*. At the carapace size of 3.5 mm, the megalopa metamorphoses into either a male or primary female. At 15 mm size, the adult dwarf male changes sex to a secondary female. In *Pandalus platyceros*, the planktonic larva metamorphoses into a male at the carapace size of ~ 3 cm; between 3.5 and 4.2 cm carapace size, the transition from male to female occurs. Incidentally some males of the protandric decapods do not undergo sex change and remain as males (e.g. pandalids). Six species belonging to the genus *Pandalus* are reported to display protandric digyny. In some protandrics, various proportions in a population metamorphose into primary females, bypassing the male phase (e.g. Crangonidae; Frechette et al., 1970, Pandalidae; Butler, 1964, Processidae; Boddeke et al., 1991). In *P. borealis*, adult males changes sex to the secondary females but the female 'headroses' can also change sex to male. Interestingly, the fluvial populations in *Samastacus spinifrons* are protandric but the lake populations are not (see Vogt et al., 2004). Long term laboratory experiments have shown that 2% *Crangon franciscorum* (Gavio et al., 2006) and 5% *C. crangon* (Schatte and Saborowski, 2006) undergo sex change from male to female, i.e. only a small fraction in population of the shrimps are protandric. Figure 3.16 shows the ontogenetic pathways of sex change in selected sequentials. Obviously, sexuality and sex change in these protandrics remain highly flexible.

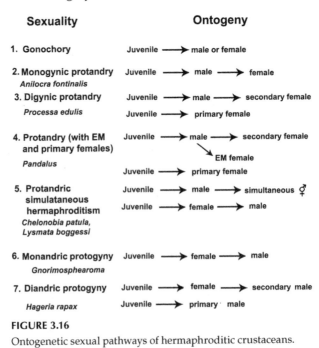

FIGURE 3.16

Ontogenetic sexual pathways of hermaphroditic crustaceans.

Protandric Simultaneous Hermaphroditism (PSH) Protandrics inclusive of Protandric Simultaneous Hermaphrodites (PSHts) have been reported from 2.8% of ~ 1,800 species of caridean shrimps (Bauer, 1986). Females are marked by the loss of appendixes masculina and modified endopodites of the first and second pleopods. Primary males are identified by the presence of the appendixes masculina and prehensile third pereiopods. The gonads of males are ovotestes with well developed paired testes but underdeveloped ovaries, and the shrimp reproduces as a male. Eventually, the ovarian zones in the ovotestes also mature. Unlike protandrics, which lose the male secondary sexual characteristics during the ovarian development, PSHts maintain most male anatomical characteristics but lose the coupling hooks cincinnuli (Fig. 3.15B) and appendixes masculina on the first and second pleopods (Baeza et al., 2007). As a result, these PSHts cannot self-fertilize and have to necessarily outcross. Shrimps acting as females prefer smaller over larger shrimps as male mating partners; the male mating ability is also greater for smaller than larger hermaphrodites. However, the ability to acquire the resources for mating increases markedly with increasing body (Baeza et al., 2007).

The credit for reporting the existence of PSH in a caridean shrimp goes to an Indian (Kagwade, 1981). However, R. T. Bauer and A. J. Baeza must be credited for elaborating the reproductive strategy of PSH in 18 caridean species (Table 3.8). In comparison to the density of protandrics (e.g. 22 shrimps/m^2 in *Thor manningi*, Bauer, 1986), that for PSHts is far low and ranges from 1 shrimp/km^2 in *Exhippolysmata oplophoroides* Baeza et al., 2010) to 2.5-13.1 shrimps/km^2 in *Lysmata boggessi*, Baeza et al., 2014). Surveying large areas (~ 100,000 km^2) in the central Mediterranean Sea, Spano et al. (2013) investigated the abundance of 294 malacostracans. Limiting the abundance to 55 species, they reported the density to range from 1 indivi/km^2 for *Homola barbata* to 1,645 indivi/km^2 for *Pagurus alatus*. Hence, the encounter rate among con-specifics is expected to be extreme in these low-density species (Charnov, 1982, Baeza et al., 2010). As a consequence of these low densities, grouping and crowding are imposed on some PSHts (Table 3.10). A second imposition is on skewing sex ratio in favor of hermaphrodites. A third one is the highest allocation of time and resources for female reproduction. Testing sex-dependent energy and time costs as well as sex-dependent mortality of reproductive output in *L. wurdemanni*, Baeza (2006) found that 1. Males grew faster than PSHts of the same size and age, indicating that the energy cost of reproduction was higher in PSHts than in males. In fact, growth rate of an individual decreased faster, when reproducing as a female than as a male. 2. Time cost of reproduction during the female phase was longer than that for males 3. Predation related mortality of smaller PSHts was not greater than that for smaller males, indicating that sex-dependent mortality may have an insignificant role in reproduction. Baeza (2007b) went a step ahead to test (i) how resources are optimally allocated to reproductive function during male and female phases and (ii) what determines the shift in optimal allocation with advancing age and increasing body size. He found that the resource allocation for the female gonad of *L. wurdemanni* PSHts was 118-folds higher than that for the male gonad. This is perhaps the most extreme case for female-biased allocation. Secondly, the proportion of resource allocation to female function was higher in small than in larger hermaphrodites. In terms of both energy and time, egg production is costlier than sperm. Male phase hermaphrodites replenish their sperm reservoirs within 2 d after matings, as males. But female phase hermaphrodites require 11 d or longer to refill their ovaries after spawning as females. The time required to refill ovaries is also long in gonochorics. For example, it is 9-15 d in *Callinectes sapidus* and ~ 12 d in penaeids (see Baeza, 2007b).

TABLE 3.10

Socioecology and PSHt sex ratio in temperate and tropical protandric simultaneous hermaphroditic carideans

Species	Socioecology	Sex Ratio of PSHt	Zone	Reference
L. wurdemanni	Crowd	0:14-0.88	Temperate	Baeza (2007b)
L. californica	Crowd	Biased	Temperate	Bauer and Newman (2004)
E. oplophoroides	Crowd	0.27	Subtropical	Baeza et al. (2010)
L. boggessi	Groups	0.84	Subtropical	Baeza (2009)
L. hochi	Groups	0.23	Tropical	Baeza and Anker (2008)
L. holthuisi	Crowd	0.61	Tropical	Anker et al. (2009)
L. nayaritensis	Crowd	0.92	Tropical	Baeza et al. (2007)
L. galapagensis	Crowd	0.98	Tropical	Baeza (2009)

3.6 Intersexuality

All simultaneous hermaphrodites are intersexes but some intersexes are not simultaneous hermaphrodites. The essential difference is the tolerance between the two opposing sexual tendencies in simultaneous hermaphroditism. Intersex is the product of interference between the two opposing sexual tendencies; consequently, neither of them can express decisively and may remain non-functional (see Pandian, 2012, p 8). Intersexuality includes (i) gynandromorphism and (ii) intersexes. Intersexuals, especially gynandromorphs can be useful as models for research on sex determination, developmental pathways and mechanisms by which parasites affect sex differentiation in their hosts (Sassaman et al., 1997, Olmstead and LeBlanc, 2007).

3.6.1 Gynandromorphism

These are individuals with clearly defined areas of female and male characteristics. The distribution of these areas may be bilateral (left/right), axial (anterior/posterior) or mosaic (patchily distributed). Bilateral and mosaic phenotypic gynandromorphs are attributed to random mutations such as chromosomal non-disjunction (see Krumm, 2013). For example, a few rare individuals of *Nephrops norvegicus* display complete bilateral separation: half of the body has internal and external characters of male and remaining half has female characters (Farmer, 1972). Axial gynandromorphs (male anterior and female posterior) have been reported from many branchiopods such as *Artemia franciscana*, *Branchinecta lindahli*, *B. mackini*, *B. packardi* and *B. tolli* (see Krumm, 2013) as well as chydorid cladocera *Pleuroxus* (Frey, 1965). All axial gynandromorphs possess male secondary antennae and female genitalia, while the mosaics display bilateral gynandromorphism in the secondary antennae and normal female thoracic appendages and genitalia. In *B. lindahli,* the frequency of gynandromorphs is 0.002-0.003% in laboratory-reared populations (Krumm, 2013). Through quantitative morphological measurements of the endopods, Krumm (2013) found that there was no significant difference between gynandromorph endopods and female endopods and thereby provided evidence for the absence of intersexuality in *B. lindahli*. Incidentally, her results support the hypothesis that sex determination in *B. lindahli,* and possibly in other anostracans is cell-autonomous.

3.6.2 Intersexes

These are individuals with phenotypes having intermediate characteristics between normal females and males. They are reported to occur across crustaceans (e.g. Copepoda: Gusamao and McKinnon, 2012, Isopoda: Rigaud and Juchault, 1998) but mostly from crayfish and crabs. Permanent intersexes with female and male gonoducts and gonopores but with unisexual gonads, either ovaries or testes are reported in the burrowing crayfish *Parastacus defossus, P. pilimanus, P. pugnax, P. saffordi* and *P. varicosus* (Rudoph et al., 2001) and in the non-burrowing *P. brasiliensis* (see Vogt et al., 2004), parthenogenic *Procambarus fallax* (Martin and Scholtz, 2012) and *Cherax quadricarinatus* as well (Sagi et al., 1996). In Cambaridae, Turner (1935) listed over a few thousands aberrant sexual characteristics. But there are other reports indicating that they are not so prevalent: there was only one female with male-like pleonic appendages among 10,000 cambarid individuals, two among 1,800 specimens belonging to two species of *Procambarus* and seven among 30,000 *P. clarkii* (see Martin and Scholtz, 2012). In intersexes of *C. quadricarinatus*, male and female genital openings are persistent in both 7- and 19-m old individuals (Sagi et al., 1996). The hermit crabs display a range of intersexualism. In *Clibanarius vittatus*, the intersexes possess open male and female gonopores. Despite their presence in *C. antillensis*, they are not open. In *C. selopetarius*, the male gonopores open on the coxae of the 5th pleopods but those of female on the coxae of the 3rd pleopods (Turra, 2004). Of 1,114 specimens of a crayfish *Samatstascus spinifrons* examined, 41 were intersexes. All the intersexes had gonoducts of both sexes, but 30 were with testes, four with ovaries and seven with ovotestes (Rudoph, 1999). In isopods, intersexes vary greatly from fertile females bearing small male characters to sterile males with female genital orifices (Rigaud and Juchault, 1998). Understandably, *Gammarus minus* intersex females brood slightly a fewer but larger embryo than those of normal females. Hence, maternal investment during brooding also does not differ significantly between intersex and normal female (Glazier et al., 2012).

3.7 Mating Systems

Crustaceans display all the recognized mating systems. Monogamy is restricted to burrow-inhabiting caprellid amphipods and alpheid carideans. In this mating system, heterosexual pairs share a microhabitat for a longer duration than reproductive cycles and usually copulate only with the same partner (Correa and Thiel, 2003). Within polygamy, polygyny is more prevalent. In *Eulimnadia texana*, a single male may inseminate as many as 12 hermaphrodites in a day (Weeks et al., 2006c). In a process of convenience polygamy, a female rock shrimp *Rhynchocinetes typus* may allow mating and transfer of spermatophores from subordinate to aggressive males. However, the females may remove the spermatophores of subordinate males before receiving spermatophores from dominant ones. Using molecular markers, multiple paternity has been demonstrated. For example, of 15 females, 4 (27%), 5 (33%), 2 (13%) and 4 (27%) have fertilized their eggs using spermatophores received from 1, 2, 3 and 4 males, respectively (Bailie et al., 2014).

With regard to size selection among mating partners, three different patterns seem to emerge. In monogamous mating system, the pairs are of almost equal size (Fig. 3.17 A). For example, the mean size of paired males and females of *Alpheus estuariensis* was 8.2 and 8.4 mm CL, respectively (Costa-Souza et al., 2014). In *Gammarus insensibilis* too, males select females of their own size (Thomas et al., 1996a). In the polygynic systems, males were always larger than females (Fig. 3.17B). A female stone crab *Menippe* sp (e.g. 80 mm CW) was paired with larger males ranging from 85 to 115 mm CW

(Wilber, 1989). In the polyandric system, females pair with preferably smaller males (Fig. 3.17C). In the protandric simultaneous hermaphrodite *Lysmata wurdemanni* during its female phase prefer smaller males over larger ones (Baeza, 2007a). More information is required before this generalization can be confirmed.

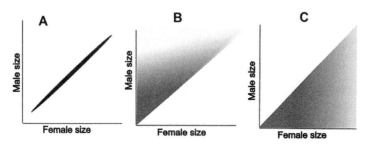

FIGURE 3.17
Patterns of mating pair size selection of males by females A. monogamous mating systems: the mating partners are almost equal size (drawn using data reported for *Alpheus estuariensis,* Costa-Souza et al., 2014). B. In polygynic mating systems, female selects larger males (drawn using data reported for *Menippe* sp, Wilber, 1989). C. Polyandric mating system: female selects smaller males e.g. *Lysmata wurdemanni* (Baeza, 2007a).

Pairing and Mating Positions   In consonance with the diverse morphological diversity of crustaceans, pairing and mating positions are expected to vary widely. Johnson et al. (2001) have summarized six different mating positions in peracarids, a taxon representing crustaceans, which are not limited within bivalve shells. A detailed description of the entire sequence of mating events is reported in the bivalve spinicaudate *Lynceus brachyurus* (Patton, 2014). In the bivalve conchostracans, six different mating positions are described (Cohen and Morin, 1990).

3.7.1 Male's Presence Needed

The presence of males is required to (i) prolong the life span of mating partners (ii) accelerate vitellogenesis, (iii) initiate ovulation, (iv) continue egg production and (v) reduce infertile eggs.

Life Span   In general, males have a relatively shorter life span, as they spent more time on energy consuming activities like mate searching and so on. For example, the time spent on energy consuming activities by the clam shrimp *Eulimnadia texana* is 88 and 9% in males and hermaphrodites, respectively (Zucker et al., 2001). Understandably, the life span of male shrimp is 25 d, when the male is reared alone. However, it is 27 and 31 d, when reared with two and 15 hermaphrodites, respectively (Zucker et al., 2001). On the other hand, the inseminated female copepods *Acartia tonsa* has a longevity of 35.5 d, in comparison to 15.8 d for the unmated female (Parrish and Wilson, 1978).

The placement of spermatophore (s) on the female's body accelerated vitellogenesis in the red shrimp *Aristeus antennatus* (Carbonell et al., 2006). When females were isolated and thereby prevented from mating in anostracan *Streptocephalus dicotomus*, the eggs did not descend from the oviduct into the median uterus and no secretion was poured into the uterus for the purpose of shell formation (Munuswamy and Subramoniam, 1985c). In the clam shrimp *Lynceus brachyurus*, following mating and insemination, a male abandoned a depleted female to mate with another having eggs ready for fertilization The number of eggs deposited decreased to zero within 4-5 d, when all the males were removed from the group of ovulating females (Patton, 2014). In standardized culture of *A. tonsa*, fecundity was 44, 46 and 17 eggs/♀/d in (i) a mated female with continued presence of male, (ii) mated female but with no male and (iii) unmated female, respectively. The fraction of infertile eggs decreased from 30% in the first group to different levels in the second and third groups (Wilson and Parrish, 1971).

3.7.2 Induction of Parthenogenesis

With continued absence of males, a rare paracambarid combarid *Orconectes limosus* switches from sexual to facultative parthenogenesis. *O. limosus* mate twice a year, in autumn and spring (see p 39, 93), although, they lay eggs only in spring (Hamr, 2002). Buric et al. (2011) reared them in mated and unmated groups of 30 females each. The first group was allowed to mate freely with 15 males kept in the same tank. In the second group, the females were, however, separated from males by a screen, which allowed visual and chemical communications between 30 females and 15 males but physical contact for mating and spermatophore placement were not allowed. Both the groups were reared (from August) for a period of 10 m (till May). Surprisingly, the unmated females produced 96 offspring/female, while the mated females 1.5 time more (139) offspring/female. Analysis of seven variable microsatellite loci of unmated females and their respective juveniles revealed that the offspring produced by the unmated females had multilocus genotypes identical to the mothers. Incidentally, Yue et al. (2008) also reported the occurrence of natural clones in *Procambarus clarkii*, a close relative of *O. limosus*, indicating the tendency of these cambarids to switch to parthenogenesis. From their experiments complemented by genetic analysis, Buric et al. (2011) have proved that at least one species of cambarid crayfish is capable of facultative parthenogenesis.

3.7.3 Social Induction

The sex changing peppermint shrimp *Lysmata wurdemanni* provides a rare opportunity to test the role played by sex and size in a mating system on the course of sex change with increasing body size. The shrimp begins with Male

Phase (MP) but changes to Female Phase (FP) in protandric simultaneous hermaphrodites. Newly molted FP is inseminated as female by other FP serving as males. Hence, FPs are outcrossing simultaneous hermaphrodites, while MPs mate as males only. The MPs are capable of sex change to simultaneous hermaphrodites on attaining a size of 6 to 8 mm CL, depending on abiotic (Baldwin and Bauer, 2003) and/or social environment. To test the effects of sexual and (small or equal) body size on the fraction of five MP candidates (focal MPs), four different combinations were selected (Table 3.11). In small scale experiments, sex change up to the 50th d was estimated by rearing a single MP alone or with one or another individual of different sizes and sexual phases. In these combination experiments, the results obtained by Baeza and Bauer (2004) are briefly summarized: 1. Sex change in the large scale combinations, the difference between rearing five focal MPs were with five small MPs or five MPs of similar size was not significantly different, indicating that the size difference of MPs has no social role to accelerate sex change. 2. Similarly, there was no significant difference in the combinations of five FPs or four MPs + one FP, indicating the presence of MPs does not alter sex changes. 3. The combinations of focal MPs with five similar sized MPs changed sex earlier than focal MPs reared with five FPs. Clearly, it is the presence of a large number of MPs rather than the presence of single FP accelerated sex change. 4. More than 95% of the solitary MPs changed sex earlier indicating that the MPs were readily awaiting sex change, once they attain a size of 6 mm. 5. The fraction of sex changes decreased to 90, 84 and 80% in the other combinations showing that in the presence of FP, the MP delayed sex change and opted to grow up to 8 mm size; sex changed FP in simultaneous hermaphrodite at the largest size may produce more eggs confirming the hypothesis of the size advantage model.

3.7.4 Hybridizing Systems

In 1999, Nakatani described an albino strain of the crayfish *Procambarus clarkii*. The wild crayfish (WT), when mated as female or male with albino (AM), the F_1 offspring were WT, indicating that WT is dominant and AM recessive. However, F_2 progenies were in the Mendelian ratio of 0.75 WT: 0.25 AM, i.e. 0.25 carried WT genes alone and 0.50 harbored both WT and AM genes.

Begum et al. (2010) undertook further investigations. In the first series, four different crosses between WT and AM were made. When the cross involved the dominant WT females and males, fecundity was three times (143 eggs/♀) more than that (48 eggs/♀) produced by a cross between the recessive AM ♀ X AM ♂. Likewise, the cross between WT ♀ and AM ♂ produced two-times (99 eggs/♀) than that between AM ♀ X AM ♂. Clearly, the dominant WT is more fecund either as female or male as a mating partner.

TABLE 3.11
Social induction of sex change from protandric male phase to simultaneous hermaphroditic female phase in *Lysmata wurdemanni* (compiled from Baeza and Bauer, 2004). †approximate values only

Social or Size Combinations	Sex Change on 50th Day (%)
Large Scale Experiment	
5 focal MPs + 5 small MPs	85[a]
5 focal MPs + 5 MPs	95[a]
5 focal MPs + 4 MPs + 1 FP	75[b]
5 focal MPs + 5 FPs	65[b]
Small Scale Experiment	
1 solitary focal MP	96
1 Focal MP + 1 small MP	90
1 Focal MP + 1 MP	84
1 Focal MP + 1 FP	80

Not significant within a and b groups

In the second multiple mating series, AM female was mated first by AM male and WT male as second. The ratio between WT and AM juveniles was 0.56 WT : 0.44 AM and was not significantly varied from 0.5 : 0.5 ratio. However, when AM female was mated with WT male as first and then AM males, the ratio 0.85 WT : 0.15 AM was significantly different. In the first group, the AM females fertilized their eggs equally with spermatophores received from AM and WT males, despite WT males being the second to mate. But in the second group, the AM females selectively fertilized their eggs from spermatophores received from WT males, although AM males belonged to their own strain. As fecundity of AM females fertilized by WT males was significantly more (76 eggs/♀) than AM females fertilized by AM males (46 eggs/♀), the choice was for the dominant WT males.

Geographic strains of the European and American copepods *Tisbe battaglia*, *T. bulbisetosa*, *T. clodiensis*, *T. holothuriae* and *T. lagunaris* are intercrossable. Their F_1 'hybrids' are fertile and give rise to variable number of offspring, ranging from 29% in the Banyuls ♀ X Beaufort ♂ in *T. clodiensis* to 63% in Helgoland ♀ X Beaufort ♂ in *T. holothuriae* (Battaglia and Volkmann-Rocco, 1973). The DNA sequence analysis of the cosmopolitan cladoceran *Moina micrura* has shown that two clones have originated one from Europe and other from Australia (Petrusek et al., 2004).

3.7.5 Alternative Mating Strategy (AMS)

In the polygynous mating system, competition for mate becomes more and more intense. Males of lower competitive ability may adopt AMS 'to make use of a bad situation' (Taborsky, 2001) and pave ways for the evolution of two or more mating morphs (Gadgil, 1972). However, " the struggle between the male 'morphs' for possession of females...results not in death of unsuccessful competitors but in a few or no offspring" (Darwin, 1859). Oliveria (2006) classified 140 species belonging to 28 families of fishes (Taborsky, 1998) that display AMS into three patterns: these patterns may also be adopted to crustaceans as well. Accordingly, 1. Plastic reversibles, which switch reversibly in either direction back and forth from one morph to another during their life time (e.g. *Orconectes immunis*, Fig. 3.18). *O. immunis* switches from copulatory to non-copulatory morph during different seasons. This has already been explained elsewhere (p 39). 2. Plastic transformants, which irreversibly switch from one morph to another in a single direction during their life time (e.g. *Macrobrachium rosenbergii*). 3. Fixed alternative sex linked morphs, adopting a specific irreversible strategy for life time (e.g. *Euterpina acutifrons*).

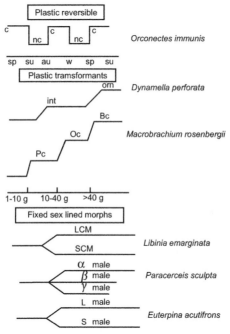

FIGURE 3.18
Schematic illustration for plastic reversible, plastic transformants and fixed sex linked morphs of crustaceans. sp = spring, su = summer, au = autumn, w = winter, c = copulatory morphs, nc = non-copulatory morphs. Int = intermediate and orn = ornamental morphs, LCM = Large Chelate Morph, SEM = Small Chelate Morph, L male = Large male, S morphs = Small morphs. Pc = Pink chelate morphs, Oc = Orange chelate morphs, Bc = Blue chelate morphs

Most crustaceans are polygynous. Understandably, they too display AMS. But the available information on AMS of crustaceans is incomplete and/or scattered widely indicating the need for more information. For the first time, an attempt was made to extend Oliveria's (2006) classification to crustaceans. Table 3.12 briefly summarizes selected characteristic features of some crustaceans that display AMS. Notably, there are more representatives for the presence of two morphs, as predicted by Gadgil (1972). However, representatives for the presence of three morphs are also present. A brief account is described on the (i) representative advantage of competitive morphs over 'sneaking' morphs in *M. rosenbergii,* (ii) endocrine control of morphism in *Libinia emarginata* and (iii) genetic mechanism of morphism in *E. acutifrons.*

TABLE 3.12

Characteristic features of three patterns of mating morphs that display Alternative Mating Strategy (AMS) in crustaceans

Species and References	Features	Morph 1	Morph 2	Morph 3
		Plastic Reversibles		
Orconectes immunis Decapoda (see Pandian, 1994, p 77)	Occurrence I pleopods	Spring & summer Corneous, hard, sculptured	Summer & winter Soft, uniramous, unsculptured	
		Plastic Transformants		
Dynamella perforata Isopoda, Glynn (1968)	Size Penis Appendixes - masculina Uropod Telson	Intermediate Absent Absent Not enlarged Not elaborate	Ornamented Present Present Enlarged Elaborate	
Macrobrachium rosenbergii Decapoda, Ra'anan and Sagi (1985)	Size Relative chela size (cm) Claw color Courtship	Small 1-10 g 0.5-0.7 Pink Submissive, not courting but sneaking	Medium 10-40 1.0-1.5 Orange Aggressive, but not courting	Large > 40 g 1.5-2.0 Blue Aggressive, courting and guarding
		Fixed Sex linked Morphs		
Euterpina acutifrons Copepoda, Haq (1973)	Size (mm) Active Adopted Sex ratio of F_1	Small, 0.5-0.6 Yes, warm 85% small + 15% large	Large 0.65-0.70 No, cold 94% large + 6% small	
Libinia emarginata Crab, Sagi et al. (1992)	Claw Abrasion Gonad Behavior	Small Abraded Small Not aggressive, Sneak and do not guard	Large Unabraded Large Aggressive, Court and guard	
Paracerceis sculpta Isopoda, Shuster (1987), Shuster and Sassaman (1997)	Male Life span (d) Length (mm) Uropod Telson Harem	'sneaker' 68 18 Not elaborate Unsculptured Non-haremic	'satellite' 63 11 Intermediate Intermediate Non-haremic	'territorial' 92 21 Elaborate Sculptured Haremic

M. rosenbergii is a commercially important freshwater palaemonid prawn. In it, the presence of pink (PC), orange (OC) and blue color chelate (BC) morphs are described (Ra'anan and Sagi, 1985). Its sex ratio is 0.5 ♀ : 0.5 ♂. However, the morph ratio within males is 0.25 PC : 0.20 OC : 0.05 BC. By stocking these morphs in different ratios with females (unfortunately, the number of females in each case was not reported), Ra'anan and Sagi (1985) reported that when 3 BC + females were reared, 93% females were fertilized, of which 97% (i.e. 90.2% cf Carbonell et al., 2006 indicating 10% oocytes remaining undeveloped) were berried. The corresponding values for 3 OCs and 6 PCs are 33 and 38%, respectively. Of them, 42 and 44% females were fertilized and berried. Of the total females, 14, 17 and 90% were berried, when fertilized by 6PCs, 3OCs and 3BCs, respectively. Clearly, the BC have a reproductive advantage over the OCs and PCs. It is possible to seal the gonopore with a drop of adhesive super glue (cyanoacrilate), which blocks the release of sperm but does not prevent aggressiveness and mate guarding in males. When one BC was stocked with 2 OCs or 5 PCs, 17 or 33% females were fertilized. In a third combination series, one BC + 2 OCs or one BC + 5 PCs were stocked with females. In the presence of BC, OC could not fertilize even a single female. However, 4% females were fertilized by sneaking five PCs. Clearly, by sheer number and sneaking behavior, the PCs have an advantage over the OCs.

Information available on mating morphism in spider crab *L. emarginata* relates to hormone levels and mating behavior. In this crab, there are two mating morphs: (i) Large clawed, unabraded (with intact pubescent epicuticle covering exoskeleton), post-autumn-molted morph (LCM) and (ii) small clawed, abraded, unmolted (for a year) morph (SCM). LCM has a larger reproductive system. Mandibular Organs (MOs) of LCM are also two-six times more active. The LCM aggressively competes for, mates with, carry and guard the receptive female (Sagi et al., 1992). Contrastingly, the SCM displays a non-competitive 'sneaking mating' strategy (comparable to sneakers in fishes, see Pandian, 2013,) with a solitary female or a group of females, and quickly inseminate, when LCM is distracted. It neither carries nor guards the female (Sagi et al., 1992).

In *E. acutifrons*, there are (i) small, warm-adapted and (ii) large, cold-adapted morphs. Haq (1973) crossed each of the large and small morphs with virgin females. The morph ratio was estimated in these crosses for two successive broods at 16°C and 20°C. The ratios of the large and small morphs among the progenies of these crosses strongly suggest the presence of a genetic mechanism, in which genes responsible for the morphism are linked to sex determination and a role played by autosomes in producing 6% smaller morphs and 15% large morphs, sired by large and smaller morphs, respectively.

4

Asexual Reproduction and Regeneration

Introduction

A taxonomic survey on the modes of reproduction in aquatic invertebrates shows that the hemocoelomates namely Arthropoda and Mollusca do not reproduce asexually (Table 1.1). Hence, they may not retain Embryonic Stem Cells (ESCs) to clone the whole animal and reproduce asexually. With ongoing differentiation during early development, the stemness of stem cells progressively decreases from totipotency to pluripotency, multipotency, oligopotency and finally to unipotency. As they have the capacity to regenerate one or more of their appendages, crustaceans seem to have retained adequate mass of Oligopotent Stem Cells (OlSCs) only. To achieve cloning by asexual reproduction in cnidarians, turbellarians and colonial ascidians, the minimum mass of Pluripotent Stem Cells (PSCs) required is estimated to range between 100 and 300 (Y. Rinkevich et al., 2009). It has, however, been claimed that it can be achieved with as small as 10-15 PSCs in parasitic colonial rhizocephalans (Isaeva, 2010). In fact, Shukalyuk et al. (2005, 2007) have brought some evidence in support of their claim that these rhizocephalans possess adequate mass of stem cells to reproduce asexually.

4.1 Parasitic Colonial Rhizocephalans

In their life cycle, the parasitic rhizocephalans include free living larval stages and parasitic adult stage. Subsequent to the nauplius-like planktonic stage, the female cypris larva passes through kentogon and vermogon stages. The vermogon injects a cluster of cells into the body of the host crab. These cells develop into interna, from which an externa is developed on the body surface of the host. Morphologically, the interna consists of roots and stolons that branch out into the hemocoelic cavity of the crab. As dense ring or muff, it attaches to the host's gut. Thus, the parasitic colonial rhizocephalans *Polyascus* (*Sacculina*) *polygenea* and *Peltogasterella gracilis* consist of two systems: a trophic system with numerous dentritic roots in the interna and a reproductive system with many Primordial Germ Cells (PGCs) (?) in the externa at different stages of development. Undifferentiated stem-like cells, present inside the stolons, emerge out to form asexual buds of externae and play a role in the morphogenesis of the earliest buds and subsequently migrate into the ovary as Oogonial Stem Cells (OSCs) (see Shukalyuk et al., 2007). In fact, Isaeva et al. (2009) consider that "the stem cells are cultured in the host's hemocoelomic medium". Non-colonial rhizocephalans manifest only a single externa on the host's body surface, whereas the colonial forms reproduce asexually to produce multiple externae, whose number may range from a dozen to hundreds (Glenner et al., 2003). The capability of these parasitic colonial rhizocephalan females to reproduce asexually has been earlier reported by many authors (e.g. Hoeg and Lutzen, 1995, Takahashi and Lutzen, 1998, Liu and Lutzen, 2000, Glenner et al., 2003). Incidentally, the cypris-like male larvae transform into trichogen and finally settle on the receptacles of the externa and commence spermatogenesis (Shukalyuk et al., 2007).

However, it must be noted that the injected PSCs by vermogon may also be exhausted with increasing number of asexually produced externae. During molting of the host crabs, the externae may be lost leaving a scar. However, regeneration of the lost externa occurs only rarely (Lutzen, 1981), i.e. as long as the interna is still left with PSCs. Kristensen et al. (2012) have noted that the loss of externa usually leads to the death of interna. As the limited number of PSCs present in the interna and their migration, following differentiation of OSCs, into externa(e), the interna seems to be left with no more PSCs at some point of time. Hence, the loss of externa leads to the death of interna and the parasite itself.

In 2005, Shukalyuk et al. reported that the stem-like cells in the parasitic colonial rhizocephalans selectively express alkaline phosphatase activity, a histochemical marker for the presence of stem cells. Further, these stem-like cells in *P. gracilis* selectively express Proliferating Cell Nucleus Antigen (PCNA), a cellular marker for cell reproduction. To confirm molecular nature of the stemness of these stem-like cells, Shukalyuk et al. (2007) found the expression of *vasa*-related genes and DEAD-box RNA helicase gene products in the stem-like cells of *P. polygenea* and thereby claimed that these cells are stem cells. However, *P. polygenea* male and the non-colonial rhizocephalan *Clistosaccus paguri*, which do not reproduce asexually, also express the *vasa*-related *PpVLG* (*P. polygenea*'s *vasa*-like gene). Hence, the evidence thus far reported by Shukalyuk et al. (2005, 2007) may not be adequate, albeit the fact that the parasitic colonial rhizocephalan females reproduce asexually remains unquestionable. However, isolation and transplantation of the claimed 10-15 'stem cells' to infect an uninfected crab *Hemigrapsus sanguiensis* remains to be demonstrated.

4.2 Autotomy and Regeneration

In general, autotomy refers to the breaking of an animal's body into two or more pieces, as a mode of asexual reproduction in annelids and echinoderms. In crustaceans, which do not reproduce asexually, autotomy means the reflex of severing of one or more appendage(s); it is an adaptation to escape from predators and limits the wound. While it provides immediate survival benefits, the loss of appendage (s) and the consequent 'regenerative load' cost long term dysfunction, and resources and energy (Juanes and Smith, 1995), which otherwise could have been channeled for growth and/or reproduction (e.g. Barria and Gonzalez, 2008). Many authors have described the process of regeneration at the cost of somatic growth during successive molts; however, no one has thus far described the effect of natural/artificial loss of spermatophore-transferring appendages on reproduction. Barring Vogt (2010), no one has described the effect of autotomy of one or more appendages on the reproductive performance especially of females.

TABLE 4.1

Appendage loss (%) in field populations of decapods (from Mariappan, 2000, modified, permission by C. Balasundaram)

Species	Sex	Loss (%)
Prawn		
Macrobrachium nobilii	Juvenile	11
	Male	15
	Female	22
Lobsters		
Panulirus argus		40
Nephrops norvegicus	Male	62
	Female	41
Homarus americanus	Male	40-44
	Female	30-61
Crabs		
*Atergatis floridus**	Male	41
	Female	18
Callinectes sapidus		25
Carcinus maenas	Juvenile	2†
	Male	18††
	Female	55
Cancer magister		45
C. pagurus	Male	13
	Female	10
Chionoecetes bairdi	Juvenile	35
	Male	43
	Female	23
Cyrtograpsus angulatus		80
Necora puber	Juvenile	23
	Male	33
	Female	29
Paralithodes camtschatica	Juvenile	29
	Male	15
	Female	20

*Poisonous crab, †small ~ 27 mm CW, ††large ~ 73 mm CW

Autotomy is common in natural populations of decapods (Table 4.1). Fishing practices are responsible for substantial limb loss in decapods. Muthuvelu et al. (2013) have described injuries caused by prawn fishing trawlers in the Pondicherry coast of India. The proportion of limb loss ranges from 2% in small males of *Cancer maenas* to 80% in *Cyrotograpsus angulatus* and within a species, it increases from 2% in juveniles to 55% in large females of *C. maenas*. In sexually dimorphic decapods, the cheliped is lost more frequently. The importance of cheliped in feeding, sexual display and mating can be understood from its size. The chelae of crayfish *Orconectes rusticus* are important chemosensory appendages (see Belanger and Moor, 2013) in perception and discrimination of female odors (Belanger et al., 2008). The size of chelipeds constitutes 10-26% of the body weight of *Macrobrachium nobilii*, 20% in *C. maenas* and 50% in *Menippe mercenaria* (see Mariappan et al., 2000).

Following a molt, maximum growth of regenerating chelipeds may be equivalent to 12-15% of the weight of a pre-molt decapods, however, at the cost of about one third reduction in somatic growth. Regeneration of a lost limb to its original size is dependent on size of the crabs at the time of claw loss (McLain and Pratt, 2011) and duration of a given molt cycle (Barria and Gonzalez, 2008). In cheliped regenerating anomuran *Petrolisthes laevigatus*, Barria and Gonzalez (2008) found that more than sub-optimal feeding, the decreased intermolt duration reduces somatic growth. Incidentally, anecdysic crabs do not molt and regenerate the claws and chelipeds after the terminal molt. Regeneration of the autotomized parts may require from two successive molts in *Callinectes sapidus* to 4.7 molts in *Paralithodes camtschatica* (Juanes and Smith, 1995). Strikingly, the occurrence of complete regeneration of chelae/chelipods indicates the presence of stem cells within muscles of claw/cheliped. Secondly it also explains 70% survival of the autotransplanted (transplantation within the same individual) crab (see Table 4.3). The sand fiddler crab *Uca pugilator* takes four molts for complete regeneration of its claws (McLain and Pratt, 2011). While regeneration may be accomplished within a year in juveniles, it may require as long as 7 y in adults (Juanes and Smith, 1995). Interestingly, the transition of chelae from pink to blue in *M. rosenbergii* is not accompanied by allometric changes in segments (Kuris at al., 1987). The pink males transform faster into blue ones in the absence of blue chelate morph or when reared in isolation (Ra'anan and Cohen, 1985).

From careful observations in *Procambarus fallax*, Vogt (2010) recorded that autotomy and subsequent regeneration of the last limb did not shorten the life span, albeit reduced the number of spawning. On 850 d of its life, an autotomized B3 female had a body weight of 8.7 g and three spawnings (see Table 3.5), in comparison to 16.6 g and five spawnings in normal A1 female. Another female, which had a heavy 'regenerative load' by having lost both chelipeds and six walking legs at the age of 402 d, weighed 14.3 g and spawned only twice. Clearly, autotomy reduces not only somatic growth but also reproductive output.

In decapods, autotomy occurs at a pre-formed fracture plane, which is located at the basis-ischium segment of the pereopods (Fig. 4.1). Intense stress on a limb causes quick abandoning of the distal end of the limb by a specific mode of action on musculature of the proximal limb (Vogt, 2010). Following autotomy, the fractured plane is immediately closed by a pre-existing membrane and thereby the loss of hemocoel is arrested. Regeneration of the limb is subsequently initiated by an emerging blastema, consisting of mitotically active cells. These cells have the ability to differentiate into four different tissues namely (i) epidermis, (ii) nerve cells, (iii) connective tissues and (iv) muscle cells (Hopkins et al., 1999). Notably, the first two are ectodermal derivatives and the last two are mesodermal derivatives.

4.3 Claw Tissue Transplantation

This is an important area of research with potential to reveal the presence of stem cells in claw tissue and their role in regeneration of autotomized appendages. Unfortunately, only a few publications are available; they have also not received their due attention, as indicated by less than six-seven citations/publication. A reason for this may be that the authors have not looked at their contributions from the point of stem cells. The following narrative is an attempt to highlight their contributions from the angle of stem cells.

Taking advantage of the availability of wild and albino strains in *Procambarus clarkii*, Mittenthal (1980, 1981) and Nakatani (2000) made reciprocal heterospecific transplantations of cheliped tissues or tissues from pollex or dactyl (Fig. 4.1) into the stump of autotomized walking leg or eyestalk. The claws were regenerated. The morphology of the claws was determined by the donor but the color by recipient. Clearly, the claw tissues contained the 'stem cells', from which a claw could be regenerated and in that process both the tissues of the donor and recipient participated. But their work, especially that of Nakatani, suffered from poor (7-8%) survival of the regenerated crayfish (Table 4.3).

In a series of publications, Kao and Chang (1996, 1997, 1999) made better planned and more meaningful transplantations and ensured a higher survival of the transplanted crabs *Cancer anthonyi, C. gracilis* and *C. productus*. Firstly, they showed that autotransplantation (transplantation within the same crab) of claw tissues of *C. gracilis* (32 mm CW) into the autotomized stump of the 4th walking leg induced regeneration of a complete claw. However, frozen claw tissues or tissues from walking legs failed to induce regeneration. Notably, tissues from walking legs from the relatively old crab failed to induce regeneration. Also, the stem cells underwent irrevocable damage or death, when frozen. Centro-lateral transplantation of claw muscle tissue into the autotomized stump of the 4th walking leg regenerated a claw with normal handedness. Most regenerated claws displayed a combination of characteristics of both the claw and leg, suggesting that donor tissue induce regeneration, to which the recipient was a responding field in regeneration of the crab. For example, tissues present in eyestalk do not induce regeneration but provide a responding field (Kao and Chang, 1996). Clearly, live claw muscle tissues alone induce regeneration of the claw. Hence, the claw tissues possess the Oligopotent Stem Cells (OlSCs) that are capable of differentiation into the epidermal derivatives namely (i) epidermis and (ii) nerve cells including dactyl and chemoreception sensory cells as well as mesodermal derivatives namely the (iii) connective tissues and (iv) muscle cells.

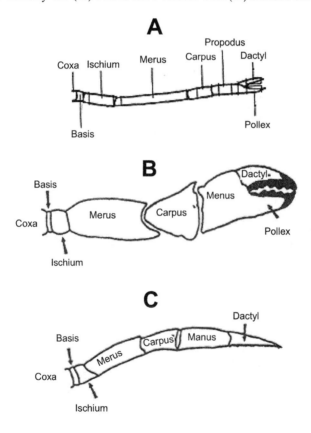

FIGURE 4.1
Free hand drawings to show A. the segments in the second cheliped of *Procambarus clarkii* (from Nakatani, 2000) and segments of claw B. and walking leg C. of the crab *Cancer gracilis* (from Kao and Chang, 1999).

In their second publication, Kao and Chang (1997) showed that autotransplantation of tissues drawn from the 4th walking leg of relatively young crab *C. gracilis* (12 mm CW) into the eye sockets regenerated complete walking legs in the eye sockets. Moreover, the transplantation of tissues of claw's digit either from the distal dactyl or pollex or from the proximal claw segments ischium and merus (Fig. 4.1) regenerated complete claws in the eye sockets within two-three molts. When the donor's tissues were from the claw digits, the most distal claw segments regenerated first followed by the proximal claw segments in the subsequent molts. Thus tissues from the distal portions of the claw are also capable of regenerating the proximal portions of the claw in the eye sockets. Notably, tissues drawn from the walking leg of an older (32 mm CW) crab could not induce regeneration but the same drawn from an younger crab (12 mm CW) did induce regeneration of the claw as well as a complete walking leg. Understandably, the functional OlSCs in the young crab are perhaps lost as age advanced and the anecdysic crab undertook terminal molt at the size of ~ 30 mm CW.

Autotransplantations of the limb tissues, drawn from the 4th walking leg of *C. gracilis* (14 mm CW) into the claw stumps regenerated (i) walking leg-like limb, (ii) bifurcated limbs, (iii) a composite limb with claw and walking characteristics or (iv) a normal claw. Clearly, functional OlSCs are also present in the 4th walking leg, until the crab attains a body size of 14 mm CW and OlSCs are capable of regeneration. Reciprocal interspecific transplantations of claw tissues of the slender clawed *C. gracilis* (14 mm CW) and stout clawed *C. productus* (30.5 mm CW) were also undertaken. The regeneration resulted in the production of a claw-like limb, (ii) a walking leg with toothed dactyl, (iii) a walking leg with 2 dactyl, (iv) two pollexes or (v) a normal walking leg. Most regenerates had characteristics of both the donor and recipient, which confirmed that the donor tissues induce regeneration, to which the recipient provided a responding field. However, it may be noted that irrespective of age, functional OlSCs are present within the tissues of the 4th walking leg but are functional only up to the terminal molt undertaken by the anecdysic crab. Notably, autotransplantation resulted in regeneration of 70% of the normal claw but interspecific (reciprocal) transplantations regenerated 54% of normal walking legs. With regard to the number of organizers and limb fields, and the involvement of Distal-less protein in bitamous limb formation, the differences reported between transplantation studies in the crabs on one hand and cockroach/amphibians on the other, the available information is briefly summarized in Table 4.2.

Understandably, autotransplantation resulted in 33-71% survival in *C. gracilis* and 81% in *C. productus* (Table 4.3). But that involving multiple tissues from either pollex or dactyl or ischium to the stump of the 4th walking leg resulted in low survival of 35% in *C. gracilis* but 73% in *C. productus*. Notably, transplantations of claw or leg tissues from a relatively old donor *C. productus* to the claw stump to the relatively young recipient *C. gracilis* resulted in survival from 50-73% . Unusually, the reciprocal interspecific transplantation from a young donor *C. gracilis* to an old *C. productus* produced 100% survival. With limited information, it is difficult to generalize whether the observed wide variations are due to (i) inherent difference in number and/or stemness or stem cells (ii) the level of immune-tolerance to different tissues from the same individual or an individual belonging to another strain and species (iii) young vs old recipient and/or (iv) difference in thickness of cheliped between *C. gracilis* and *C. productus*.

TABLE 4.2

Comparison of transplantations of the crabs with those of cockroach and amphibians

Crabs	Cockroach /Amphibians
Donor tissues are from fully differentiated crabs	Donor tissues are from regenerated blastema or limb buds
The regenerated two limbs are characterized by variable morphology and branching points along the proximo-distal axis suggesting that the two distal organizers and limb fields are derived one each from donor and recipient tissues	Autotransplantations of distal portion of limbs to limb stump with reversed antero-posterior or dorso-ventral axis produce identical triplicate legs and all of them generate at the same proximo-distal level. Three distal organizers generate three identical limb fields, regardless of antero-posterior polarity. The organizers include one from donor and two from recipient.
The generation of bifurgated limb supports that Distal-less protein is involved in crustacean biramous limb formation	Expression of Distal-less proteins in small isolated group of cells in an imaginal leg disc induces the formation of the second leg

TABLE 4.3

Survival of successfully regenerated crayfish and crabs (compiled from Nakatani, 2000, Kao and Chang, 1997, 1999)

Species	Donor Tissue	Recipient Stump	Survival (%)
Cancer gracilis	Claw	Claw	71
	4 W	Claw	33
C. productus	4 W	Claw	81
C. gracilis	4 W	Claw††	100
C. productus	4 W	Claw†	73
	Claw	4 W†	50-60
*C. productus**	Claw	4 W	78
C. gracilis	Claw	4 W	35
Procambarus clarkii[1,2]	Claw or Pollex or dactyl	4 W[2,1]	7-8

† = *C. gracilis*, †† = *C. productus*[1] albino[2], wild strain of *P. clarkii*,
* = multiple tissue autotransplantation

Incidentally, 7-8% of the regenerated crayfish alone survive following inter-strain transplantations. It is not clear whether the crayfish has low immune-tolerance. In vertebrates, amputation of a part of unpaired fins and tail of fishes is followed by regeneration (see Sheela et al., 1999). However, the regenerative capacity is progressively decreased in higher vertebrates. For example, amputation of a segment of a finger or part or whole of appendage is no more followed by regeneration. Amazingly, thanks to the presence of stem cells amidst the muscles of claw/leg of crustaceans regenerate a full complement of the claw and cheliped, as well as to tolerate the graft from the same individual or individual belonging to another strain or species by immuno-modulations. Transplantation research in crustaceans is sure to reward stem cell researchers and immunologists.

From the descriptions on transplantation and regeneration studies, the following may be inferred: 1. Functional OlSCs are present within the muscle tissues of the claw and 4th walking legs. Irrespective of age/size, the OlSCs are present and functional in the claw. The OlSCs are also present possibly at the pre-formed fracture plane, located at the basis-ischium segment of the pereiopods and functional throughout the life span of diecdysic decapods. However, they are perhaps lost, when anecdysic crabs undertake the terminal molt and are no longer harbored within the muscle tissues of the claw and walking leg. 2. The potency of OlSCs is limited to differentiation of ectodermal derivatives (i) epidermis, (ii) nerve cells and mesodermal derivatives of (iii) connective tissues and (iv) muscle cells, 3. In the parasitic colonial rhizocephalan females too, the roots and stolons in the interna as well as the tissues of externa comprise of (i) epidermis, (ii) nerve cells, (iii) connective tissues and (iv) muscle cells. Of course, the 'stem cells' are also capable of differentiation into oogonia. However, OlSCs are not capable of differentiation into endoderm and its derivatives. Notably, asexual reproduction in parasitic colonial rhizocephalan females and regeneration of claw and walking leg in decapods share many common features of differentiation, which is limited to a few tissues alone. Not surprisingly, the parasitic colonial rhizocephalan females with limited number of tissues reproduce asexually.

5

Cysts and Resting Stages

Introduction

In diecdysic crustaceans, every molt may not be followed by spawning and brooding, as a consequence of considerable energy drained on oogenesis and spawning. The release of neonates/offspring is well synchronized with favorable environmental conditions. When conditions are not favorable, entomostracans produce cysts, but malacastracans undergo a resting phase. Besides providing an opportunity for recombination, cysts serve as a 'seed pool' rendering survival of a population in periodically stressful or unpredictable environments.

5.1 Ovarian Diapause

As indicated, females that have been drained of considerable energy and nutrient resources due to oogenesis, spawning and brooding, may prolong the interspawning period or undergo ovarian diapause. Tropical cladocerans undertake 15 molts during the adult stage. Of them, 20% molts are not followed by brooding, i.e. a resting stage is included following every four successive adult molts (see p 34-35). In *Macrobrachium nobilii* too, 39% of the adult molts are not ovigerous (Pandian and Balasundaram, 1982), i.e. a resting stage is included following every three successive adult molts.

Embryos of some decapods cease to develop beyond gastrula stage, as they undergo diapause, while being brooded by *Maja squinado* for six w, *Cancer pagurus* eight w *Corystes casselanrus* 14 w and *Hyas coarcticus* 16 w (Wear, 1974). The presence of diapausing embryos on the brooding sites inhibits subsequent molt (see Hubschman and Broad, 1974) and thereby provides a resting phase in these females in the decapods.

In temperate crustaceans, the resting stage is enforced due to the absence of one or another structure related to fertilization and/or brooding. The ornamentation on the spermatophore-transferring appendages are absent during summer and winter in crayfish *Orconectes immunis* (see p 39) and thereby resting stages are included. In others, setae in egg carrying oostegites are absent to provide a resting stage for a shorter or longer period. For example, the oostegite setae (e.g. Amphipoda: *Amphiporeia lawrenciana*, Downer and D.H. Steele, 1979) or oostegites themselves (e.g. Isopoda: *Jaera ischiosetosa* D.H. Steele and V.J. Steele, 1970) are absent to provide these peracarid females to have a resting phase. In decapods, egg-carrying setae are absent in *Palaemon squilla*, *Crangon crangon* and *Panulirus longipes cygnus* to provide a resting phase in these decapod females. The Gonad Inhibiting Hormone (GIH) arrests not only vitellogenesis but also causes the disappearance of one or another structure related to egg brooding and (cf Le Roux, 1931 a, b) thereby provide a resting phase to the females across all malacastracans (see Pandian, 1994, p 130-131).

5.2 Diapausing Cysts

The branchiopods namely tadpole shrimps (Notostraca) and clam shrimps (Spinicaudata) produce cysts alone (Table 5.1). The non-large branchiopods namely ostracods also produce cysts. However, anostracans are capable of producing both cysts and subitaneous eggs. For example, females of *Artemia* spp as well as temperate anostracans *Cheirocephalus diaphanus* and *C. stagnalis* (see Brendonck, 1996) produce subitaneous eggs and dormant cysts. Some ostracods are also capable of producing both subitaneous eggs and cysts. For example, the ostracod *Heterocypris incongruens* produce subitaneous eggs and diapausing cysts (see Delorme, 2009). Rozzi et al. (1991) have shown that the proportion of cyst production depends on the genotype of ostracod, temperature and photoperiod.

Diapausing cysts and ephippia can be produced sexually or parthenogenically in *Artemia* spp and daphnids. Sexual and parthenogenic *A. parthenogenetica* females[*], for example, produce nauplii ovoviviparously and cysts oviparously. Table 5.2 lists the effects of these oviparous and ovoviviparous modes of reproduction on selected life history traits. Ovoviviparity is not uncommon among ostracods (e.g. *Darwinula*, *Cyprideis*, Delorme, 2009). Incidentally, cladoceran females produce eggs parthenogenically or ephippia sexually, from of which neonates are released. However, some calanoids like *Limnocalanus macrurus* produce resting eggs only in a habitat but subitaneous and resting eggs in another (Williamson and Reid, 2009). Notably, malacastracan females may pass through resting stage but do not produce cysts.

---

[*] Golubev et al. (2001) identified it as *A. salina*. However, the conference, in which the paper was presented, recommended the name as *A. parthenogenetica*

TABLE 5.1

Sexually and/or parthenogenically produced entomostracan cysts

Taxon	Reported Observations
Anostraca	Bisexuals obligately produce oviparous cysts e.g. *Streptocephalus dichotomus*. Expectedly, *Artemias* produce cysts both sexually and parthenogenically. They are also capable of producing directly nauplius ovoviviparously. *Cheirocephalus diaphanus* and *C. stagnalis* also produce subitaneous eggs and diapausing cysts
Notostraca	Sexually produce oviparous cysts, e.g. *Triops cancriformis*
Spinicaudata	Sexual reproduction obligatory produce cysts. e.g. *Eulimnadia texana*
Cladocera	Obligate parthenogenics produce ephippial e.g. *Daphnia cucullata* neonates or ephippia. Facultative parthenognics are capable of producing neonates or sexual ephippia cysts. e.g. *Daphnia pulex*. Some produce ephippia only through sexual reproduction, e.g. *Daphnia ephemeralis*, while others parthenogenic ephippia only e.g. Spanish lineage of *Daphnia pulicaria* and *D. middendorffiana*
Ostracoda	Bisexuals produce oviparous cysts. e.g. *Eucyspris virens*. Parthenogenics (e.g. Darwinoids) also produce cysts. *Heterocypris incongruens* produce both subitaneous eggs and dormant cysts
Calanoida	Bisexuals capable of producing sibitaneous eggs or dormant resting eggs

TABLE 5.2

Life span and reproductive traits of parthenogenic and bisexual families of Ukrine *Artemia parthenogenetica* (compiled from Golubev et al., 2001)

Parameters	Parthenogenic	Bisexual
Life span (d)	73.6	65.0
Generation time (d)	40.9	29.9
Generation time as% of life span	56	46
Number of broods	5.1	3.4
Broods with cysts (no.)	4.4	0.9
Absolute fecundity eggs + nauplii (no.)	180.9	94.3
Absolute fecundity of cysts (no.)	153.7	26.5
Absolute fecundity of nauplii (no.)	27.1	67.1
Fecundity at I brood (no.)	29.1	34.1
Larvae produced as% of absolute fecundity	15	72

Adaptations to Unpredictable Habitats Cysts producing entomostracans inhabit temporary aquatic habitats. These astatic aquatic habitats dry out or freeze periodically, or are subjected to striking changes in water level. Some of them remain dry for several years; others are inundated for a few weeks or days only. Some of them are flooded several times a year; others have only a single wet phase (Brendonck et al., 1996). Hence, these aquatic bodies are transient, hazardous and unpredictable but provide predator-free 'biological vacuum'. Having chosen to inhabit these aquatic bodies since Cambrian, these entomostracans have developed excellent adaptive features to survive and flourish in them. Firstly, they grow very fast and attain sexual maturity within a few days or weeks. For example, notostracans (1-2 cm long) require just 2-3 w to develop from an egg to attain sexual maturity (Weeks et al., 2014 b). Anostracans (1-2 cm long) do it in 3 w at 15°C but within 45 d at 4-17°C. The spinicaudatan *Eulimnadia diversa* (3-4 mm) requires only 4-11 d to develop from egg to adult (Dodson and Frey, 2009). On the fast growing potential of Anostraca, Dumont and Munuswamy (1997) commented that they "have a biomass multiplication factor of up to 10,000 between hatching and maturity" (see also Atashbar et al., 2012). Secondly, they also have one of the highest reproductive potential coupled with capacity to produce cysts. For example, the life time fecundity of *Streptocephalus proboscideus* is up to 4,000 cysts. Table 5.3 shows that the reproductive features of some anostracans are characterized by reproductive life span ranging from 49% in *Branchinecta orientalis* to 79% in *S. torvicornis* of their respective LS and life time cysts production from 262 in *Tanymastigites perrieri* to 2,640 in *S. torvicornis* (Beladjal et al., 2003a, b).

TABLE 5.3

Reproductive traits of selected anostracans reared at 25°C

Parameter	*Branchipus schaefferi* Beladjal et al. (2003a)	*Streptocephalus torvicornis* Beladjal et al. (2003a)	*Tanymastigites perrieri* Beladjal et al.(2003b)	*Branchinectes orientalis†* Atashbar et al. (2012)
Reproductive period (d)	17.6	36.7	12.0	16.5
Reproductive period as proportion of LS (%)	73	79	69	49
Clutch (no.)	13.3	8.2	15.0	5.5
Clutch (no. of cysts)	107.6	221.8	–	51.5
Cysts production (no./♀)	1734	2640	262	306

†at 24°C

Entomostracan cysts withstand desiccation and freezing up to -18°C (e.g. *Heterocypris incongruens* (Delorme, 2009). Interestingly, both subitaneous eggs and diapausing cysts/ephippia resist digestion and remain hatchable (Table 5.4). Small proportions (13-18%) of subitaneous eggs of anostracans, collected from feces of birds, are hatchable but of copepods are higher (50%). Subitaneous eggs are less resistant to digestion than resting eggs of copepods.

TABLE 5.4

Resistance to digestion and hatchability of subitaneous eggs and diapausing cysts of entomostracans

Species and Reference	Passage Via Digestion	Hatchability (%)
	Anostraca	
Subitaneous eggs of	Fecal soil	2.8-12.5
Branchinella mesovallensis	Kidder† feces	12.4
Branchinella lindhali	Mallard† feces	17.6
Rogers (2014)		
	Cladocera	
Daphnid ephippia	Predators	Resist digestion
Mellors (1975)		
	Copepoda	
Labidocera aestiva	Polychaete predators:	
Marcus (1984)	Feces of *Capitella* sp	66
	Subitaneous eggs	84
	Diapausing cysts	
	Feces of *Strebdospio benedicti*	50
	Subitaneous eggs	84
	Diapausing cysts	
	Ostracoda	
Cyprinodopsis viduu,	Feces of goldfish and	Hatchable
Cyprinotus incongruens	swordfish, wild and domestic	
Physocypris sp, *Pomotocypris* sp	ducks	
Delorme (2009)		
Elpidium	Feces of tadpole *Scinaxax*	Resists digestion
Lopez et al. (2002)	*perpusillus* and mouse	

5.3 Cyst and Ornamentation

The cyst morphology is a promising feature in taxonomic identification. The cyst's shape ranges from cylindrical (*Eulimnadia cylindrova*, Fig. 5.1D) to spherical (*Artemia* spp. Fig. 5.2A) and to a twisted form. Cysts of many entomostracans are ornamented, which provides protection against predators (Dumont et al., 2002). Understandably, the cysts of *Artemia* spp are smooth and devoid of ornamentation (Fig. 5.2A), as they float on supersaline waters with no predators; hence, ornamentation may be needless, albeit the cysts of Puthalam (Tamil Nadu, India) strain of *A. parthenogenetica* have a rough surface (Sugumar and Munuswamy, 2006). The ornamentation includes crests, ridges, furrows and large depressions (Fig. 5.1A, 5.2D). In *E. magdalensis*, the spherical cyst is ornamented by a large (25-33) depression with a flat bottom (Fig. 5.1A). The *E. cylindrova* cyst is cylindrical with wide and domed extremities, resulting in polygonal shape (Fig. 5.1D). However, these details of ornamentation are severely affected by the combined effects of abrasion and filling sediments. Due to abrasion, the ridges are eroded, and rows and depressions are filled by fine sediments. Consequently, the cyst surface can significantly be altered, according to the time spent in sediment (Fig. 5.1B, C, 5.1E, F). Still, the cylindrical cysts are more readily identifiable than spherical ones (Rabet et al., 2014). As nauplii can be hatched from un-crushed cysts abraded to different levels, egg viability also remains unaffected by reduced shell thickness. Cyst crushing can be more harmful than abrasion during the time spent in the soil. With larger alveolar layer (see Fig. 5.2 C1, 3), the strength required to crush *Artemia* cysts is fairly low (Hathaway et al., 1996).

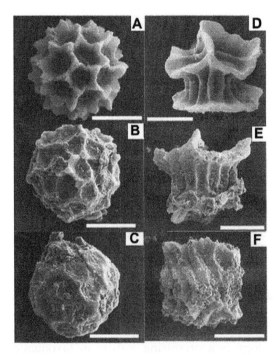

FIGURE 5.1
Effects of abrasion and sedimentation on the morphology of (A, B, C) *Eulimnadia magdalensis* and (D, E, F) *E. cylindrova*, scale: bar = 100 μm (from Rabet et al., 2014, permission by N. Rabet).

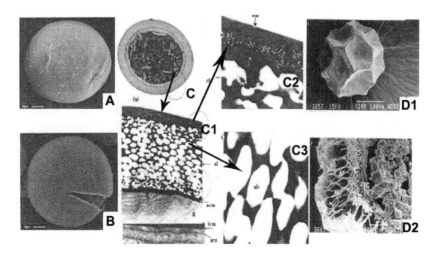

FIGURE 5.2
Scanning electron micrographs of hydrated cysts of A. *Artemia franciscana* and B. *A. parthenogenetica* showing smooth surface. C. Transmission electron micrographs of C showing a cross section of *A. parthenogenetica* cyst, C1 = Cyst membranes, C2. Thick cortex, C3. Alveolar layer (from Sivagnanam, 2005, permission by N. Munuswamy), D. Scanning electron micrographs of *Streptocephalus dichotomus*, D1. Cyst and D2. Fibrillar alveolar layer of ~ 10 μm thickness (from Munuswamy et al., 2009, permission by N. Munuswamy).

The entomostracan cysts are of different sizes. In *Eulimnadia*, for example, the maximum diameter of the cyst ranges from 139 μm in *E. cylindrova* to 145 μm in *E. colombiensis* (Rabet et al., 2014). Using density difference between cyst and sediment, cysts are recovered from sediments. The alveolar layer size in the outer cover of the cyst (Fig. 5.2 C3, D2) diminishes with increasing density of cysts and is as an important factor in cyst floatation. The supernatant containing floating cysts is filtered through progressively smaller sized sieves with mesh size ranging from 900 down to 80 μm. Through this technique, five-10 cysts of *Eulimnadia* spp are obtained from 200 g soil (Rabet et al., 2014). Clumping of cysts by the mother – an adaptation to stabilize the egg bank – reduces labor in systematic sieving technique. In Taiwanese Siangtian pond, dried soil samples (20 cm × 20 cm × 5 cm) typically contain intact cyst clumps of fairy shrimp and clam shrimps. The cyst can directly be counted, although the values are underestimates (Wang et al., 2014). Direct staining of 18 m y old sediments with calcofluor, a fluorescent optical brightener, revealed the presence of much smaller gonidium. Immunofluorescence using polyclonal antibodies raised against the deep-sea isolate of *Aspergillus terreus* confirmed its presence in the form of 18 m y old gonidium in the sediment sub-section, from which it was isolated (Raghukumar et al., 2004, Damara et al., 2006). A possibility of coupling it with flow cytometric separation of the calcofluor

-positive cyst from calcofluor -negative sediment (see also Pandian, 2011, p. 137) is to be explored.

Cyst Structure *Streptocephalus dichotomous* cyst consists of a thick outer cortex (5-6 μm), the middle thin alveolar layer (~ 10 μm, see DeWalsche et al., 1991), characterized by an alveolar mesh and an inner most cuticular layer (2μm) (Fig. 5.2 D2). In *Branchinella thailandensis*, the thick cortex measures ~ 20 μm, the thin middle alveolar layer and an inner most cuticular layer of 5 μm (Plodsomboon et al., 2013). The consistently spherical shape of the alveoli *of Artemia parthenogenetica* suggests that they are filled with or formed by gas bubbles, which assist the cyst to float (Sivagnanam et al., 2013, see also Sugumar and Munuswamy, 2006). The well-developed alveolar layer considerably diminishes the cyst density after desiccation and is an important floating mechanism. The floatability is important in spatial dispersal (Caceres et al., 2007) and harvesting of the cysts at a commercial level. Interestingly, subitaneous egg of copepod *Simodiaptomus* (*Rhinediaptomus*) *indicus* is surrounded by a thin three-layered outer chorion but the diapausing egg is covered by highly complex thick and four-layered outer chorion (Dharani and Altaff, 2004b).

Shell Glands In *S. dichotomous*, there are two pairs of shell glands, the anterior and posterior pairs are located anterior and posterior to the ovisac/ uterus. The glands are connected to the ovisac through fine ducts (Fig. 3.2A). The secretary cycle of these glands is synchronized with that of ovarian cycle. In their structure and function, the glands are similar to those of *Artemia* (Munuswamy and Subramoniam, 1985b). However, the thickness of the outer cortical layer shows considerable variations between bisexual and parthenogenic strains of *Artemia*. The subitaneous eggs have a thinner cortex than that of cysts (Sivagnanam et al., 2013).

5.4 Commercial Potential of Cysts

Table 5.5 lists the abundance of cysts resting eggs of a few entomostracans from sediments. Diapausing cysts/resting eggs are banked in densities ranging from 10^3 to 10^6/m² of sediments in both freshwater and seawater. The nauplii hatched from some of these cysts can serve as live 'fodder' in aquaculture, especially to rear delicate larval stages of fishes and prawns. Hence, these cysts amidst sediments or on soils are indeed 'biological gold mines'. Nevertheless, 'mining' them may prove costlier than their actual cost.

TABLE 5.5

Abundance of entomostracan cysts (from Pandian, 1994, updated)

Taxon	Abundance (No./m^2)
Freshwater	
Anostraca	
Hoplopedium gibbereum[5]	~170,000
Branchinella kugenumaensis[2]	29,514
Streptocephalus vitreus[5]	~16,000
S. dichotomus[6]	25†
Spinicaudata	
Eulimnadia braueriana[2]	31,969
Lynceus biformis[2]	2,41,184
Cladocera	
Ceriodaphnia pulchella[5]	~520,000
D. pulex[3]	40,000
Daphnia dentifera[3]	30,000
Bosmina longirostris[5]	~20,000
D. ephemeralis[3]	11,000
D. pulicaria[3]	10,000
D. galeata mendotae[4]	350
Copepoda	
Diaptomus pallidus[5]	~650,000
D. sanguineus[5]	~45,000
Sea Water	
Anostraca	
Artemia monica[1]	7,000,000
Cladocera	
Penilia avirostris	122,000
Evadne tergestina	2,000
Podon polyphemoides	4,000
Copepoda	
Acartia clause	3,200,000
Eurytemora affinis	3,200,000
A.erythreae	780,000
A. tonsa[5]	~370,000
Centropages abdominalis	148,000
Tortanus forcipatus	62,000
Labroides aestiva[5]	~43,400
A.steueri[5]	~42,000
Ce. hematus[5]	~36,500
Calanopia thompsoni	24,000
L.wollastoni[5]	~20,500
Labidocera bipinnata	3,500

1. Anon (1993), 2. Wang et al. (2014), 3. Caceres and Tessier (2004a, b), 4. Caceres (1998), 5. Hairston et al. (2000), 6. Dumont and Munuswamy (1997), †mg/m^2, ~ mean values

The just described high reproductive potential of some anostracans too is not as high as that of *Artemia*. By virtue of its being one of the highest reproductive potentials, *Artemia* can produce cysts in such quantities required for industrial operation. Among the entomostracans, cysts of *Artemia* alone float; the floating cysts render the collection of them easier at the cheapest cost. Cysts of some entomostracans tend to float initially (e.g. *Daphnia* ephippium, Dodson and Frey, 2009). However, all of them other than *Artemia* immediately or subsequently sink to the bottom in both freshwater and denser seawater. The sunken cysts make their harvest a laborious and costlier job. For example, employing low-salaried, unskilled labor to sieve the pond bottom, measuring 2,170 m², 25 mg cysts of *Streptocephalus dichotomus*/m² alone could be harvested. Hence, the labor cost of harvesting the cysts from natural pond is in the range of US $ 300/50 g cysts (Dumont and Munuswamy, 1997). This is costlier, as importing *Artemia* cysts costs < US $ 600/kg.

An alternative is to aquaculture of some of these entomostracans, especially the fairy shrimps. Anostraca comprises of 258 species and is divided into 21 genera. Of 240 freshwater species, a dozen have been cultured at the laboratory scale (Munuswamy, 2005). Some fairy shrimps such as *S. dichotomus* are nutritionally adequate and prove to be a potential source of live feed. In an excellent analysis, Dumont and Munuswamy (1997) showed that aquaculture is also laborious and costlier in terms of both water and labor. Their analysis is summarized below: 1. Water cost of rearing the fairy shrimps is very high. The life time (~ 90 d) fecundity of *S. proboscideus* is ~ 10 mg cysts. Incidentally, *S. proboscideus* cysts may float (Brendock et al., 1996). Population density of the fairy shrimp ranges from 2/l in Kenya to 5-9/l in ponds of Tamil Nadu, India. Rearing even at the highest density of 50 females/l, the quantity of water required to produce 1 kg cysts/d (i.e. 0.3 ton/y) is 200 m³ or a pond measuring 20 m length × 10 m breadth holding water to a depth of 1 m. 2. *Artemia* thrives in such large aquatic systems, in which natural cysts productivity is in the range of 10^3-10^5 ton/y and reaches the industrial requirement. The cyst productivity of fairy shrimps in aquaculture systems is 0.3 ton/y and measures far below the industrial requirements.

5.5 Brine Shrimps

The family Artemiidae comprises of 89 species belonging to the genus *Artemia*. Currently, they are reported to inhabit around the world: *Artemia franciscana* (America, part of Europe), *A. monica* (North America), *A. persimilis* (Argentina), *A. parthenogenetica* (Asia, Africa, Europe, Australia), *A. sinica* (Central Asia, China), *A. tibetiana* (Tibet), *A. tunisiana* (Europe and North Africa), *Artemia* sp? (Kazakhstan) and *A. urmiana* (Iran). *Artemia* spp are a biological oddity in every sense of the word (Mohammed et al., 2010). They are one of the few crustaceans, which have led to extensive investigations by ecologists, reproductive biologists, cytogeneticists, biochemists and molecular biologists. In *Artemia* spp, some races are parthenogenics but others are bisexuals. Thin shelled subitaneous and/or thick shelled dormant cysts are produced through parthenogenic and sexual modes of reproduction; the thin shelled eggs are retained in the brood pouch and hatched as nauplii after a few days.

A. *salina*, for example, is diploid (2n = 42 chromosomes), bisexual in America but parthenogenic in France. Triploid and tetraploid parthenogenics occur in India and Italy, respectively. The *Artemia* complex has global distribution, except in Antarctica (Munoz and Pacios, 2010). *A. parthenogenetica* is recorded from the Old World namely Africa, Asia and Europe, especially from the Mediterranean Basin but not from Australia and Americas. Of 127 records of the African *Artemia* complex, 50% are *A. salina*, 38% *A. parthenogenetica* and 12% *A. franciscana* with an isolated distribution from Morroco, Egypt, Nambia, South Africa and Madagascar (Kaiser et al., 2006). In Algerian (sebkhas) saltmarshes, bisexual *A. salina* is present but frequently co-occurs with diploid and tetraploid parthenogenic strains (Ghomari et al., 2011). Molecular evidence suggests that parthenogenesis in *Artemia* is relatively ancient with a single parthenogenic lineage branching from an Old World sexual ancestor of ~ 5 m y ago. Incidentally, 200,000 y-old fecal pellets from Lake Urmia, Iran (Djamali et al., 2010) and 27,000 y-old cysts of *Artemia* from Great Salt Lake, Utah, USA (Clegg and Jackson, 1997) have been found. Automictic recombination, that can occur in diploid but not in polyploids, seems to have played a key role in long term maintenance of the parthenogenic lineage. The occurrence of *A. parthenogenitica* is reported in the ploidy status of triploid (India, Turkey, Browne et al., 1984) tetraploid (Spain, Italy, Browne et al., 1984) and pentaploid (China, Zhang and King, 1993).

Artemia produce cysts that can survive over an obligate, intense period of desiccation and successfully hatch, when water with optimum level of oxygen is available. Hentig (1971) reared *A. salina* at different temperature (10, 15, 20°C) and salinity (5, 15, 32, 70 g/l) combinations and estimated the production of offspring. The results reported by him (in German) are briefly summarized (Table 5.6): 1. At low salinity of 5 g/l and certain combinations of low salinity and low temperature, neither eggs nor cysts are produced. 2. The lowest temperature-salinity combinations, at which offspring are released, are at 15°C and 32 g/l salinity or 20°C and 15 g/l salinity. 3. The release of offspring in terms of absolute fecundity (110 to 473) or clutch number (1-3.6) heightens with increases in temperature and salinity 4. At salinity of 70 g/l, the proportion of cysts produced increases from 8% at 15°C to 23 and 39% at 20 and 30°C, respectively. The corresponding values are 25, 98 and 184 cysts/female at 30°C. The proportion of cysts produced at 30°C also increases from 14 to 39% 38 cysts/female at 15 g/l salinity and from 14% to 184 cysts/female at 70 g/l salinity.

TABLE 5.6

Comparison of reproductive life span, performance and cyst production by parthenogenic and sexual *Artemia* spp (source: Browne and Wanigasekera, 2000, Hentig, 1971*)

Species	Salinity (g/l)	Reproductive Life Span (d)	Offspring (No./♀)	Cyst (%)
Parthenogenic				
A. parthenogenetica	120	111	603	17
	180	113	542	82
Sexual				
A. franciscana	60	49	66	0
	120	58	470	22
A. sinica	120	52	287	29
	180	45	121	80
A. persimilis	120	62	427	92
	180	96	482	93
A. salina	15*	–	270	14
	32*	–	470	29
	70*	–	473	39
	120	39	131	100
	180	71	260	100

Both sexual and parthenogenic species of *Artemia* adopt a hybrid strategy of simultaneously producing subitaneous eggs and dormant cysts (Table 5.6). This strategy may ensure rapid colonization and sustenance in the new favorable but unpredictable habitats. Bearing the costs of sex and meiosis as well as mate-searching, sexually reproducing *Artemia* spp have almost half the reproductive life span as that of *A. parthenogenetica* (Table 5.6; see also Table 2.3). Consequently, *A. parthenogenetica* produces about one and a half to twice more offspring, as those of sexually reproducing *Artemia* spp. This is also true of cladocerans; parthenogenic cladocerans produce twice more number of eggs, as those of the sexual females (Glossener and Tilman, 1978). At salinities > 120 and 180 g/l, *Artemia* spp produce higher proportions of their offspring as cysts, irrespective of sexual or parthenogenic modes of reproduction. Clearly, hypersalinity, as a stressor, is an inducer of cyst production. In fact, Browne (1982) has suggested that decrease in food supply and/or low oxygen level may also induce cyst production, as is the case with cladocerans. However, experimental evidence goes to show the genetic control of this highly adaptive reproductive strategy of cyst production. Selecting six polymorphic enzyme loci of *A. franciscana* of the Great Salt Lake of Utah, USA, Gajarodo and Beardmore (1989) showed that the number of cyst production increases from about 100 to 520 in females characterized by increasing heterozygosity from 0 to 3 but at an increasing energy cost from ~ 600 to 1,000 j/cyst. Besides, they have also noted that the more heterozygous females attain early maturity and produce offspring more frequently and expensive cysts. Incidentally, polyploidy within asexual lineages of many branchiopods is associated with an increase in heterozygosity (see Zhang and King, 1993). Not surprisingly, most branchiopods are characterized by polyploidy; the consequent heterozygosity facilitates cyst production.

The results reported by Browne (1980) clearly indicate that *A. salina* is a typical geographic parthenogen. Two of its Indian strains namely Chennai (MA) and Kutch (KU) reproduce by obligate parthenogenesis; three of its American strains are obligate dioecious outbreeders (Table 5.7). In cyclic parthenogenic cladocerans like *Daphnia magna*, Methyl Farnesoate (MF) secreted by mandibular organs silences the meiosis inhibitor gene and thereby produces sexual females and males (Olmstead and LeBlanc, 2002) as well as induces meiotic maturation of gametes. It is not clear whether the mandibular organs of *A. salina* secrete MF. There are excellent opportunities to induce producing males not only in the Indian parthenogenic strains of *A. parthenogenetica* but also in many other parthenogenic lower crustaceans by exposing them to MF or its analogs like fenoxy carb.

In *A. salina*, life time production of viable cyst by a single female belonging to MA (88 cysts) and KU (nine cysts) strains lay nearly on the opposite ends of the five strain spectrum (Table 5.7). The same is true for its reproductive output. Encountering unpredictable and stressful habitats, both sexual and parthenogenic strains of *Artemia* have undergone diverse course of evolution that no consistent pattern is apparent on the reproductive performance of different strains within a single species or among different species of *Artemia*.

TABLE 5.7

Reproductive performance of obligate parthenogenic Chennai (MA) and Kutch (KU) strains as well as obligate dioecious outbreeding San Francisco (SF), Lake Chaplain, California (CH) and Cabo Rojo, Puerto Rico (CR) strains of *Artemia salina* (compiled from Browne, 1980)

Parameter	MA	KU	SF	CH	CR
Life time fecundity (no./♀)	1442	1532	1627	1057	969
Mature (F_1) adult (no./♀)	1063	1356	1619	660	569
Reproductive output (mg dry nauplii mass/♀)	4.6	2.5	4	2.5	3.4
Zygotes encysted (no./♀)	193	46	453	193	210
Viable cyst (no./♀)	88	9	313	19	46

Incidentally, the larger nauplii hatched from the large cyst of 3n *A. parthenogenetica* of Chennai is too large (~ 500 µm length in comparison to 430 µm length of *A. franciscana*, Mohammed et al., 2010) to be used as food for the protozoea stage 17 of penaeids cultured on a large scale in India. Consequently, many Asian countries import cysts of diploid bisexual *A. franciscana*, which produce smaller nauplii. For example, the alien nauplii serve as live feed for 85% of the cultured finfish and shellfish in India. Investigating the extend of invasion and naturalization of exotic *A. franciscana*, Sivagnanam et al. (2011) and Vikas et al. (2012) have found the disappearance of indigenous *A. parthenogenetica* and massive invasions by alien *A. franciscana* into hypersaline habitats of India, especially in the states of Tamil Nadu and Gujarat, where extreme conditions prevail. Life history differences between diploids and polyploids originate from allelic divergence, increase heterozygosity and DNA contents. In general, polyploids have higher levels of heterozygosity and produce larger cysts, which can withstand and survive greater stress.

5.6 Daphnids

At this juncture, the difference between a dormant cyst and a resting egg may have to be defined, as these terms are indiscriminately used one for the other, especially by molecular biologists. Cladocerans and copepods produce ephippia and resting eggs, respectively. Water content of the resting eggs is around 70% (Clark et al., 2012). Conversely, the cysts of *Artemia* undergo desiccation and their water content is as low as 0.7% (Clegg, 1978). In *Artemia* cysts, metabolism is almost totally suppressed but similar experiments have not been undertaken for hydrated diapausing eggs of cladocera to know at what level their metabolism is suppressed. Investigations on long term survival of these hydrated diapausing embryos of cladocerans and copepods may provide some clues for similar preservation of aeronauts on space travel. Further, the production of cysts/resting eggs is not without risk. These cysts and/eggs may not survive the dormant period (Table 5.10), may not receive appropriate cue and hence, may fail to hatch or may be transported to an unsuitable habitat (Gyllstrom and Hansson, 2004). Despite all these high costs of production and risk undertaken by cysts and resting eggs in highly unpredictable habitat, the process of cysts/resting eggs producing entomostracans have successfully survived since the Cambrian.

The diapausing cladoceran resting eggs are encased in an ephippium that either floats at the surface or sinks to the bottom. Although many ephippia are denser than water and sink quickly, spines, gas chambers and lipid content of the resting eggs may reduce sinking rate and/or increase buoyancy. Some cladoceran females deposit their ephippia in the surface film and as the ephippia case is hydrophobic, these ephippia float, when trapped there, and air penetrate into the ephippia (see Caceres et al., 2007). The floating and sinking ephippia represent different strategies between spatial and temporal dispersal of offspring, respectively. The sunken ephippia may not leave the habitat, in which they have been produced; hence, they function as temporal dispers. Transported by wind, and water as well as birds and mammals (Mellors, 1975, Michels et al., 2001, Caceres and Soluk, 2002), the buoyant ephippia serve as spatial dispers. From their field and experimental observations, Caceres et al. (2007) found that (i) populations and genotypes of *Daphnia pulicaria* produce a mixture of buoyant and non-buoyant ephippia and (ii) the proportion of buoyant ephippia ranges from 0 to 40% in different populations.

Table 5.8 is a representative example for the vertical distribution of resting eggs of daphnid species. An investigation of the sediment in combination with industrial pollution provides a powerful avenue to understand the pollution-induced changes in species composition of a lake like Onondaga. Industrial dumping of soda ash into the lake was commenced in 1930 but terminated in 1986. During this period, the ephippial accumulation by *D. pulicaria*, *D. ambigna* and *D. galeata mendotae* was completely or almost completely eliminated. For example, there were no *D. pulicaria* eggs in the sediments between 5 and 60 cm depth, corresponding to the period from 1930 to 1986. However, *D. pulicaria*, *D. ambigna* and *D. galeata mendotae* were replaced by *D. exilis* and *D. curvirostris* during the said period. The resting eggs of *D. exilis* were abundant in the sediment below 15 but above ~ 60 cm depth, corresponding to the years 1925 and 1990 (Hairston et al., 2005).

TABLE 5.8

Vertical distribution of dormant eggs of daphnid species in sediments of Onondaga Lake (New York, USA) (compiled from Hairston et al., 2005)

Species	Sediment Depth (cm)	Dormant egg Density ($10^3/m^2$)
Daphnia exilis	15	40
	25	280
	40	120
	65	30
D. curvirostris	~30	~75
D. pulicaria	2-3	50
	70	30
	100	75
D. ambigna	~0-5	20
	100	25
D. galeata mendotae	~0-5	100

The environmental cues, that trigger production of males and sexual females, have been traced to dropping temperature, decreasing photoperiod, declining food supply and/or crowding. Depending on the prevailing risk, the 'responsive diapausing' (see Slusarczyk, 2004) cladocerans may have to evolve optimal strategy to suit different habitats. As production of males and cysts is expensive (Gajarodo and Beadmore, 1989) and risky (Gyllstrom and Hansson, 2004), different cladoceran populations may respond differently to different environmental cues or different strength of the same cue. Not surprisingly, many studies experimenting to identify the cues triggering sexual reproduction have reported variable results and found differences between species (Larsson, 1991, Caceres, 1998) and clones (Fitzimmons and Innes, 2006).

With risky and expensive investment on cysts production, the daphnids, for example, select the optimum level of investment, after assessing the time of investment and permanency of their habitats. All populations of *Daphnia ephemeralis* produce ephippia in early April, but *D. pulex* and *D. pulicaria* during May-June and *D. dentifera* in autumn. In less permanent ponds, both *D. dentifera* and *D. pulicaria* invest on 100% production of ephippia. However, the proportion of their investment in November ranges between 90 and 100% in *D. dentifera* but 3 and 75% in *D. pulicaria* (Caceres and Tessier, 2004a, b). In general, populations with low minimum densities (high risk) invest more on production of diapausing ephippia, although permanency of aquatic system may not be the sole cue to trigger production of diapausing ephippia. Exposing *D. magna* to kairomone containing rearing water, in which a single fish was previously kept for a day, Sakwinska (1998) showed earlier and higher investment on offspring production.

Prey organisms are known to sense the predator-derived kairomones, which may induce cysts/ephippia production in planktonic fooder animals (Lass and Spaak, 2003, see also Hairston, 1987). To explore the source of chemical cue that induces ephippia production in *D. magna*, Slusarczyk and colleagues designed a series of simple but meaningful experiments. As nature of the cue namely the kairomone and its source was not known, Slusarczyk and Rygielska (2004) designed an experiment to know the source of the chemical cue from the following media: 1. Control water, 2. Fish water, in which a planktivorous fish was kept previously in 10 l water, 3. Fish feces water, 4. Combination water containing fish water + fish feces water and 5. *Daphnia* water containing extract of homogenized *D. magna*. All the females that were exposed to the fish feces water or fish feces water + fish water produced ephippia. This observation clearly indicates that the major source of the chemical cue arises from fish feces. In a second series, Slusarczyk and Rybicka (2011) prepared 'kairamone' medium, mixing an aliquot of feces of rudd and crucian carp in water. The effects of the kairomone water on selected reproductive parameters are summarized in Table 5.9. Briefly, the kairomone present in fish feces advanced size and age at sexual maturity, induced ephippial production in more females and increased both clutch size and neonate production in *D. magna* at 18, 22 and 26°C. Essentially, the kairamone arising from the fish feces induced an earlier and higher investment on reproduction.

TABLE 5.9

Effect of fish kairomone medium on reproductive parameters of *Daphnia magna* at different temperatures (compiled from Slusarczyk and Rybicka, 2011)

Parameter	With no fish			With fish		
	18°C	22°C	26°C	18°C	22°C	26°C
Size at maturity (mm)	2.65	2.52	2.45	2.58	2.45	2.46
Sexual maturity (d)	195	145	128	192	138	125
Proportion of ephippia carrying ♀	0.0	0.0	0.0	0.59	0.28	0.0
Clutch size (no./clutch)	2.7	2.4	2.2	3.1	2.5	2.5
Neonate productivity (no./♀/d)	0.16	0.28	0.28	0.20	0.30	0.42

Slusarczyk (2001) performed two more series of experiments to test the combined effects of feed (*Scenedesmus oblicuus*) density (0.4, 0.5, 0.7, 1.1 and 1.6 mg C/l) and in the presence/absence of fish cue containing water, in which predatory fishes rudd and crucian carp were previously kept for a day. In the absence of fish cue, *D. magna* produced no resting eggs at any tested feed densities. However, the frequency of ephippium carrying females increased to various levels, when fed at the lowest density of 0.4 mg C/l but in the presence of fish cue. When offered feed at high density, the frequency of ephippial female decreased and the females also postponed the formation of resting eggs to later broods. At high feed density of 1.6 mg C/l and in the presence of fish cues, only 17% females produced ephippia. Hence, *D. magna* seems to trade off between chances of survival and successful reproduction with awaited predation and food availability. It selects to produce diapausing eggs, when the assessed gains are more.

Recent research has shown that chemically induced-predator defense in morphology, life history and behavior play a major role (Lass and Spaak, 2003). For example, neckteeth, formed against *Chaoborus* kairomones, increase the survival of daphnids; crusted *Daphnia* is less susceptible to *Notonecta* predation and spined *Daphnia limholdzi* are not preferred by predatory fishes. Some copepods and cladocerans can detect kairomones of *Salamander*. Behavioral changes in swarming, Diel Vertical Migration (DVM) and swimming patterns facilitate cladocerans to escape from predators. The previously non-migrating copepods initiate DVM within 4 h following the introduction of *Chaoborus*.

The oldest ephippium found is estimated as 3,300 y old. With advent of DNA technology, it has become possible to extract DNA from ephippia of daphnids like *Daphnia longispina* as old as 200 y (Limburg and Weider, 2002). Investigations on this line of 'witnessing the past' (Lampert, 2011) indicate that the dormant eggs could hatch even after 50 y of the egg banking (Mergeay et al., 2007). For more information on this aspect of palaeolimnology, Lampert (2011) may be consulted.

5.7 Copepods

In copepods, Uye (1985) recognized diapausing eggs as different from quiescent eggs. Reviewing 150 publications on the hatching phenology of branchiopods, Brendonck et al., 1996) also recognized these two types of eggs. In general, the diapause is a state of arrested development, which is genetically controlled. *Acartia californiensis, A. tonsa, Labroides aestiva, Pontella meadi, P. mediterranea* and *Tortanus forcipes* are copepods, which produce diapausing eggs. Conversely, quiescence is a state of retarded development induced by adverse environmental conditions. Hence, the quiescent eggs are almost equal to subitaneous eggs. *A. clause* and *A. steueri* are examples for copepods producing quiescent eggs (Uye, 1985).

Some calanoids produce both subitaneous and diapausing eggs. For example, *A. clause* produces several generations stemming from subitaneous eggs during the period from January to May (5-26°C), as 80% of eggs hatched within 4-5 h of incubation at 10°C. However, more and more eggs produced from June were diapausing eggs, as they did not hatch even after 14-h of incubation at 15°C (Uye, 1985). As many as 23 copepod species belonging to five families that produce diapausing eggs are listed by Uye (1985). Table 5.5 summarizes available information on abundance of diapausing eggs of entamostracans including copepods in sediments. In estuarine sediment, viability of the eggs is > 2 y, although those of *Diaptomus sanguineus* are reported to survive for 332 y (Hairston, 1996).

5.8 Diapausing Larvae

In many cyclopoid and calanoid copepods as well as ostracods, the larvae secrete an organic cyst-like covering and undergo diapause during unfavorable conditions of drought or extreme cold. Such cysts, buried in mud, are adapted to withstand desiccation and enable the larvae to survive.

Freshwater cyclopoid larvae entering diapause at the late copepodid stage may not undertake extended period of diapause in permanent water bodies but can survive for several years in dry sediments of temporary ponds (Wyngaard et al., 1991). Pandian (1994, p 131-132) summarized the occurrence of diapausing cysts in eight species of cyclopoids: *Canthocamptus staphylinoides, C. staphylinus, Cyclops strenuus strenuus, C. leucarkti, C. vicinus, Cyclopoida* sp, *Diacyclops bicuspidatus* and *D. navus*. The encysted fourth copepodids of *D. bicuspidatus* survive anaerobiosis. Although all larval stages of *D. navus* are obligate aerobes, oxygen uptake of encysted copepodid is scarcely measurable.

Distribution of diapausing *Tortanus discaudatus* copepodid[1] ranged from 6,2710/m² up to 5-6 cm depth and then decreased to 904/m² at 11-12 cm depth. The values for the copepodid[2] also ranged from 33,457/m² up to 6-7 cm depth and then decreased to 1,659/m² at 11-12 cm depth. But no eggs of *T. discaudatus* survived below 2 cm depth. Decrease in density of diapausing copepodids with increasing depth was probably due to the older (deeper) cysts being exposed to anoxia and H$_2$S for a much longer period of time (Marcus, et al., 1994). Notably, diapausing copepodid 'cysts' withstood progressively decreasing oxygen levels in the sediments but the eggs could not.

Following seven molts during larval stages, an ostracod attains sexual maturity. The diapausing seventh instar of *Condona rawsoni* has been recovered from frozen sediment of a pond, which had gone dry the previous autumn (Delorme, 2009).

5.9 Cyst Viability and Metabolism

Cyst size ranges from 139 μm in clam shrimp *Eulimnadia colombiensis* (Rabet et al., 2014) to 300 μm in *Artemia parthenogenetica* (Sivagnanam et al., 2013). Despite the small size, a viable period of the copepod cysts ranges from > 2 y in *Tortanus discaudatus* to > 332 y in *Diaptomus sanguineus* (Table 5.10). The period for the freshwater cladocerans (> 125 y) is longer than the marine cladoceran (5 y). Cysts of spinicaudates (Weeks et al., 2006c) and ostracods (Delorme, 2009) are viable for many decades; however, experimental data are not yet available. Diapausing cyclopoids survive and remain hatchable for long periods, although the period remains to be estimated. Survival of diapausing eggs and copepods decreases with increasing sediment depth; at which they may be exposed to anoxia (Marcus et al., 1994). The cladoceran epihippia too survive more for a season alone (Caceres and Tessier, 2003). Amazingly, *Artemia franciscana* cysts withstand 4 y exposures to anoxia (Clegg, 1997).

TABLE 5.10

Viability of crustacean cysts and resting stages

Species	Viable Period (y)	Reference
	Anostraca	
Artemia salina	4	Clegg (1997)
	Cladocera	
Daphnia longirostris[F]	50	Mergeay et al. (2007)
D. galeata mendotae[F]	>125	Caceres (1998)
D. pulicaria[F]	> 125	Caceres (1998)
Ceriodaphnia pulchella[F]	14	Moritz (1987)
Bosmina longirostris[F]	35	Moritz (1987)
Podon polyphemoides	5	Marcus et al. (1994)
	Copepoda	
Diaptomus minutus[F]	30	Hairston et al. (1995)
D. oregonensis[F]	30	
D. sanguineus[F]	332	Hairston et al. (1995)
Tortanus discaudatus	<2	Marcus et al. (1994)
'copepodid[1]'	5	Marcus et al. (1994)
'copepodid[2]'	14	Marcus et al. (1994)

F = freshwater crustaceans

Covered by a complex shell (Fig. 5.2C) with chorion thickness ranging from 1.2 μm in *A. urmiana* to 10.24 μm in *A. persimilis* (Mohammed et al., 2010), the encysted *Artemia* enters a period of obligate dormancy and arrested development at the gastrula stage. The dormant embryos, comprised of a syncytium of ~ 4,000 nuclei, are released from ovisac to the environment, in which they undergo dehydration. The artemian strategy of embryonic development seems not to enter into energy demanding differentiation process prior to diapause. But to do it so fast that the 4,000 nuclei containing syncytial embryo completes the entire sequence of differentiation to produce a nauplius within 8-12 h of hydration (cf Late Embryogenis Proteins [LEP]). The abilities of the dried cysts are amazing: 1. They can withstand total desiccation with only 'residual water' remaining (0.7 μg/g dried cysts). 2. When the cysts are placed over strong desiccant or under pressure, they resist temperature exceeding 100°C for over an hour and can be exposed to near absolute zero, apparently for an indefinite period. 3. They exhibit considerable resistance to bombardment by various forms of high energy radiation. 4. They are able to withstand distortion and mechanical damage and 5. The durable nature of the shell is revealed by the fact that the cyst can be soaked for a month or longer in acetone, a wide range of alcohols or other organic solvents with no decrease in viability (Pandian, 1994, p 134). Not surprisingly, a volume of literature has been accumulated on the puzzling metabolic capabilities of *Artemia* cysts, although a few publications on metabolism of other anostracan cyst (e.g. Munuswamy et al., 2009) and resting eggs (e.g. Pauwels et al., 2007) are available.

Cyst Metabolism The credit for pioneering research on the biochemical process of dormancy in *Artemia* must go to J. S. Clegg. Dormancy is characterized by abundance of glycerol, trehalose, chaperones, antioxidants and LEP. Glycerol is present (2-6% of dry weight) in *Artemia* cysts but in no other stage of its life cycle (Clegg, 1962, 1965). With its physico-chemical properties, glycerol alluviates many problems related to dehydration. 1. Its relatively high electric constant has considerable solvent properties for macromolecules and inorganic electrolytes, 2. Its ability to stabilize a variety of proteins and to prevent their aggregation and precipitation protect many enzymes and other proteins against denaturation. 3. Its least compressibility can bear the enormous stress of desiccation on structure, whose dimensions are at the Amstrong level (see Pandian, 1994, p 135) and 4. By virtue of its structure, glycerol may play the role of water as a substitute in the 'glycerated cysts' (Crowe, 1971).

Trehalose The encysted *Artemia* embryo contains 17% of its dry weight as trehalose (cf 12% trehalose in *Streptocephalus dichotomus* cysts, Munuswamy et al., 2009), which does not occur in any other stage of *Artemia*. Clegg (1965) demonstrated the absence of trehalose in the hemolymph of female *Artemia*, fed on yeast containing 12% of its dry weight as trehalose. Trehalose is synthesized at the expense of glycogen but it is rapidly metabolized at the termination of dormancy. The trehalose-glycogen inter-conversion in the dormant egg may reduce glucose to form a variety of insoluble 'melanoidens'. Besides, trehalose is not susceptible to aminolyses. Taken together with these properties, large quantities of substrates for energy metabolism and biosynthesis can be stored as trehalose in the dried embryo with greater stability and integrity (see Pandian, 1994, p 135-136).

Under aerobic conditions, trehalose is metabolized at 15 µg/mg dry mass of *A. franciscana*. However, no change in its content was detectable in cysts during 4 y of uninterrupted exposure to anoxia. From experimental analysis, Clegg (1997) estimated the following: 1. Trehalose content of *Artemia* embryos is 165 µg/mg. 2. With lower limit of detection, 10 µg trehalose/mg/4 y is equivalent to the loss of 0.3 ng trehalose/mg/h on metabolism. Amazingly, this means that the encysted *Artemia* embryo undergoes a Metabolic Depression Rate (MDR) of 500,000 times lower than the aerobic rate. Incidentally, lactate dehydrogenase activity, an index of aerobic metabolism, is depressed to < 1.6% of the aerobic rate in *A. parthenogenetica* cyst (Hemamalini and Munuswamy, 1994). 3. The heat loss measured in anoxic embryos is < 10% of the minuscule (Hand, 1995). Hence, it may require 18 y for *Artemia* embryos exposed to uninterrupted anoxia to exhaust the available 165 µg trehalose/mg (Clegg, 1997).

The diapause-destined *Artemia* eggs arrest development at the gastrula stage, encyst and enter diapause, characterized by depression to the lowest metabolic rate (Clegg, 1997). In fact, virtually all metabolism and DNA, RNA and protein syntheses are arrested. Intracellular pH decreases to 6.4-6.6 signals (Hardewig, et al., 1996) the initiation of global arrest of protein synthesis down to < 3% (Hoffmann and Hand, 1994). For example, radioactive proteins remain around 3,400-3,700 disunits/mg wet biomass/min from d 1 to 4 y of encystment (Clegg, 1997). This confirms that during diapause, there is neither protein synthesis nor degradation. Encysted embryos tolerate uninterrupted anoxia for an extended period of 4 y (Clegg, 1997) and dehydration up to 0.7 µg/g cyst (Clegg, 1978) and temperature fluctuation exceptionally well. The p26 heat shock protein and the ferritin homolog artemin, synthesized in diapause-destined embryos, prevent heat induced denaturation of citrate synthesis and confer thermo-tolerance. Artemin has arisen from ferritin by gene duplication; subsequent divergence has retained chaperone activity intrinsic to ferritin but has eliminated a role in iron homeostasis. Accumulation of artimin inhibits heat induced protein denaturation, an activity characteristic of molecular chaperones (Chen et al., 2007). Incidentally, Shunmugasundaram et al. (1996) reported that sulfohydrol group may play an indirect role in protecting the reversibly denatured molecules from deleterious effects of peroxides. The presence of reduced oxidants like ascorbic acid and glutathione in the desiccated cysts suggests that oxidants and glutathione mitigate the influence of free radicals in the encysted embryos.

Transcription and translation studies have shown that dormant embryos contain significant amount of mRNA. And there was no net degradation of mRNA pools during dormancy. "Cytochrome c in mitochondria is essential for oxidative phosphorylation but its release into cytoplasm initiates assembly of the apoptosome, the molecular machinery activating the caspase 9 pathway programmed cell death" (Clark et al., 2012). However, *Artemia* caspase activation is not dependent on cytochrome c but is regulated by nucleotide concentrations and thereby diapausing stress leading to programmed cell death and apoptosis is eliminated in the cyst.

The abundance of p26 has been documented in many *Artemia* species (*A. franciscana*, *A. monica*, USA) *A. parthenogenetica* (Siberia, France), *A. sinica* (China), *A. urimiana* (Iran) and *A. salina* (Tunisia). However, p26 or p26-like proteins have not been detected in other than *Artemia* (e.g. *Branchinecta* spp, *Triops* spp, *Daphnia* spp, Clegg et al., 1999). It is in this context, the publication by Munuswamy et al. (2009) reporting the presence of p26 and artemin-like proteins in the cysts of the Indian fairy shrimp *S. dichotomus* is significant. Western blot analysis of the cysts showed the presence of p26 and artemin-like proteins with similar electrophoretic pattern and cross reactivity with polyclonal antibodies raised against p26 and artemin. Hence, the presence of p26 and artemin-like proteins is documented for the first time in *S. dichotomus* other than *Artemia* cysts. Further, p26 was found localized in the nuclear fraction of *S. dichotomus*, as it is in *A. franciscana*.

5.10 Hatching Phenology

Hatching in batches is a trait of crustaceans. The brooded embryos attached to the setosed appendages of decapods are hatched in three-22 batches (see Pandian, 1994, p 64). Hatching phenology of entomostracan cysts and emergence of resting eggs of cladocerans and copepods are asynchronous and varies from population to population and year to year. Ephippia of different populations of *Daphnia pulicaria* maintained under the same temperature-photoperiod that occurs in nature, ephippial hatching fraction ranged between 6 and 50% only (Caceres and Tessieer, 2003). On exposure to different salinities (5-60 g/l), hatching success of *Artemia parthenogenitica* cysts is 91% at 30 g/l salinity but decreases to 55-58% at the tested extremes of salinity (Ramasubramanian and Munuswamy, 1993). The tropical entomostracan cysts hatch only after a period of drying, especially when a dried pond is refilled with rain water and diluted (specific conduction < 80 μmole/cm^2, *Streptocephalus dichotomus*: Bernice, 1972, Sam and Krishnaswamy, 1979) and after repeated wetting and drying in many branchiopods (*S. vitreus, S. proboscideus, S. sudanicus, Branchinecta mackini, Triops spp, Eulimnadia antlei*, Brendonck, 1996). In temperate forms, a period of freezing may be obligatory (e.g. *Eubranchipus vernalis*, Mozley, 1932). Table 5.11 summarizes the factors like osmotic pressure, oxygen level, lower pH, elevated free pCO_2, temperature and photoperiod that are reported to trigger hatching in entomostracan cysts.

Other than these environmental factors, application of morphogen Retinoic Acid (RA) and Calcium Ionosphere A23 187 can accelerate the process of hatching in anostracans (Dumont et al., 1992). Hatching of *S. dichotomus* cysts increased from 31 to 50% at 10^{-6} M RA, but remained around 43% only at 10^{-8} M RA. With Calcium Ionosphere, it increased from 50-55% and a combination of RA and Calcium Ionosphere, it increased to about 60%.

Hatching success of resting eggs/emergence of larvae decreased with increasing sediment depth. For example, it decreased from 12.0% at 0-1 cm depth to 3.1% at 2-3 cm depth in copepodid[1]. The larvae present between 3 and 12 cm depth failed to emerge. Conversely, emergence of copepodid[2] increased from 6.7% at 1-2 cm depth to 21.4% at 4-6 cm depth. Notably, the success of resting eggs of *Podon polyphemoides* was 5.3% only at 1-2 cm depth (Marcus et al., 1994). Evidently, resting copepodids withstood the reduced oxygen levels better than the resting eggs of *P. polyphemoides*.

TABLE 5.11

Environmental factors that trigger hatching in cysts/resting eggs entomostracans (compiled from Brendonck et al., 1996, Pandian, 1994, p 126)

Reported Observations

Osmotic Pressure

Anostraca: *Branchinecta gigas, B. mackini, Cheirocephalus diaphanus, Streptocephalus dichotomus, S. seali*

High Oxygen Level

Anostraca: *Eubranchipus holmani, Linderiella occidentalis, Siphonophanes grubei, S. seali,*
Notostraca: *Triops longicaudatus, Limnadia stanleyana*

Low pH

Anostraca: *E. vernalis, Si. grubei, T. longicaudatus, S. proboscideus*

Elevated Free CO_2

Anostraca: *Si.grubei*

Temperature

Anostraca: *B. paludosa* (5°C), *L. occidentalis* (10°C), *B. lindahli* (15°C), *S. macrourus* (14-20°C), *S. dorothae* (15-30°C), *S. texanus* (22-30°C), *S. dichotomus* (30°C), *Lepidurus packardi* (15°C), *Le. couesii* (20°C)

Notostraca: *T. longicaudatus* (22-30°C), *L. stanleyana* (20°C), *Caenestheriella setosa* (25°C), *C. gynecia* (22-30°C)

Copepoda: *Centropages abdominalis, C. yamada* (10°C, 15°C), *Labroides aestiva* (11°C), *Acartia tonsa* (14°C), *Tortanus forcipatus, Labidocerca bipinnata, Calanopia thompsoni* (15°C), *A. erythreae, A. clause* (20°C)

Photoperiod

Anostraca: *S. macrourus*

Notostraca: *T. longicaudatus, T. cancriformis, E. antlei, L. stanleyana*

The dynamics of emergence from resting eggs of two daphnids during spring in Oneida Lake (New York, USA) is described by Caceres (1998), indicating that the emerging frequency varies widely from species to species, location to location and year to year. The emergence of *D. galeata mendotae* was 1,200, 100, 100 and 0/m² during the years 1992, 1993, 1994 and 1995, respectively. These values, however, were 3,200, 13,700, 2,400 and 2,200/ m² in *D. pulicaria*. In the lake, the frequency was < 2/m² at depths but 15.6/ m² in shallow area. Further, the cumulative resting egg production during 1992-1995 was 1,500/m² in *D. galeata mendotae* and 20,000/m² in *D. pulicaria*. The emergence frequency was 1.2% for for *D. galeata mendotae* and 1.1% in *D. pulicaria* (Caceres, 1998). In Lake Constance (Europe), vertical distribution of *D. galeata* resting eggs increased from ~2-3 × 10⁴ of < 5 y old eggs/m² to

~5 × 10⁴ of 15 y old eggs/m² and subsequently decreased to $< 1 \times 10^4$ of 35 y old eggs/m². Almost parallel looping trends (Fig. 5.3) were apparent for the relationships between (i) egg density and sedimented egg age, (ii) hatching success and age of resting eggs. This clearly indicates that younger the age of resting eggs, the greater the hatching success. The eggs older than 35 y may not hatch at all (Weider et al., 1997, see also Table 5.9).

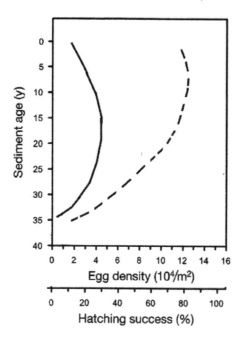

FIGURE 5.3
Simplified freehand drawing to show the relationships between sediment age and (broken line) density of resting eggs and (continuous lines) hatching success of *Daphnia galeata* in Lake Constance (redrawn from Weider et al., 1997)

To survive on a long term basis in unpredictable habitats, not all cysts/ resting eggs hatch during any given wetting/favorable period. This unsynchronized hatching phenology results in the development of egg sediments. Cysts/resting eggs are accumulated at densities of 10^3-10^6/m² (see Table 5.5). Yet the hatching frequency ranges from 1.2% in the *D. galeata mendotae* (Caceres, 1998) to 2.8% in a fairy shrimp (Simovich and Hathaway, 1997). The variable hatching is recognized as a bet-hedging strategy, in which the emergence from the cysts/ hatching of resting eggs is spread over several favorable/wetting periods. One or other environmental factor is reported to play a role in hatching/emergence (Table 5.11). However, it is not clear, how some cysts/eggs alone are triggered (by one or another factor) to hatch/ emerge, while others continue dormancy (see Benvenutta et al., 2009).

However, many authors have attempted to identify the factor or combination of factors that trigger hatching of cysts/emergence of resting eggs. For example, Benvenutta et al. (2009) attempted to identify the factor responsible for hatching the cysts of clam shrimp *Limnadia badia*. Of the tested physico-chemical factors, temperature, oxygen level, pH, salinity and total dissolved solids, the factor responsible for multiple hatching of larvae even within a clutch of cysts could not be identified. It is for the molecular biologists to resolve the differences between cysts/resting eggs within a clutch and identify the genetic mechanism responsible for the asynchronous hatching of entomostracan cysts/resting eggs.

6

Sex Determination

Introduction

The very presence of gonochorics, sequential hermaphrodites and parthenogenics within Crustacea suggests the operation of diverse mechanisms of sex determination. The determination ranges from genetic, as in some penaeids, to primarily genetic with autosomal modulations, perhaps at the downstream level of genetic cascade of differentiation, as in many sessile and parasitic crustaceans. Genes involved in sex determination may be located on a single sex chromosome with or without autosomal influence on sex determination or restricted to a single gene locus (e.g. *androgenic gland gene, LAC*). The polygenic sex determination system, when strongly influenced by one or other environmental factor, induces wide flexibility in sex ratios. As in aquatic animals like fishes (see Pandian, 2014, p 1), the reproductive strategy of crustaceans seems to sustain and increase genetic diversity and/or reproductive success.

6.1 Genome and Sequencing

In crustaceans, genome size (C-value) ranges from 0.20-0.36 pg in daphnid cladocerans (Vergilino et al., 2009) to 0.94-2.81 pg in gammarid amphipods (Libertini et al., 2008) and to 0.78 – 5.71 pg in calanoid copepods (Gregory et al., 2000) (Table 6.1). The estimated mean of 0.23 pg genome size of *Daphnia pulex* is close to that (0.204 pg) arrived at by genome sequencing, where the genome contains 199×10^6 bp (see Vergilino et al., 2009). Whereas the variations in genome size are narrow in daphnids, but is wider, i.e. twofold in gammarids and seven-fold in calanoids. In fishes too, the genome size ranges from 0.4 pg in bufferfish to 7.2 pg in green sturgeon. Fishes with fastest population doubling time have smaller genome (1.2 pg) than those (1.65 pg) growing slowly (Pandian, 2011, p 20). Interestingly, the fastest population doubling daphnids have the smallest (0.2 pg) genome in comparison to amphipods (~ 2.0 pg). Another interesting feature is the direct relation between genome size and body size of calanoid copepods (Fig. 6.1). A third interesting feature is the direct correlation between genome size and quantum of C- heterochromotic DNA in amphipods (Table 6.1).

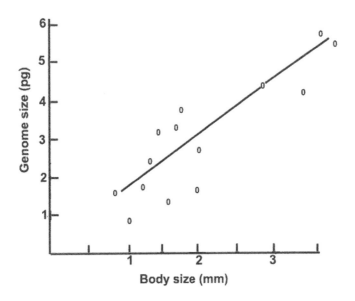

FIGURE 6.1

Relationship between genome size and body size in copepods (data are drawn from Table 6.1).

Mitochondrial genome of many crustaceans has been partially (e.g. *Homarus americanus*, Boore and Brown, 1995) or completely sequenced (e.g. *Tigriopus japonicus*, Machida et al., 2002, *Penaeus monodon*, Wilson et al., 2000). It contains 15, 333 bp in *D. pulex*, 15,822 bp in *Artemia franciscana*, (see Wilson et al., 2000) and 16,026 bp in *Portunus trituberculatus* (Yamauchi et al., 2003). The arrangement of genes in *P. monodon* is almost identical with that of *D. pulex*. However, sequences of mitochondrial genome may be more important to phylogenetic analysis (Ji et al., 2014). Another area, in which mitochondrial sequence can be of use, is hybridization. For example, mitochondrial gene sequence has shown that all putative natural hybrids between the red snow crab *Chionoectes japonicus* and snow crab *C. opilio* in Korean waters have sequences identical to *C. japonicus*. Clearly, *C. japonicus* serves as maternal partner in all the natural hybridization events (Kim et al., 2012).

TABLE 6.1

Genome size (C-value in pg) of some crustaceans (compiled from Gregory et al., 2000, Libertini et al., 2008, Vergilino et al., 2009)

Species	n	3n/Body Size (mm)
Cladocera		
Daphnia pulex	0.22–0.24	0.20–0.23*
D. pulicaria	0.22–0.28	–
D. carinata	0.24	–
D. middendorffiana	–	0.21–0.25*
D. trenbosa	0.27-0.31	0.24–0.36*
Copepoda		
Eurytemora composita	0.78	1.16†
Lepidodiaptomus tyrrelli	1.33	1.60†
Limnocalanus macrurus	1.63	2.00†
Diaptomus sicilis	1.72	1.20†
D. insularis	1.61	0.90†
Osphranticum labronectum	2.45	1.25†
Di. leptopus	2.77	2.08†
Hesperodiaptomus shoshone	3.11	–
Le. wilsonae	3.25	1.26†
Di. nodus	3.33	1.60†
Di. forbesii	3.81	1.54†
Hes. victoriaensis	4.37	2.70†
D. arcticus	4.67	3.23†
Heterocope septentrionalis	5.45	3.35†
Hes. n. sp	5.54	3.23†
Hes. nevadensis	5.71	3.15†
Amphipoda		
Apohyale crassipes	0.94	0.53††
Protohyale schmidti	1.05	–
Orchestia montagui	1.71	1.00††
A. prevostii	1.71	1.19††
O. cavimana	1.77	1.18††
Platorchestia platensis	1.86	1.26††
Talitrus salator	2.20	1.43††
O. mediterranea	2.28	1.51††
Sardorchestia pelecaniformis	2.68	1.83††
O. gammarellus	2.81	1.82††

*3n, † = body length (mm), †† = quantity of C- heterochromatic DNA

The first crustacean genome that has been sequenced is that of *Daphnia pulex* (Colbourne et al., 2005, 2011). The genome is compact with 200 Mbp with 31,000 protein coding genes in comparison to those (180 Mbp, 13,700 protein coding genes) of *Drosophila*. Over 36% of *D. pulex* genes have no homology to any other species. The *Daphnia* Genome Consortium, established at Indiana University, Bloomington, U.S.A, with 375 scientists from 19 countries participation, aims to support annotation and functional characterization of these genes. These include $_c$DNA libraries, a library of > 200,000 Expressed Sequence Tags (ESTs), genetic linkage maps, microarrays for expression studies and a protein sequence database. Prior to the commencement of its genome analysis in 2008, enormous data have been accumulated on the central role played by *Daphnia* ecology and trophic dynamics have been accumulated. Hence, Lampert (2011) has rightly considered that "*Daphnia* does not provide 'just another genome' but a prime candidate for ecogenomics". Recent advances in ecological genomics and microarray analysis of *Daphnia* have opened the doors to understand the genetic basis of parthenogenesis and its consequences for genes and genomic evolution (Eads et al., 2008, Kato et al., 2010). However, the mitochondrial and nuclear genomic sequences have not so far identified the sex chromosomes and sex determining genes, as well as genes responsible for induction of parthenogenesis and diapause.

6.2 Sex Specific Markers

In view of their economic importance, sex specific molecular markers have been identified and characterized in three penaeids *Fenneropenaeus chinensis*, *Penaeus japonicus* and *P. monodon* (Table 6.2). From analysis of ESTs, Wu et al. (2009) have considered 28 genes may be involved in reproduction and development of *Macrobrachium nipponense*. In general, a penaeid marker has identified the commencement of sex differentiation rather than identifying sex determining molecular marker. Only one contig out of 48 is male specific in *F. chinensis*. Similarly, the expression of *pMTST-1* is limited to males in *P. monodon*. Others like the identified ESTs in *P. japonicus* indicate that sex differentiation commences, when mysis metamorphoses into post larva. Incidentally, Jerry et al. (2006) employed seven highly polymorphic microsatellite markers for parentage identification. This has provided a base to develop sex linked molecular markers.

Sex linked molecular markers are specific to one or other sex chromosomes. Many such markers are known for fishes. For example, using *Sry* specific primers, Kirankumar et al. (2003) identified XY specific products with 333 and 588 bp in XY males of golden rosy barb *Puntius conchoinus* but 200 bp in XX females. Genome of tetra *Hemigrammus caudovittatus* amplified by the *DMRT1* specific primers produced 237 and 300 bp products in XY males but 100 bp in XX females (David and Pandian, 2006). In crustaceans, sex linked molecular markers have been identified in *Penaeus monodon* (Staelens et al., 2008) and *Macrobrachium rosenbergii* (Ventura et al., 2011a) (Table 6.2). In *P. monodon*, the AFLP marker alleles associated with sex in a panel of 52 unrelated shrimps examined, two AFLP markers were sex linked. Of them, only the sequence of AFLP marker E06M45M3470 provided adequate DNA sequence length to design primers for a locus specific assay. Indel-4 and indel-5 primers amplified a female specific 82 bp allele and a non-sex (male) specific 76 bp allele (Fig. 6.2). Radhika et al. (1998) opened an avenue to identify sex linked molecular markers in anostracans. Phenol oxidase activity in the hemolymph of *Streptocephalus dichotomus* male is only a third of that of females. Moreover, electrophoretic analysis revealed that male plasma has a single fraction of tyrosine hydroxylase but the female has three isozymes showing diphenol oxidase activity.

Interestingly, the *DMRT* (*doublesex-and mab-3-related transcription* factor) is a major sex determining gene in Nile tilapia *Oreochromis niloticus*. In this Nile tilapia, *tDMRT1* and *tDMRTO* are reported to express in testis and ovary, respectively (Guan et al., 2000). The expression of *DMRT1* is also known to be upregulated in testis of mice, although it is expressed in the genital ridge of both sexes initially (Raymond et al., 2000). In crustaceans, the *DMRT* families of genes are involved in sex differentiation rather than sex determination. Farazmand et al. (2010) reported the expression of *DMRT*-related genes in differentiation of female gonads of sexually reproducing *Artemia urmiana* and parthenogenically reproducing *A. parthenogenetica*. In *Daphnia* gonads, *DMRT93B* mRNA alone were detected in testis but *DMRT1 1E* and *DMRT 99B* in ovary (Kato et al., 2008). Unlike in vertebrates, *DMRT* genes are involved in sex differentiation in anostracans.

The Primordial Germ Cells (PGCs) are progenitors of germ cell lineage and possess the ability to differentiate into either oogonia or spermatogonia (Wylie, 2000). Hence, they carry heritable information to the next generation. The presence of PGCs has been reported in *Pandalus platyceros* (Hoffman, 1972) and *M. rosenbergii* (Damrongphol and Jaroensastraraks, 2001). In *M. rosenbergii*, PGCs with prominent nuclei and granulated cytoplasm have been detected in 2 d–old embryo. Their positions on the 16th and 15th d-old embryos suggest their migration to the gonadal primordium. Their spermatogonial cells show resemblance to PGCs. The *vasa* gene is the first molecular marker to identify the PGCs. It has been isolated, cloned and proved that the gene is specifically expressed in germ cells (Olsen et al., 1997). Hence, *vasa* are the most reliable markers for studies on germ cells. Feng et al. (2011) traced the origin and migration of the *vasa*-gene. The expression of *vasa* from two-cell embryo to the blastula indicates that *Fc-vasa*-like mRNAs are maternally inherited, as in goldfish *Carassius auratus* (Otani et al., 2002). Subsequently, the gene in *F. chinensis* is centralized gradually and is restricted to five cell clusters as primordium cells of PGCs at the limb bud stage. The clusters are located at the base of antennule, antenna and cephalic lobe. The PGCs then migrate dorso-laterally in naupliar and zoeal stages. They enter into genital ridge at mysis stage 1. These events are similar to those in fishes (Pandian,

2011, p 131-133). In *F. chinensis,* the gonad is first detected histologically at post larva 1 stage and Germ Stem Cells (GSCs) are distinguished as clusters in juveniles of 30 mm body length. Not surprisingly, *Transformer-2C* commences expression at mysis stage of female *F. chinensis* and the male specific *ESTs* begin to express strongly at post-larval stage of *P. japonicus* (Feng et al., 2011) (Table 6.2).

TABLE 6.2

Representative examples of molecular markers to identify sex, sex determining locus and time of its expression in shrimps

Species and Reference	Reported Observations
	Sex Specific Markers
Fenneropenaeus chinensis	
Xie et al. (2010)	The differential expression of genes like *heat shock protein* 90 is expressed at a higher level during development
S. Li et al. (2012)	*Transformer-2C,* known to play an important role in sex differentiation, is expressed at a higher level in the ovary than in other tissues. Its expression is suddenly increased in mysis stage of females.
S. Li et al. (2013)	Of 48 contigs identified from cDNA from male specific 4th and 5th periopods including androgenic glands and spermatophore sac, one contig is specifically expressed in male shrimp
Penaeus monodon	
Preechaphol et al. (2007)	Of expression pattern of 14 out of 25 sex related gene homologs in ovary and testis, *ovarian lipoprotein receptor* homolog is expressed in ovaries only. A homolog of *ubiquiton specific protenase 9 X chromosome (Usp9X)* is expressed preferably in ovaries in 4-m old shrimp
Leelatanawit et al. (2009)	Characterization of 5 full length cDNA, *P. monodon testis specific transcript-1 (pMTST1)* is expressed in testes but not ovaries
P. japonicus	
Callahan et al. (2010)	Determination of 24 Expressed Sequence Tags (ESTs) has revealed that 6 of them are expressed in ovaries only and 5 others in testes alone. Expression of these ESTs are low and inconsistent until PL 110, i.e. 110 d following metamorphosis from mysis to post-larva
Macrobrachium nipponense	
Wu et al. (2009)	Of 3, 294 sequencing reactions, 3,256 ESTs longer than 100 bp are obtained. Cluster and assembly analysis has yielded 1,514 unique sequences. Of them, 28 genes may be involved in reproduction and development.
	Sex linked Markers
P. monodon Staelens et al. (2008)	PCR based genotyping of the set of 52 unrelated shrimps for the sex specific indel polymorphism showed 82 bp female specific allele and 72 bp non-sex (male) specific allele.
M. rosenbergii Ventura et al. (2011a)	Of a total of 1,795 AFLP bands screened, 65 were specific to males and 85 to females; on conversion of 16 and 12 bands to SCAR markers ~ 140 bp product was amplified only in females.

6.3 Karyotypic Analysis

It is a basic tool by which information on the number, size and morphology of chromosomes is deduced. Karyotyping replaced the older technology that determined nuclear DNA content. In Crustacea, diverse chromosome systems and relatively undifferentiated chromosomes are present. The chromosome size, for example, measures 1-3 μm in shrimp *Atyaepyra desmarestii* (Anastasiadou and Leonardos, 2010). The 2n number of chromosomes ranges from four in *Acanthocyclops robustus* (Yang et al., 2009) to 254 in *Eupagurus ochotensis* (Niiyama, 1959). The range is limited from 70 in *Trachypenaeus curvirostris* to 90 in *Penaeus semisulcatus* within Penaeidae but from 56 in *Palaemon serratus* to 118 in *Macrobrachium rosenbergii* in Palaemonidae (see Gonzalez-Tizon et al., 2013). In *Acanthocyclops vernalis*, the 2n number is eight for the American population but 10 for the European populations. The values reported for *A. americanus* are six in Chechoslovia but six, eight and 10 in Ukraine. The Ukraine populations harboring different numbers of chromosomes are reproductively isolated (Kochina, 1987). A reason for this variation in 2n number may be due to the presence of one or more B chromosomes (cf Beladjal et al., 2002). In *Nephrops norvegicus* too, the chromosome complement shows a great variablility in 2n (131-140) and n (72-74) values within and among individuals. C-banding and *DdeI* staining have revealed the presence of eight numbers of tightly condensed and strongly stained asynaptic chromosomes. They are also randomly distributed between the daughter cell. Deiana et al. (1996) identified them as B chromosomes. Neither the light microscopic resolution (3,700 time magnification, Niiyama, 1959) aided by differential staining techniques nor the different chromosome C banding (e.g. Deiana et al., 1996) and FISH (e.g. Mlinarec et al., 2011) techniques have not yielded information related to sex chromosomes. For more information on cytological aspects of chromosomes Hedgecock et al. (1982) may be referred.

6.4 Heterogametism

"Depending on the nature of the genetic factor that has gained dominations in the sex determination process, either male specific Y chromosome or female specific W chromosome emerges. The isolation of such chromosomes in one sex leads to the remarkable cytogenetic transformation, i.e. the establishment of sex chromosomes, and heterogamety and homogamety. Thus sex is determined by gene (s) located on a certain chromosome and genes on the other chromosome may not have any effect in the monogenic system" (Devlin and Nagahama, 2002) (see Pandian, 2011, p 20). Sex is decisively determined by a single *Sry* gene located on a morphologically distinguishable chromosome in mammals. In crustaceans, it seems to be determined, however, by many genes located on different not readily distinguishable chromosomes. As many as seven to 10 sex determination systems are thus far recognized in crustaceans. They are XY, XnY, XO, XnO, ZW and ZW_1W_2 (Table 6.3). Interestingly, ostracods have explored employing different systems within XnO. As many as five such systems are known: X_2O, X_3Y, X_4Y, X_6Y and $X_{4-7}Y$ are employed by *Notodromas monacha, Cypria compacta, C. exsculpta, Heterocypris incongruens* and *Cyclocypris ovum*, respectively (Niiyama, 1959). In *Cypria* sp, males have the highest number of X chromosomes ($X_{11}Y$) (see Mazurova et al., 2008). In *Cyprinotus incongruens*, females have six pairs of X chromosomes + four pairs of autosomes, whereas male has one set of six X chromosomes + four pairs of autosomes + one Y chromosome. The male produces two types of sperm – one with 10 (XXXXXXAAAA) chromosomes and the other with five (AAAAY) chromosomes (Bauer, 1940).

From a series of karyotypic studies, Niiyama (1959) deduced the existence of male heterogametism in eight out of 26 decapods and one out of six isopods. Unfortunately, he has provided karyotype figures only for different stages of spermatogenesis. Nevertheless, the presence of heterogametism in many other crustaceans has been deduced from (i) sex-linked genes and (ii) cross breeding (Ginsburger-Vogel and Charniaux-Cotton, 1982) as well as (iii) hybridization, (iv) crossing females with masculanized neofemales and (v) ploidy induction (Table 6.3).

As early as in 1957, Anders recognized the existence of sex linked traits r^+, R^2 and r in the XY sex determination system of *Gammarus pulex*. Male ratios were 24 , 68 and 98% in the presence of recessive r^+r^+, semidominant r^+R^2 and dominant $R^2 R^2$ alleles, respectively. From a series of publications, Battaglia (1961) deduced genetic polymorphism controlled by at least three alleles in a harpacticoid copepod *Tisbe reticulata,* of which two (*violacea*, v and *macula, m*) were dominant on a common recessive. Female ratio was higher among *violacea-macula* heterozygotes of either gene. Inbreeding among pure strains led to homozygosity within the fourth or fifth generation and resulted in complete elimination of heterozygous females carrying *m* or *v* allele.

Voorouw et al. (2005) reported the paternal transmission of male biased sex ratio during three generations of another harpacticoid copepod *Tigriopus californicus*. In this copepod, neither temperature (Voorouw and Anholt, 2002) nor crowding (Voorouw, unpubli.) play a significant role in structuring sex ratio. Voorouw et al. found that the (i) mean primary sex ratio was skewed toward males, (ii) observed skewing was larger than binomial expectation and (iii) extra-binomial skewing was paternally transmitted from F_1 to F_2. In this context, the findings by Beladjal et al. (2002) are relevant. They found that the sperm of fairy shrimp *Branchipus schaefferi* contains 10 autosomes and one to three B chromosomes. The paternally transmitted sex ratio is associated with the presence of B chromosome. They suspected that the B chromosomes carry the sex determining gene so that all individuals with one or more B chromosome develop into males, and those that lacked it into females. Fathers, whose sperm containing two or three B chromosomes transmit a higher load of sex determining genes and consequently sire more sons. The reported observations of Battaglia, Voorouw et al. and Beladjal et al. indicate that males seem to play a role in skewing male ratios in offspring, whatever the means adopted.

TABLE 6.3

Heterogametism in crustaceans (compiled from Niiyama, 1959, Ginsburger-Vogel and Charniaux-Cotton, 1982, modified, added and updated)

Anostraca

XY: *Cheirocephalus nankinensis*

XO: *Branchipus vernalis*

ZW: *Artemia salina*

Spinicaudata

ZW: *Eulimnadia texana*

Ostracoda

XY: *Cyclocypris laevis*

XnY: *Cypria compacta, C. deitzei, C. exsculpta, C. fordiens, C. whitei, C. ophtalmica, Cyc ovum, Heterocypris incongruens, Cypria sp, Cyprinotus incongruens*

XO: *Cyc. globosa*

XnO: *Notodromas monacha, Physocypris kliei, Platycypris baueri, Scottia browniana*

Copepoda

XY: *Acartiella gravelyi, Acartia keralensis, Tortanus barbatus, T. forcipatus, T. gracilis*

XO: *Acaratia centura, A. negligens, A. plumosa, A. spinicauda, Centropages typicus, Ectocyclops strenzki*

ZW: *Acanthocyclops vernalis, Eucyclops serrulatus, Megacyclops viridis, Cyclops viridis*

Isopoda

XY: *Anisogammarus anandalei*

XO: *Tecticeps japonicus*

ZW: *Paracerceis sculpta, Idotea balthica*

ZW$_1$W$_2$: *Jaera albifrons, J. albifrons forsmani, J. ischiosetosa, J. praehirsuta, J. syei*

Amphipoda

XY: *Gammarus anandalei*

XO: *G. pulex, Marinogammarus marinus, M. pirloti*

ZW: *Armadillidium vulgare, Naesa bidentata, Orchestia cavimana, O. gammarellus*

Decapoda

XY: *Eriocheir japonicus, Gaetice depressus, Hemigrapsus penicilatus, H. sanguieneus, Pachygrapsus crassipes, Plagusia dentipes*

XO: *Ovalipes punctatus*

X$_1$X$_2$Y: *Cervimunida priniceps*

ZW: *Macrobrachium rosenbergii, Penaeus monodon, Fenneropenaeus chinensis, P. japonicus, Cherax quadricarinatus, P. vannamei*

Experimental studies on hybridization and triploidization have revealed female heterogamety in penaeid shrimps. Hybridizing *Penaeus monodon* ♀ and *P. esculentus* ♂, Benzie et al. (1995) found male biased sex ratio. According to Haldane rule, a higher mortality in such crosses is experienced by the heterogametic sex. Hence, Benzei et al. concluded that penaeid females are heterogametic. Sex ratio of a few penaeid triploids was female biased. For example, the ratios were 0.67 ♀ : 0.33 ♂ (in comparison to 0.37 ♀ : 0.61 ♂ in 2n) in *P. monodon* (Pongtippatte et al., 2009), 0.8 ♀ : 0.2 ♂ in *Fenneropenaeus chinensis* (Xiang et al., 2006) and 1.0 ♀ : 0.0 ♂ in *P. japonicus* (Coman et al., 2008). These penaeids were considered as female heterogametic.

Among crayfish (Parastacidae), intersexuality is known from a number of genera, for example, *Procambarus* (see p 113). Parnes et al. (2003) recorded that the F_1 progeny from a normal sire of *Cherax quadricarinatus* segregated into 49% females, 47% males and 3.7% intersexes. These intersex individuals are functional males but are genetic females. On crossing with normal females, it was found that of F_1 progeny sired by the intersex male, there were 70% females, 25% males and 5.5% intersexes. These observations reveal that C. *quadricarinatus* is female heterogametic and intersexes carry ZW chromosomes

Another example for the existence of female heterogamety in a marine isopod *Paracerceis sculpta* hails from 12 y field observations and hybridization experiments on the inheritance of autosomal and sex-linked cuticular pigment patterns. Shuster et al. (2014) recognized three pigment patterns: 1. *Laterals-2-red, L2r* having pigmentation on lateral margins of a six-seven body segments, 2. *Three red strips, 3rs,* having red pigmentation zones running through the body length and 3. *cephalon dark-cd,* having black head capsules. 98% of *cd* were females. Marked and unmarked parents were crossed in all possible combinations. Progenies ratio of *L2r* and *3rs* were of the Mendelian pattern. Hence, *L2r* and *3rs* were controlled by dominant autosomal alleles. However, unmarked males crossed with *cd* females sired 0.5 ♀ : 0.5 ♂ and with *cd* daughters and no *cd* sons. Sons from these families never sired *cd* daughters. Clearly, the sex-linked expression of *cd* suggests female heterogamety in *P. sculpta.*

There are indications on the existence of sex determination systems in many cirripedes. The presence of sexually dimorphic cyprids in *Ulophysema oersundense* indicates the presence of genetic sex determination system (Melander, 1950). Documenting the existence of genetic sex determination system in *Balanus galeatus*, Gomez (1975) reported their sex ratio as 0.74 ♀: 0.26 ♂. From an experimental study, Svane (1986) documented the presence of genetic sex determining system in *Scalpellum scalpellum*, as only a maximum of 50% cyprids settle as males, when provided surplus hermaphrodites with empty receptacles. In parasitic colonial rhizocephalans, female cypris larva passes through the kentogen and vermogen stages, while the male cyprid transform into trichogen. Clearly, sex is irrevocably determined at cypris stage itself in these maxillipods. Hence, a search for sex specific molecular markers, sex determining genes and sex chromosomes in these maxillopods may be rewarding.

6.5 Ploidy Induction

Barring 557 viviparous fishes, all other 25,500 and odd teleostean fishes are oviparous (Pandian, 2013, p 37). Their eggs can readily be experimentally fertilized at a desired but fixed time. As ploidy induction involves manipulations within a few minutes following fertilization, i.e. during the first and second meiotic divisions, fishes are readily amenable to studies on ploidy induction (Pandian, 2011). In majority of crustaceans, fertilization is internal (see Table 2.8). More than 96% crustaceans also brood their eggs on one or the other site of their body (Table 2.10). Moreover, their eggs do not survive, when removed from the brooding site, e.g. attached to the setae of abdominal appendages (see Table 2.14). Consequently, most crustaceans are not accessible to ploidy induction. However, calanoid copepods, penaeids and some euphausiids shed their eggs freely. To meet the demand by aquaculture farms, penaeid eggs are artificially fertilized. Hence, the eggs of these egg-shedding crustaceans are accessible to ploidy induction. Notably, no ploidy induction study on calanoid is thus far undertaken.

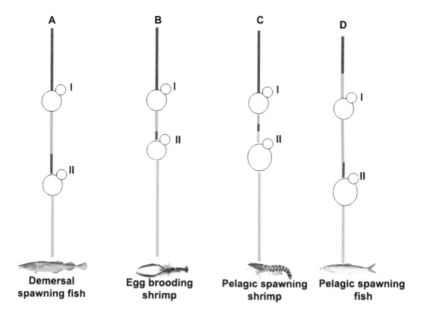

FIGURE 6.2

Events of meiosis (as indicated by I and II), spawning (green), fertilization (red) and hydration (blue) in egg brooding and pelagic spawning shrimps and fishes. Note the following: (i) the duration between release of polar body I and II is shorter in shrimps (ii) polar body I is released prior to spawning in fishes but after in shrimps and (iii) hydration occurs prior to spawning in pelagic spawning fishes but after spawning in penaeid shrimps. Note the large size of hydrated pelagic eggs of shrimp and fish.

Hydration is an important event that facilitates floatation of pelagic oocytes/eggs. Interestingly, there seems to be remarkable differences between the timing of the hydration event in pelagic eggs of fishes and crustaceans. In fishes, the ovulated oocytes are hydrated within the oviduct of female prior to spawning. For example, the oocyte of pelagic spawning cod is so heavily hydrated that an oocyte is 8.2 times larger in volume and 7.2 times heavier than the pre-hydrated oocytes (see Pandian 2014, p 44). Conversely, hydration seems to occur immediately following spawning in penaeids. No information is yet available in the process of hydration of oocytes in pelagic spawning crustaceans. Of course, hydration occurs during embryonic development in demersal eggs of fishes and brooded eggs of crustaceans. Hydration of oocytes within oviduct of fishes causes severe mortality of spawners (see Pandian 2014, p 44). But the hydration during embryonic development of brooded eggs causes the loss of developing embryos. Unlike fishes, embryo brooding crustacean mothers do not suffer mortality owing to hydration during embryogenesis.

In comparison to fishes and shellfishes, the progress made on research in ploidy induction of crustaceans is slow. This may be due to: (i) in microlecithal eggs of penaeids, the release of the first and second polar bodies is in quick succession. For example, they are released 5 and 15 min following spawning and fertilization in *Penaeus monodon* (Pongtippattee et al., 2005), 16 and 20 min in *Litopenaeus vannamei* (Garnica-Rivera, 2004), 8 and 20 min at 19°C in *Fenneropenaeus chinensis* (Li et al., 2003a) and 15 and 35 min at 24.5°C in *P. japonicus* (Norris et al., 2005).

Unlike fishery biologists, crustacean biologists have made a special effort to introduce higher heterozygosity by inducing Pb I triploids. However, the time lag from the release of polar bodies to fertilization is shorter in crustaceans than in fishes. Triploids that retain polar body I (PbI or MI) receive maternal two chromosomal sets plus a third set from the paternal source. Polar body II (PbII or MII) triploids receive one set of maternal chromosomes, the second maternal chromosome set, an exact copy of the same plus the third set from the paternal source. As a result, Pb I triploids are expected to display higher heterozygosity than Pb I triploids. However, of eight microsatellite loci, Sellars et al. (2009) found that CSPjO12 was the only marker that showed significant difference in heterozygosity between Pb I and Pb II triploids of *P. japonicus*. Unexpectedly, the Pb I triploids were less heterozygous than Pb II triploids. Moreover, Sellers et al. have not found any consistent trends in heterozygosity between Pb I and Pb II triploids. Li et al. (2003b) also did not note any difference between Pb I and Pb II triploids of *F. chinensis*.

TABLE 6.4

Summary of polyploid inductions in penaeids

Species and Reference	Threshold for Induction	Observations
	Triploid Induction	
L. vannamei		
Campos-Ramos (1997)	10-12 min old eggs to 10°C for 10, 15 or 20 min	62-88% 3n blastula and gastrula
Sellars et al. (2012a)	1 min old eggs to 200-μm 6-DMAP for 6-8 min	29-100% unhatched embryos, 50% MI 3n nauplii
	7.5 min old eggs to 200-μm 6-DMAP for 10 min	65-100% unhatched embryos, 52-100% M II 3n nauplii.
	8 min old eggs to 11.7-13.3°C for 8 min	MI 3n nauplii, 5-100% unhatched embryos; no larvae survived for > 7 d
P. japonicus		
Norris et al. (2005)	15 min old eggs to 150 μmol/16-DMAP for 15 min	41% MII 3n embryos
Coman et al. (2008)	Embryos to 150 μmol/16-DMAP for 10 min	73-93% MII 3n nauplii. No males. On 300 d, growth varied from families to family and MI to MII
P. monodon		
Pongtippatte et al. (2012) see also Sellars et al. (2012b)	15 min old eggs to 1 mg cytochalastin-B/l for 10 min	60% hatching and survival but 4% embryos with 3 pronuclei
	15 min old eggs to 1 mg 6-DMAP/l for 10 min	60% hatching and survival but 4% embryos with 3 pronuclei
	15 min old eggs to 8°C for 10 min	40% hatching and survival but 38% embryos with 3 pronuclei, 3n grew faster than 2n. 3n sex ratio: 0.67 ♀ : 0.32 ♂ in comparison to 2n : 0.39 ♀ : 0.61 ♂
P. indicus		
Sellars et al. (2009)	200 ml 1 m M 6-DMAP to 1-2 min for 6 min	No difference in survival between treated and control. No consistent trend in growth of post-larvae from different families. no observable development of ovary
	200 ml 1 m M 6-DMAP to 1-2 min for 8-12 min	
Wood et al. (2011)	15 min old eggs to 9°C for 6 min	0-76% MII 3n nauplius

Table 6.4 Contd.

Species and reference	Threshold for induction	Observations
F. chinensis		
Li et al. (2003a)	Time span for release of PbI ranges from 8 min at 19°C to 25 min at 15°C. For PbII, these values were 20 and 42 min	Hatching ranged from 30-80%, of which 50-100% were 3n. 3n decreased from 80% at Embryo and 60% at Naupliar, 55% at Zoea 1, 40% in Mysis and 10% in PL stages
Li et al. (2003b)	18-20 min old eggs to 29-32°C for 10 min	High mortality during metamorphosis and requires longer duration. 3n grow slowly, OSI: 2.3 in 2n female but 0.7 in 3n female, TSI: 0.6 in both 2n and 3n ♂, sperm like spermatid with no spike
Xiang et al.(2006)		At 10 cm size, 3n weighed 14 g but 2n weighed 13 g, ovary with primary oocytes alone, sperm and spermatophore in vas deferens, 3n sex ratio: 0.8 ♀ : 0.2 ♂
Li et al. (2006)		No significant difference between MI and MII 3n
Tetraploid Induction		
P. japonicus		
Sellars et al. (2006)	25 min old eggs shocked at 35 or 36°C or 5°C for min 150 μM 6-DMAP	8-98%, 13-61% and 15% 4n embryos for those shocked at 35, 36 and 5°C, respectively No induction, none survived,
Sellars et al. (2009)	9-114 min old embryos to 32-33.5°C for 3-4 min	Effectively inhibits cleavage of first mitotic division. Survival : 58% embryos and 38% nauplii
P. monodon		
Foote et al. (2012)	18, 20 and 22 min old eggs to 2 or 14°C	No nauplii hatched, shocking at 8-14°C produced nauplii but not tetraploids

The demand is high for quality penaeid female brooders. With mostly female heterogamety (e.g. *P. japonicus, F. chinensis*), induction of gynogenesis in penaeid shrimps may provide 75% females only i.e. the sex ratio of ZZZ 25% ♂ : ZZW 50% ♀ : ZWW 25% ♀. Not surprisingly, no publication is yet available on induction of gynogenesis. Triploidy is also known to induce sterility. However, triploidy has been induced in *L. vannamei, F. chinensis, P. japonicus, P. monodon* and *P. indicus* during the last 20 years (Table 6.3), in expectation that the triploids may grow faster than diploids.

Table 6.3 summarizes available information on triploidy induction in five penaeid species. The following may be inferred: 1. Viable triploids were successfully induced in *P. japonicus, P. monodon, P. indicus* and *F. chinensis* but not in *L. vannamei*. However, attempts to induce tetraploidy in *P. japonicus* and *P. monodon* were also not successful and did not produce even 4n nauplius. 2. The triploids suffer heavy mortality during metamorphosis from one larval stage to the next (*P. monodon*). 3. Larval durations were also extended. Growth of triploids varied widely from family to family in *P. monodon* and their triploids were reported to grow 9% faster than diploids. 4. Female ratio in triploids ranged from 0.67 in *P. monodon* to 1.0 in *P. japonicus*, suggesting female heterogamety in penaeids. 5. The triploids were sterile. Ovarian development was more severely affected than testicular differentiation. In triploids, sperm-like cells without spike and spermatophores were noted in vas deferens of *F. chinensis* males.

The chemical inducer cytochalasin-B is toxic. But the purine analogue 6-dimethylaminopurine (6-DMAP) is identified as a safer and more effective chemical alternative to cytochalasin-B (Desrosiers et al., 1993). However the use of 6-DMAP to retain the first or second polar body by the Australian workers in *P. vannamei* has not yielded success. In fact, Pongtippatte et al. (2012), who have attempted to induce triploidy in *P. monodon*, have succeeded with cold shock but not with either cytochalasin-B or 6-DMAP. The chemical induction has resulted in the production of 3-16% embryos with 3-pronuclei, albeit their induction has produced 60% hatchlings. Notably, hatching success of cold shocked embryos is low (40%) but 38% of the embryos harbor 3 pronuclei. Using confocal microscopy coupled with immunofluorescence, Morelli and Aquacop (2003) have made cytological analysis of triploid *P. indicus* induced by heat shock. The pronuclear migration is sensitive to micro-tubule depolymerizing heat shock but amphimixy remains unaffected. Heat shock inhibits or retards cytokinesis but has no effect on centrosome duplication. At the second cleavage, the uncleaved eggs divide directly into four cells, as a consequence of the presence of four centromeres, but some divide into three, while others into two. A similar effect has also been described in *P. indicus* after being treated with 6-DMAP (Peeters, 1996). But it is not clear why the same does not occur in *L. vannamei, P. japonicus* and *P. monodon*. In triploid and tetraploid *L. vannamei* embryos, the abnormalities have been traced to the abnormal syngamy and consequent differences in the number of pronuclei in daughter cells at the first and second cleavages (see Fig 5, 6 of Zuniga-Panduro et al., 2014).

7

Sex Differentiation

Introduction

Sex determination and sex differentiation are highly diverse processes that have evolved independently a number of times (Hodgkin, 1990). Unlike in fishes, sex differentiation in crustaceans is not a labile and protracted process. Further, it is a little different and more complicated in crustaceans than in fishes. Unlike vertebrates, endocrine and gametogenic functions are clearly separated into distinct organs, the Androgenic Gland (AG) and the testes, respectively (Sagi et al., 1997). All hermaphroditic fishes are potential hermaphrodites but are functional bisexuals at a given point of time (Pandian, 2013, p 213). However, there are functional and self-fertilizing hermaphrodites in crustaceans. The process of sex differentiation is disrupted by pollutants in both crustaceans and fishes. Unlike in fishes, it is also disrupted by parasites and food. In gonochoric and protogynic fishes, administration of androgen, especially 17β-methyltestosterone (MT) successfully induces masculinization. However, crustaceans are not amenable to hormonal sex reversal. Vogt (2007), the only publication available on this aspect, indicates that the relatively more flexible crayfish *Procambarus alleni* exposed to MT from egg to adult stage was not masculanized. On the other hand, administration of Juvenile Hormone (JH) or its analog induced parthenogenic production of males and females in daphnids. JH has so far not been used to induce sex change or alter larval stage in fishes.

7.1 Crustacean Hormones

Hormones and neuroendocrines act as chemical messengers of genetic cascade that realizes sexualization and maintenance of sex in an individual. In decapod crustaceans, reproduction is a highly complex process requiring precise coordination. Crustacean endocrinology has moved from classical approach of ablation and implantation of endocrine organs to modern techniques involving immunological (e.g. Nithya and Munuswamy, 2002) and recombinant technology (Sagi et al., 2013). Table 7.1 lists endocrines and neuroendocrines of crustaceans with proven physiological actions related to reproduction (Nagaraju, 2011). Notably, decapod gonads also possess steroids that are recognized with vertebrates (Swetha et al., 2011). Figure 7.1 shows the approximate locations of the endocrine glands in a decapod shrimp. At least seven organs are identified to produce a dozen and odd hormones and neuroendocrines. Their actions are targeted on as many as 10 organs namely, gonads, hepatopancreas (HP), *secondary sexual chraracteristics (SSCs)*, Y-organ, eyestalk, brain, Thoracic Ganglion (TG), ovaries and hemolymph. Strikingly, the gonad maturation is under the dynamic regulation of as many as six hormones (AH, GIH, Ecdysteroids, MF and CHH) and neuroendocrine Gonad Stimulating Factor (GSF). The ovarian maturation seems to be regulated by estrogens, FSH and LH. This may indicate how complex reproduction is and the need for precise regulation and coordination of each event in the reproductive sequence culminationg in gamete spawning/milting. Ever since Panouse (1943) showed the presence of gonad inhibitory factor in *Palaemon serratus*, publications on endocrine role on reproduction have accumulated. They have been reviewed from time to time (e.g. K. G. Adiyodi, 1985, Fingerman, 1987, 1997a, Huberman, 2000, Tsukimura, 2001, Nagaraju, 2011).

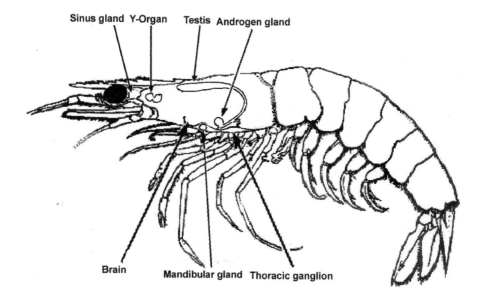

FIGURE 7.1

Approximate locations of the major endocrine glands in a male shrimp. In the place of testis, an ovary is located in a female.

TABLE 7.1

Endocrine and neuroendocrine hormones of crustaceans (compiled and simplified from Kusk and Wollenberger, 2007, Nagaraju, 2011, Swetha et al., 2011)

Production Site	Hormone	Target Organ (s)	Targeted Action (s)
Androgenic gland	Androgenic hormone (AH), protein	Gonads, HP, SSCs	Masculinization of gonads and SSCs
X-organ-Sinus gland	GIH, peptide	Gonads, HP	Inhibits gonadal maturation
X-organ-Sinus gland	MIH, peptide	Y-organ	
X-organ-Sinus gland	CHH, peptide	Gonads	Inhibits molt
X-organ-Sinus gland	MOIH (neuropeptide)	Hemolymph	Regulates reproduction Regulates MF
Y-organ	Ecdysteroid,	Eyestalk, gonads	Stimulates molting and vitellogenesis
Vitellogenesis Stimulating Ovarian Hormone (VSOH)	Steroids	Ovary, SSCs	Vitellogenesis, SSCs
Mandibular organ	Methyl farnesoate (MF), terpenoid	Gonads, HP, Y-organ, brain, TG	Stimulates molting, metamorphosis, gonadal development and ecdysteroid production
Mandibular organ	Farnesoic acid	Gonads, HP	Regulates reproduction
Brain, thoracic ganglion	Gonad stimulating factor	Gonads, HP	Stimulates gonadal development
Gonads	FSH, LH, HCG	Ovaries	Stimulates ovarian maturation
Gonads	Estrogens, progesterone	Hemolymph, ovaries	Stimulates ovaries

GIH = Gonad inhibiting hormone, MIH = Molt inhibiting hormone, CHH = Crustacean hyperglycemic hormone, FSH = Follicular stimulating hormone, LH = Luteinizing hormone, HCG = Humam chorionic gonodotrophin, MOIH = Mandibular organ inhibiting hormone, HP = Hepatopancreas, TG = thoracic ganglion

7.1.1 Androgenic Gland (AG)

Cronin (1947) was the first to describe AG in the crab *Callinectes sapidus*. However, the implantation of AG into female amphipod *Orchestia gammerellus* by Charniaux-Cotton (1953, 1954, 1959) led to the dramatic discovery of masculanizing AG. Subsequently, Katakura and Hasegawa (1983) and LeGrand et al. (1987) demonstrated that an immature female implanted with AG reverts to neomale, while an immature male subjected to andrectomy by bilateral AG ablation reverts to neofemale in isopod *Armadillidium vulgare*. Andrectomy or implantation of the AG in immature/maturing decapods has resulted in complete or partial sex reversal in commercially important freshwater decapods such as *Macrobrachium rosenbergii* (Nagamine et al., 1980a, b, Sagi et al., 1990), *Procambarus clarkii* (Taketomi and Nishikawa, 1996),

Cherax destructor (Fowler and Leonard, 1999) and *C. quadricarinatus* (Khalaila et al., 2001, Barki et al., 2003). In hermaphroditic decapods too, the presence of AG is required for normal differentiation, regeneration and maintenance of gonadal male sex and SSCs (Charniaux-Cotton, 1959, 1960, 1961, Touir, 1977).

The Androgenic Hormone (AH), secreted by AG, plays a major role in male sex differentiation and development by regulating the expression of primary and secondary sexual characteristics (Sagi and Khalaila, 2001). In most crustaceans, "sex is genetically determined; the Gonad Stimulating Hormone (GSH) promotes the formation of AG, which releases AH and in turn, it induces the development of testes and male *Secondary Sexual Characteristics" (SSCs)* (Gusamao and McKinnon, 2009). In the absence of AGs, the ovaries are spontaneously developed in a neuter and the individual becomes a female. The ovarian hormone, in its turn, stimulates the development of SSCs in females (Kusk and Wollenberger, 2007). Incidentally, irregular production of AH is associated with intersexuality in the amphipod *Echinogrammus marinus* (Ford et al., 2005). Also manipulation of AGs in crayfish *C. quadricarinatus* results in production of intersexuals (Sagi and Khalaila, 2001).

In decapods, AG is located adjacent to the vas deferens (Fig. 7.1). It is attached to the terminal ejaculatory bulb called ampulla. Murakami et al. (2004) reported two kinds of AGs in *P. clarkii*, one inside the body cavity and the other inside the coxa. In AG, cells are arranged as thin, parallel and anastomosing cord or a compact lobed structure. For more details on structure and cellular organization of the AG, Alfaro-Montoya and Hernandez (2012) may be consulted.

The appearance of a triangular structure with three setae or retention of tubular structure in endopodites of the first pair of pleopods in females and males, respectively in 50 d-old post larva marks 'the point of no return' in sex differentiation of *Litopenaeus vannamei*. Exposure of older than 50 d post larvae to different temperatures (18, 27, 32°C), photoperiods (light, shadow, darkness), food densities and starvation fails to revert sexual differentiation and sex ratio (Campos-Ramos et al., 2006). In *M. rosenbergii* too, implantation of AG on immature females of 6.5-7.5 mm body length induces complete sex reversal (Àflalo et al. 2006). But that in the larger ones (>8 mm) may partially revert or may not revert at all in females larger than 28 mm body size (Nagamine et al., 1980a, Malecha et al., 1992).

For AG ablation or implantation, the giant prawn *M. rosenbergii* has been the choice of many scientists. The reasons for this choice are that: (i) it is one of the largest commercially important prawns; being large , it may withstand the stress of AG implantation or ablation, (ii) the males grow faster than females; hence all male seedlings are in great demand by aquaculturists and (iii) during its ontogeny, besides undergoing sexual differentiation, the males, also pass through three morphologically distinct stages of dominance hierarchy from pink (1-10 g), orange (10-40 g) and blue (>40 g) clawed morphotypes (see Table 3.12). In these morphotypes, the length of AG of Orange

Claw Male (OCM) is more than two times longer than that of Pink Claw Male (PCM) and the breadth of AG of Blue Claw Male (BCM) is also more than two times larger than that of OCM (Okumura and Hara, 2004). To test whether the AG ablation influences morphotypic differentiation, Sagi et al. (1990) ablated AG from PCM and OCM (Table 7.2). They found that 1. A majority (93%) of PCM transformed into OCM but could not further into BCM. Only 24% of the OCM were able to transform into BCM. Hence, the presence of AG is obligatorily required to ensure morphotypic differentiation also. 2. Following the respective transformation, the SSCs like genital papillae, sperm duct and ampulla were lost and the loss being more severe in PCM. 3. Spermatogenesis continued to occur in the absence of AGs, although the quantity and/or rate of sperm production could be affected, as the testis size was decreased. For example, Okumura and Hara (2004) reported that the spermatocytes continued to occupy a major part of testicular tubes of PCM at the intermolt stage C. but those of OCM were filled with spermatozoa. Unfortunately, Sagi et al. (1990) were silent on the induction of ovarian development in the absence of AG in PCM and OCM. However, Nagamine et al. (1980b) had earlier shown that the appearance of sex specific SSCs are reliable morphological markers of the recipient's ability to respond to the absence of AG. They found that male prawns andrectomized prior to the appearance of SSCs did not develop male specific SSCs. But those andrectomized after the appearance of male specific SSCs did not lose them subsequently, i.e. even in the absence of AGs, the male specific SSCs continued to exist.

TABLE 7.2

Effects of andrectomy on pink claw males (PCM) and orange claw males (OCM) of *Macrobrachium rosenbergii* (compiled from Sagi et al., 1990, Okumura and Hara, 2004†)

Parameter	PCM	OCM	BCM
Morphotype Transformation to			
PCM (%)	–	93	0
OCM (%)	–	–	24
SSCs			
Disappearance of genital papilla (%)	100	95	–
Atrophy of sperm duct (%)	95	57	–
Absence of ampulla (%)	100	66	–
Spermatogenesis			
Testis somatic index	0.09	0.14	0.5†
Testis with sperm (%)	73	100	100

For an easier understanding, the data reported by Nagamine et al. (1980a) on implantation of AG in *M. rosenbergii* females of different sizes are simplified in Table 7.3. Apparently, AG implantation completely reverted sex to produce neomales only in immature prawns of < 9-12 mm body length. In others, the implantation induced either partial reversal (13-28 mm females) or no reversal at all in mature females (> 28 cm body size). Notably, the AG ablation induces more readily morphotyphic differentiation than masculinization with AG implantation. The findings of Nagamine et al. may be summarized: 1. Of 16 female recipients, only 13 differentiated into neomales within one-three post-implantation molts. 2. Only the three youngest *M. rosenbergii* females (< 9-12 mm) implanted with AG developed complete vas deferens. In others, they were incompletely developed and more curled and/or crooked. 3. No neomales spawned but retained female gonopores. Neither had they developed reproductive setae. 4. Six of 13 neomales developed masculinized cheliped (7.5 mm), on attaining a body size of 28 mm CL, the CL at which mature cheliped (12 mm) is developed in normal males. Hence, the partially sex reversed neomales were unable to develop complete vas deferens and chelipeds typical of male. 5. In the inverted V shaped ovary, the middle region of the ovarian arms continued as ovaries. Primary (ovarian) vitellogenesis occurred but secondary (hepatopancreas, HP) one was inhibited. Incidentally, the ovarian regeneration with oocytes occurred in female mud crab *Scylla paramamosain*. *In vitro* incubation of ovarian tissues at secondary vitellogenesis stage in AG extract decreased amino acid uptake (Cui et al., 2005). Apparently, the AG targeted HP to inhibit secondary vitellogenesis. Spermatogonia were present in *M. rosenbergii* males but not spermatocyte and spermatid at the posterior (which is the normal position of testis in an ovotestis, see Fig. 3.3C) end of an arm or anterior end. Evidently, the recipients were too old to differentiate into fully functional testis out of the ovary. However, functional testis did originate from the ovary of a young (PL$_{30-50}$) recipient (Afalo et al., 2006). Incidentally, an inverse relationship between the level of masculinization and size at AG implantation was demonstrated in amphipods (Charniaux-Cotton, 1958, 1970) and isopods (Katakura, 1960, 1989, Le Grand-Hamelin, 1977).

TABLE 7.3

Effects of AG implantation on *Macrobrachium rosenbergii* of different body sizes (simplified from Nagamine et al., 1980a)

Sexual Maturity	CL Size (mm)	Gonopore complex	Appendixes masculina	Mature cheliped	Vas deferens	Spermatogonia in ovary
Immature I	9-12	+	+	−	+	+
Immature II	13-19	−	+	−	+	−
Mature	22-29	−	+	+	+[2]	−
Re-implanted	33	−	+[1]	+[1]	+	−

1 = absent in 29 mm sized prawn, 2 = absent in > 27 mm sized prawns

As recipient females of ~12 mm body size (Nagamine et al., 1980a) failed to differentiate functional testis from the ovary, Malecha et al. (1992) reduced the recipient size to 8-10 mm and finally to 6.5-7.5 mm size. They found that the putative neomales that were recipients at the size of 8-10 mm were unable to mate but those recipients at 6.5-7.5 mm size were able to mate and sire progenies. Fifteen matings between these neomales and normal females yielded 13,294 progenies, of which 10,128 were females and 3,166 were males. The obtained sex ratio of 0.76 ♀ : 0.24 ♂ revealed that *M. rosenbergii* is female heterogametic with ZZ/ZW sex determination mechanism and WW super females are viable. Production of complete and fully functional sex reversal can be accomplished by implanting AG from an old female into putative female of 6.5-7.5 mm size only. Eight backcrosses involving WW super females and ZZ normal males produced 11,034 females and 1,664 males with a resulting sex ratio of 0.87 ♀ : 0.13 ♂. The unexpected 13% males are likely to be phenotypic pseudomales carrying ZW genotype (see Kirankumar and Pandian, 2004, David and Pandian, 2006) or arose due to overriding autosomal factor (s) altering the course of sex differentiation process (Pandian, 2014, p 35). The pseudomale ratio ranging from one male to 1,578 females to one male to 1.35 females calls for research on sex-linked molecular markers to identify the phenotypic males like AFLP marker of Ventura et al. (2011b), as well as autosomal factors capable of altering sex differentiation process (Table 6.2).

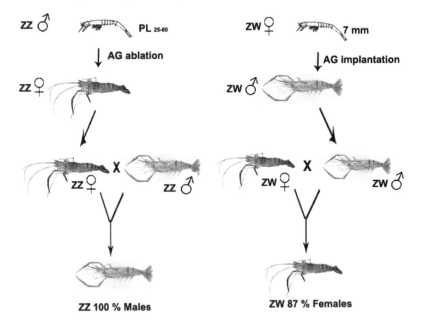

FIGURE 7.2
A two step procedure to produce all-male (left panel) and all-female (right panel) *Macrobrachium rosenbergii* progeny (information drawn from Malecha et al., 1992 and Aflalo et al., 2006).

In view of the fact that the male *M. rosenbergii* grows faster than the female and pressing demand for a technique to produce all male progenies, Aflalo et al. (2006) developed a novel two-step procedure (Fig. 7.2) for mass production of all-male seedlings. They identified putative males (PL_{25-60}, 0.15-1.0 g) by the presence of genital papillae on the 5th walking legs. Microsurgical AG ablation was performed by removing the 5th walking legs together with AG. To ensure that the AG was completely ablated a major fraction of sperm duct was also pulled out. Of 1,940 post larvae ablated, survival decreased to 80 and 66% within 24 h and 30 d after ablation, respectively. Notably, with decreasing size, at which AG is ablated, survival decreases. Nagamine et al. (1980a), who andrectomized > 12 mm sized juveniles, reported no mortality. When AG was ablated at 6.5-7.5 mm size, Malecha et al. (1992) reported 90% survival. Aflalo et al. were cautious to cull out those putative neofemales, which developed appendixes masculina on the 2nd pleopod between 30 and 90 d after ablation (at 15.6 mm size, see Malecha et al., 1992), an indicator for failed AG ablation. In view of the findings by Nagamine et al. (1980b) that male prawn, andrectomized prior to the appearance of SSCs, do not develop them subsequently, Aflalo et al. even checked by amputating the 2nd pleopod bearing appendixes masculina and their absence subsequently.

At the age of 12 m, 225 of 1,280 putative neofemales alone developed ovary, i.e. 14.6% of the total AG ablated. In this first group, 11.4% of the total AG ablated neofemales successfully mated and were berried, as against 1.9 and 1.3% that were mated and berried in the second group, in which 4,000 seedlings underwent AG ablation. All the berried females produced 100% male progenies. Incidentally, George and Pandian (1995) also developed a two step procedure to produce all male progenies in female heterogametic black molly *Poecilia sphenops*. The sex specific gonopores appear on the 5th walking legs first in PL_{25-60} after metamorphosis to post larva. The male specific appendixes masculina appear on the 2nd pleopods by 30-60 d after AG ablation (Aflalo et al., 2006) at 15.6 mm body size (range = 9.0–26.0 mm, Malecha et al., 1992). If AG implantation or ablation is performed prior to the appearance of the sex specific SSCs in the prawn, complete and functional sex reversal occurs from putative female to male or putative male to female. At this early stage, the gonad remains undifferentiated with bisexual potency. In the presence of AG and its AH action presumably during the period from 30-60 d in the post larva, the bisexual potency of the gonads is irrevocably lost in decapod males. It is likely that the oogonia are also lost at that stage. Contrastingly, both oogonia and spermatogonia drawn from mature adult fishes that have already spawned/milted continue to retain bisexual potency (see Pandian, 2011, p 146-159, 2012, p 93-102).

This detailed description on experimental endeavors to reverse sex by AG ablation/implantation clearly indicates that 1. The prawn is amenable to sex reversal, if the surgery is accomplished only during the period from PL_{25-16} and a size of < 12 mm, when the gonads remain undifferentiated. When expression of ZZ or WZ is completed during this period, the prawn reached a point of no return, i.e. the bisexual potency of Primordial Germ Cells (PGCs) is lost. Whereas gonochoric fishes have retained the bisexual potency even after milting/spawning as adults (e.g. *Oncorhynchus mykiss*, see Pandian, 2012, p 146-149, 2013, p 93-102), the crustaceans seems to have lost the potency. Of course, there are many exceptions like the simultaneous and sequential hermaphrodites, the parasites like *Ione thoracica* as well. Consequent to the loss of bipotency, partial sex reversion induced after the point of no return results in the irrevocable loss of SSCs.

According to Walker (2005), the exposure of nauplii of the rhizocephalan parasitic barnacle *Heterosaccus lunatus*, the summer-like photoperiod (LD 16: 8) induced more males but that simulating winter (LD 8:16) induced more females. Walker considered that the photoperiod-induced sex differentiation is rare, as oogenesis has to be regulated to produce eggs of two sizes either singly or together, the larger being male eggs. However, sex is irrevocably differentiated in rhizocephalans, when nauplii metamorphosed into male or female cyprid larvae, which pass through different larval stages.

Incidentally, the American alligator's eggs incubated at 33°C produced mostly in males, while those incubated at 30°C mostly females. Dr. R. Yatsu has found that a thermosensitive protein called TRPV4 present in the eggs responses to warmer temperature by inducing the male gonadal sex differentiation. It is likely that the warmer temperature rather than photoperiod prevailing during summer produced more males but low temperature simulating winter produce more females in the parasitic barnacle. However, it is to be known whether the parasitic barnacle produces eggs containing TRPV4 protein.

Purified *A. vulgare* Arvul-AH is a glycoprotein composed of two polypeptide chains that are linked by disulfide bridges (Okuno et al., 1997, Martin et al., 1999). To know the changes in protein content and protein profile of AH with morphotyphic differentiation, Sun et al. (2000) divided the OCM into early OCM and transitional orange-blue clawed male OBCM. A fourfold increase in the protein content (µg/AG) from 19 in OCM to 45 in OBCM and to 76 in BCM was observed. Two polypeptide bands (~ 18 and ~ 16 Kd) were present in OBCM and BCM but not in OCM. The putative AH proteins had approximately the same molecular weight as that of *A. vulgare* (Okuno et al., 1999).

Investigations on AG gene cloning, sequencing and characterization of its cDNA in many decapods (e.g. *M. rosenbergii*, Ventura et al., 2009, *C. quadricarinatus*, Rosen et al., 2010, *M. lar*, *Palaemon paucidens* and *P. pacificus*, Banzai et al., 2012) have shown that AH is a protein of 176 amino acids related to the insulin-like growth factor family (see Manor et al., 2007). The insulin-like peptides of many invertebrates play a variety of roles in metabolism, growth and reproduction. AG-specific insulin-like gene is expressed in males alone and has been thus far recorded in crustaceans only (Ventura et al., 2009). Further investigations by Ventura et al. (2011b) have shown that *Mr-IAG* is expressed from 20 d after metamorphosis but prior to the appearance of SSCs (cf Nagamine et al., 1980b). Its expression is stronger in reproductively active BCM and actively sneaking PCM than in OCM. In prawns, the expression of AG is combined with male sex differentiation and growth. Using RNA interference methods, it has been possible to silence the AG encoding gene

IAG; its silencing prevents differentiation of male specific characteristics and leads to the arrest of spermatogenesis in *M. rosenbergii* (Ventura et al., 2009) and *C. quadricarinatus* (Rosen et al., 2010). Ineffectiveness and/or inadequate levels of AH induced by natural mutations may account for hermaphroditism and explain a part of sex change in sequential hermaphrodites (cf. Charniaux-Cotton, 1958).

7.1.2 Eyestalk: Sinus Gland (SG)

In 1933, Hanstrom described a structure in eyestalk of crustaceans and named it as 'Sinus Gland' (SG). However, Panouse (1943) made a dramatic discovery of the existence of Gonad Inhibiting Hormone (GIH), or more precisely Vitellogenesis Inhibiting Hormone, VIH, (Nagaraju, 2011) in the SG, when he ablated the eyestalk of *Palaemon serratus* (Fig. 7.1). The SG is a cluster of axonal endings that lies on the surface of the eyestalk ganglia in close contact with hemolymph. The cluster of cells that supplies the axons of the SG, located deeper within the ganglia, is the X-organ (Hopkins, 2012). In crustaceans that do not have stalked eyes, the neurohemal release site is located adjacent to the brain (Carlisle and Knowles, 1959). Hence, the eyestalks are considered as extensions of the brain and part of the crustacean Central Nervous System (CNS). The X-organ-SG complex, located bilaterally in the eyestalks of crustaceans, is the major endocrine control center, analogous to vertebrate hypothalmo-neuro-hypophyseal complex (Swetha et al., 2011). Hormones produced by the X-organ are stored in SG and released at appropriate stimuli. The major peptide hormones released by SG are: (i) Molt Inhibiting hormone (MIH), (ii) VIH, (iii) Mandibular Gland Inhibiting Hormone (MOIH) and a host of others. These peptide hormones are collectively called Crustacean Hyperglycemic Hormone (CHH). Besides, more than 100 neuropeptides have been identified in decapods, euphausiids and copepods, suggesting greater possibilities of crustacean hormonal pathway complexity and greater possibilities for hormonal crosstalk (Hopkins, 2012).

Eyestalk ablation (ESA) induces precocious sexual maturity (e.g. *Penaeus canaliculatus*, Choy, 1987) and vitellogenesis (e.g. *Callinectes sapidus*, Zmora et al., 2009a, b) and vitellogenesis resulting in precocious spawning (e.g. *P. serratus*, Panouse, 1943). Unilateral or bilateral ESA induces precocious ovarian maturation in the Indian tiger shrimp *P. monodon* (Mohamed and Diwan, 1991). Bilateral ESA in *P. japonicus* increases Gonado Somatic Index (GSI) in immature shrimp and vitellogenin mRNA levels in the ovary (Tsutsui et al., 2005). Holding ablated female swimming crab *Portunus trituburculatus* at 20°C and providing 15 h photoperiod produce the most profound effect on the induction of ovarian maturation and spawning (Kim et al., 2010). In *Macrobrachium idella*, ESA stimulates the inhibitory principles to play a role in synchronizing the regulatory processes of molting and reproduction (Sreekumar and R.G. Adiyodi, 1983). Consequently, the ablated female has to precociously meet the costs of molting and reproduction. Not surprisingly, Choy (1987) noted that both batch and cumulative fecundities of *Penaeus canaliculatus* are higher in intact females than in ablated ones, albeit the ablated female spawns more frequently.

In non-breeding *P. vannamei* male, the ESA induces spermatogenesis, larger testicular size and spermatophore weight, and doubles mating success (Chamberlain and Lawrence, 1981). In *P. monodon*, the unilateral ESA induces higher sperm count, larger sperm and longer spike (Gomez and Primavera, 1993). Nagaraju and Borst (2008) have reported doubling of Testis Somatic Index (TSI) in *Carcinus maenas*. In fact, the hitherto described observations led Fingerman (1997a) to propose that eyestalk contains an inhibitory factor GIH.

To date VIH has been reported from crayfish (e.g. *Astacus leptodactylus*, Mosco et al., 2013), shrimps (e.g. *Rimicaris kairei*, Qian et al., 2009), prawns (e.g. *M. rosenbergii*, Yang and Rao, 2001), lobsters (e.g. *Nephrops norvegicus*, Edomi et al., 2002), crabs (e.g. *P. trituberculatus*, Kim et al., 2010), isopod (*Armadillidium vulgare*, Greve et al., 1999) and possibly in the fairy shrimp *Streptocephalus dichotomus* (Nithya and Munuswamy, 2002). These examples may indicate the ubiquitous presence of VIH across crustacean taxa. However, structural and functional characterization of VIH peptide is limited to *Homarus americanus*, *H. gammarus*, *N. norvegicus*, *Procambarus bouvieri*, *M. rosenbergii*, *R. kairei* and *A. vulgare* (see Swetha et al., 2011). In fact, all mature CHH family peptides contain 72 and 83 amino acids and are characterized by six cysteine residues, organized into three intra-molecular disulfide bridges, which maintain the VIH tertiary structure (Swetha et al., 2011, Hopkins, 2012). Expectedly, there is considerable degree of sequence similarities between VIH and MIH. VIH or its precursor of some decapods like *N. norvegicus* has been cloned. A recombinant protein encoding *GIH* has been synthesized. Using the protein, VIH synthesizing cells in the eyestalk has been localized in *N. norvegicus* (see Swetha et al., 2011). Recently, Fu et al. (2014) have estimated the expression levels of *sp-CHH* of *Scylla paramamosion* during selected reproductive stages.

The operational process of ESA is often associated with high mortality of broodstock and production of inferior quality offspring. To overcome these problems, non-surgical procedures have been developed to induce ovarian maturation and spawning. To date two procedures are available: (i) injection of double stranded RNA (ds RNA) to knock off GIH mRNA and (ii) injection of neurotransmitters (e.g. Sarojini et al., 1995) or hormones (e.g. 17 β-estradiol, Yano and Hoshino, 2006). Lugo et al. (2006). Tiu and Chan (2007) have shown the possibility of functional gene silencing in *Litopenaeus schmiditti* and *Metapenaeus ensis*, respectively. Injection of *CHH* ds RNA decreased CHH mRNA to an undetectable level within 24 h with corresponding decrease in hemolymph glucose level (see also Sagi et al., 2013). However, it were Treerattrakool et al. (2008, 2011), who demonstrated for the first time that the ovarian maturation and spawning can be induced by ds RNA, as an alternative to ESA. Silencing of *GIH* expression increased the level of vitellogenin in *P. monodon*. A single injection of 3 µg GIH-ds RNA/g body weight into pre-vitellogenic female inhibited *GIH* expression for a minimum period of 30 d. This ds RNA-mediated GIH silencing led to ovarian maturation and eventual spawning in domesticated as well as wild female brooders, with a comparable effect of ESA in wild shrimp.

In decapods, biogenic amines are conserved molecules and are involved in regulation of an array of physiological activities. Among these amines, dopamine and serotonin are present in the CNS of crustaceans (Fingerman et al., 1994). They control and regulate the release of peptide hormones from eyestalk and brain (Kuo and Yang, 1999). For example, serotonin stimulates ovarian maturation in *Pr. clarkii* but dopamine inhibits serotonin-stimulated ovarian maturation (Sarojini et al., 1995). *In vitro* culture of M-O and Y-O in the presence of melatonin significantly increases the secretion of MF and ecdysteroid, respectively in the mud crab *Scylla serrata* (Girish et al., 2015a). Injection of spiperone, a dopamine blocker, increases ovarian diameter and Ovary Somatic Index (OSI) (Rodriquez et al., 2002a). Melatonin is present in the CNS (Maciel et al., 2008) and regulates molting (Tilden et al., 2001). It may be noted that information accumulated on the role of these biogenic amines remain a piecemeal. Sainath and Reddy (2011) made a comprehensive investigation on the functions of these amines in intact and eyestalk ablated of a freshwater edible crab *Oziotelphusa senex senex* (Table 7.4). The administration of serotonin and melatonin into intact and ESA crabs induces ovarian maturation, as indicated by increased ovarian diameter, OSI and vitellogenin level. However, injection of dopamine has no effect on intact crabs but delays ovarian maturation in ESA crabs. This simple procedure is cheaper and widely practicable; aquaculturists should take advantage of it. In a more recent study, Girish et al. (2015b) have shown that the relation between the expression levels of Retinoid X Receptor (RXR), ecdysone receptor (EcR), *ecdysone inducible gene (E75)* and vitellogenesis provide an alternative molecular intervention mechanism to the traditional eyestalk ablation to induce spawning in decapods.

TABLE 7.4

Effect of administering dopamine, serotonin and melatonin on ovarian growth and vitellogenesis in the freshwater edible crab *Oziotelphusa senex senex* intact and eyestalk ablated crab received each 10 µl injection (compiled from Sainath,, 2009, permission by S. B. Sainath)

Treatment	Ovary diameter (µm)	Ovary Somatic Index	Vitellogenein (mg/g)
Control	20.4	0.37	0.07
Dopamine (10^{-6} M)	22.36	0.39	0.07
Serotonin (10^{-6} M)	69.48	3.02	0.71
Melatonin (10^{-6} M)	73.91	3.19	0.79
ESA	72.95	2.99	0.69
ESA + Dopamine (10^{-6} M)	47.49	1.73	0.30
ESA + Serotonin (10^{-6} M)	74.31	3.14	0.73
ESA + Melatonin (10^{-6} M)	79.81	3.26	0.79

Gonad Stimulating Factor (GSF) Implantation of brain or Thoracic Ganglia (TH) revealed the presence of GSF in the crab *Paratelphusa hydromous* (Gomez and Nayar, 1965). Injection of brain and TH extract induced the ovarian maturation and secondary vitellogenesis in *Uca pugilator* (Chang, 1985). During the period of spermatogenic quiescence, injection of TG extract into *Potamon koolooense* induced AG and onset of spermatogenesis with increased testicular tubules and TSI (Joshi and Khanna, 1984). With implants of *H. americanus* TG, Yano and Wyban (1992) could induce ovarian maturation in a few recipients of *P. vannamei*, indicating that GSF may not be species specific. In this context, the publication by Avarre et al. (2007) is interesting. Presenting data on the molecular architecture and phylogenetic analysis, they reported that the main egg yolk precursor protein of decapod crustaceans is a member of the Large Lipid Transfer Protein (LLTP) super family and is homologous to insect apoLp-II//I and vertebrate apoB. Hence, it is not surprising that GSF of decapods is not species specific.

7.1.3 Y-organ and MIH

In crustaceans, molting is regulated by multi-hormonal systems but is under the immediate control of molt-promoting steroid hormones, the ecdysteroids secreted by the Y-organs (YO). Located in the anterior body cavity of decapods (Fig. 7.1), the Y-organs secrete ecdysone (McConaugha and Costlow, 1981) or ecdysteroids (Nagaraju, 2011), commonly known as Molting Hormone (MH). In crustaceans, the inactive parent compound ecdysone and biologically active metabolite 20-hydroxecdysone are synthesized from dietary sterols in the Y-organ (Chang, 1985). Ecdysteroids also play an important role in the control of reproduction and embryogenesis (Subramoniam, 2000). The MH triggers the onset of proecdysis and ecdysis in crustaceans. It is secreted by the Y-organs. Synthesis of ecdysteroids by Y-organs is negatively regulated by MIH (Lu et al., 2001). Both the ovaries and testes serve as alternate sources of ecdysteroids in crustaceans like *Artemia salina, Orchestia gammerellus, Astacus leptodactylus, Cancer anthonyi, Carcinus meanas, C. magister, Callinectes sapidus, Libinia emarginata* and *Acanthonyx lunulatus*. The hormones are reported to increase DNA synthesis, cell proliferation and spermatocyte differentiation in testes. In diecdysic crustaceans, a close correlation between the hemolymph ecdysteroid levels and ovarian maturation stages has been reported (e.g. *M. nipponense*, Okumura et al., 1992, *Oniscus asellus*, Steel and Vafopoulou, 1998). However, the anecdysic brachyuran crabs, in which the entire reproductive cycle is completed within the intermolt period, during which ecdysteroid levels are already low (Demeusy et al., 1962), respond differently. Not surprisingly, there are contradictory reports on their role in vitellogenesis (see Laufer et al., 1993) in diecdysic and anecdysic decapods.

The earliest publication by Zeleny (1905) indicated that MIH is present in the eyestalks and other structures in the nervous system of shrimps. MIH in eyestalk extract or recombinant MIH inhibits the production of ecdysteroids from the Y-organs. With reference to molting, YO's switches on, while MIH switches off. The mechanism, by which the two endocrine organs are coordinated, is still unknown but their consorted coordination induces successful molting in crustaceans (Hopkins, 2012). In the eyestalked decapods, there is a 'time lag' prior to the onset of proecdysis, suggesting that there are intervening steps between the release of eyestalk peptide and onset of ecdysteroidogenesis. But, there is no such intervention of the eyestalk MIH in amphipods, as there is a rapid decline in ecdysteroids at and after ecdysis (Hopkins, 2012). Zmora et al. (2009a, b) reported that MIH stimulates vitellogenesis at advanced ovarian maturation stage in *C. sapidus*.

In the diecdysic crustaceans, characterized by long premolt and short intermolt periods, spawning is obligatorily preceded by a molt. But the anecdysic crabs and calanoids, which have short premolt and long intermolt periods, have programmed an initial reproductive and terminal growth phases within the long intermolt period. Hence, it may be interesting to know the presence and activity levels of MIH and ecdysteroids in the anecdysic crabs, which do not molt after attaining sexual maturity at the terminal molt. No information is yet available on the MIH in these crabs. But there is evidence that secretion of ecdysteroids diminishes (e.g. *L. emarginata*, Hinsch, 1972) due to the possible degeneration of Y-organs. In the snow crab *Chionoecetes opillio*, ecdysteroid level is 2.1 ng/ml hemolymph in the female that have undergone terminal molt, but its level is 3.98 and 107 ng/ml in the large and small males, respectively (Tamone et al., 2005). Ultrastructural studies have shown that Y-organs are degenerated in the isopod *Sphaeroma serratum*, following the terminal pubertal molt (Charmantier and Trilles, 1979).

In crustacean larvae, the molt is not expected to be followed by reproductive investment. Hence, ESA may have all to do with MIH only. Larval crustaceans undergo a series of molts culminating in metamorphosis. In *Macrobrachium rosenbergii*, RT-PCR studies have shown that MIH is not present in fertilized eggs. However, the presence of MIH becomes apparent during protozoea (Yao et al., 2010). Chan et al. (1998) detected the presence of MIH mRNA in pre-hatched and newly hatched stages of the crab *Charybdis feriatus*. Nevertheless, the most important morphological and anatomical changes occur during the molt from late zoeal stage to megalopa in crabs; hence, it represents the metamorphic molt.

Effects of ESA radically changes the duration of molt cycle, depending on the time, when ESA is performed. In diecdysic prawns and lobsters, the critical time is identified as the premolt 3rd zoeal stage (Table 7.5). Ablation prior to it, results in an extra molt to supernumerary 5th zoeal stage. But, ablation after this period allowed larvae to metamorphose into normal megalopa. When ESA is performed in a zoea (Z-3 instar), the molt cycle is slightly accelerated in the anecdysic crabs, possibly due to the continued presence of MIH. However, the megalopal molt cycle is significantly accelerated. Hence, the duration of total molt cycles is reduced. Exposure of the crab *Rhithropanopeus harrissi* zoea to 1-5 µg β ecdysone/ml water induced dose-dependent reduction in the molt cycles from 6.2 d in control to 5.7 d in those exposed to 5 µg/ml (Table 7.5).

TABLE 7.5

Effects of bilateral eyestalk ablation on intermolt period and molt cycle during larval stages of decapods

Species and References	Intermolt Period/Molt Cycle	Extra Larval Stage (s)
Prawns and Lobster		
Palaemonetes pugio Hubschman (1963)	No effect	No
P. macrodactylus Little (1969)	No effect	Yes
P. varians Le Roux (1984)	Reduced duration preceeding metamorphosis	Yes
Pisidia longicornis Le Roux (1980)	Reduced duration of megalopa but no effect on zoea	No ?
Homarus americanus Charmantier et al. (1988)	Reduced durations of larval and post-larval stages, when ESA was affected at stage II or III	Yes
Crabs		
Callinectes sapidus Costlow (1963)	Reduced duration of megalopa	–
Sesarma reticulatum Costlow (1966)	Reduced duration of megalopa but not zoea	Yes
Chionoecetes opillio Tamone et al. (2005)	Increased ecdysteroid level in small claw ♂ but not in large claw (> 95 mm CW) ♂ and terminally molted ♀	–
Portunus trituberculatus Dan et al. (2014)	Ablation prior to third zoea molted to super numeracy 5th zoea.	Yes
	But after 3rd zoea led to metamorphosis	No
Rithropanopeus hairrisii Freeman and Costlow (1980)	ESA during 4th zoeal or megalopa accelerated molt cycle	Yes
McConaugha and Costlow (1981) Exposure of zoea to 1.5 µg β ecdysone/ml water	Dose dependent reduction in duration of megalopa	No

The structure of MIH has been deduced from many crustaceans and the peptide is found to have very similar amino acid sequence. Two to four (e.g. Webster et al., 2012) isoforms of MIH have been identified. The recombinant

MIH-B prolongs the molt cycle in ESA shrimps but it is not effective as MIH-A (Hopkins, 2012). In *C. feriatus*, MIH genome spans 4.3 kb and consists of 3 exons and 2 introns. Exons 2 and 3 consist of coding sequences for the mature peptides, which show the highest similarity to that of *C. sapidus* (Techa and Chung, 2013). cDNA of full length MIH *C. pagurus* is shown to code 78 amino acids in the mature MIH peptide (Lu et al., 2001). The genomic DNA of MIH in *Portunus trituberculatus* is expressed in the eyestalk ganglia, TH, ovaries, testes and others (Zhu et al., 2011).

7.1.4 Mandibular Organs and MO-IH

Le Roux (1968) was the first to describe the crustacean Mandibular Organs (MOs). MOs are paired, ductless, highly vascularized ectodermally derived glands (see Nagaraju et al., 2004). In decapods, the glands are located at the base of the inner adductor muscle tendon (Borst et al., 2001). However, the location of the MO in non-decapods is not known. The unfavorable anatomical locations of the MOs have made it difficult for the classical ablation-implementation studies to know the functions of them (Borst et al., 2001, however see also Hinsch, 1980). Consequently, a 'reverse endocrinological' study (see Nagaraju, 2007) is necessitated to understand the many functions of MOs. Methyl Farnesoate (MF) is synthesized by MOs (0.07 to 1,040 pmoles/g/h and secreted into the hemolymph at nanomolar (0.002-67 ng/ml) amounts (Nagaraju, 2007). MF is a sesquterpene that is structurally similar to insect Juvenile Hormone (JH III) except for the absence of an epoxide moiety at the terminal end of MF. Farnesoic Acid (FA) and MF are immediate precursors of JH III in insects. However, JH III has not so far been found in any crustacean (Nagaraju, 2007). MF binding proteins, which transport MF from MO to target organs through hemolymph, have been characterized in *Homarus americanus* (Prestwich et al., 1990) and *Libinia emarginata* (Li and Borst, 1991).

The functions of MF are many folded: 1. At increasing levels, it regulates ovarian and testicular maturation, 2. It also has a role in molting, as its level is negatively correlated with that of MIH. Incidentally, MOs can be distinguished from Y-organs, as the former secrets MF but the latter ecdysteroids. 3. As JH, it may delay or inhibit metamorphosis to produce giant larvae and 4. It can alter sex differentiation at the pre-vitellogenic stage in oocyte of parthenogenic female cladocerans and generate males, an account of which is provided later.

Sex specific differences in MO size and MF secretion are reported from many decapods. In *Oziotephusa senex senex*, the MO mass increased at the rate of 5 mg for every g increase in body weight (Nagaraju et al., 2004). After sexual maturity, the MOs increased to an amazing size in *H. americanus* and males had larger MO than females (Borst et al., 1987). In *O. senex senex*, MF content also increased from 8 ng/MO in 8 g to 200 ng/MO in a 16 g crab. Expectedly, the MF synthesis rate decreased from 27.5 to 20 ng MF/mg MO (Nagaraju et al., 2004). Hence, the MF secretion was correlated with body size. But it must be noted that within a day, the MF levels may widely fluctuate from 10 to 100 ng/ml hemolymph in *L. emarginata* and *H. americanus* (Borst and Tsukumura, 1992).

ESA is reported to induce hypertrophy of MO in *L. emarginata*, *Palaemon varians* and *Psidia longicornis* (see Nagaraj et al., 2004). Sex specific increase in MO mass of ESA *O. senex senex* was 5 and 6 mg for every gram increase in body weight of female and male, respectively. Nagaraju et al. (2006) reported that MO secretion considerably increased from pre-vitellogenic to vitellogenic crabs. Stimulation of ovarian maturation by exogenous MF was also observed in *O. senex senex* (Reddy and Ramamurthi, 1998). A six-fold increase in MF secretion was reported in females with growing oocytes in *L. emarginata* (Laufer et al., 1987). Injection of MF increased vitelloggenin level in hemolymph of eyestalkless spider crab (Vogel and Borst, 1989). The implantation of MO of male spider crab into female stimulated the ovarian growth in immature female (Hinsch, 1980). Clearly, MF acts directly on ovarian maturation in the presence of an MF receptor on ovarian tissue (Takac et al., 1993).

MF was shown to increase spawning, fecundity, fertility and hatching success in ablated shrimps *Litopenaeus vannamei* (Laufer, 1992) and *Penaeus monodon* (Hall et al., 1999). Dietary administration of JH III at the dose of 2.5 μg/female/d for 61 d also increased Ovary Somatic Index (OSI) in crayfish *Austropotamobius pallipes* (D'Agaro et al., 2010). But the administration of 5.5 μg MF/g diet to ESA adult female (> 60-75 g) *P. monodon* for 14 d reduced the number of spawns from 3.0 in control to 1.8, and fecundity of first three spawns from 4,100 to 3,200 eggs/g shrimp. MF was shown to arrest ovarian development at stage III; > 39% oocytes were arrested in comparison to < 4% in the control (Marsden et al., 2008). It seems that the level of MF at specific ovarian maturation stage may play a positive or negative role. For example, Mak et al. (2005) showed that low FA level stimulated *Vg* gene expression in hepatopancreas but high levels inhibited it during specific stages of oocyte development in the red crab *C. feriatus*.

Methyl farnesoate also plays role in male reproduction and behavior. In the spider crab, abraded (sexually mature) males have larger MOs than unabraded males. MF treatment stimulates testicular growth in *O. senex senex* (Kalavathy et al., 1999, Reddy et al., 2004) and *Macrobrachium malcomsoni* (Nagaraju et al., 2003). Hence, MF directly induces testicular growth, as testes are known to possess MF receptors (Takac et al., 1993). However, it may be noted that the young newly molted new-shell adolescent males of *Chioneocetes opillio* with high titers of MF are reproductively less active than the non-molting old shell adult males (Zaleski and Tamone, 2014).

Early studies involving ablation and reimplantation (e.g. Hinsch, 1980) suggested that MOs might have molt-related function. In *M. rosenbergii*, the MF levels in hemolymph rose during the pre-molt stage but declined in the post-molt stage, exactly like ecdysteroid doing it (Ahl and Laufer, 1996). Administration of MO homogenate of the blue crab *Callinectes sapidus* into the white shrimp *P. sertiferus* induced molt acceleration. Similarly, that of *Procambarus clarkii* accelerated molting in *Caridina denticulata* (Taketomi et al., 1989). These observations suggest that MF is not species specific. MF may induce precocious molt by stimulating Y-organ to secrete ecdysteroids more profusely (Tamone and Chang, 1993). Injection of MF also induced molt acceleration in crayfish *Cherax quandricarinatus* (Abdu et al., 2001), *P. clarkii* (Laufer et al., 2005) and *O. senex senex* (Reddy et al., 2004). Interestingly, the injection of JH III induced molting in adult *P. clarkii* (Rodriquez et al., 2002b). Clearly, MF induces precocious and accelerated molt in adult decapods (see Nagaraju, 2007). Incidentally, the key enzyme Farnesoic Acid *O*-methyltransferase (FAMeT) converts FA into MF. The attempt by Hui et al. (2008) to prevent molting by knocking down the gene *FAMeT* by ds RNA injection in *L. vannamei* was not a success, due to 100% mortality of the treated shrimps.

Limited information available on MF/JH control of morphogenesis in crustaceans remains contradictory. Bilateral eyestalk ablation, which facilitates faster secretion of MF in decapod larvae, slightly accelerated molt but delayed metamorphosis by producing supernumerary larval stage in *Palaemon macrodactylus, H. americanus, Portunus trituberculatus, Rhithropanopeus harrissi* (Table 7.5) and *L. emarginata* (Laufer et al., 1993). In the barnacle *Elminius modestus*, analog of JH (JHA) induced an intermediate larval stage between nauplius and cypris and required an additional molt to metamorphose (Tighe-Ford, 1977). But Yamamoto et al. (1997) demonstrated that MF affected the development of *Balanus amphitrite* by inducing metamorphosis instead of the expected inhibition/delay of it. These contradictory observations may arise from the differences in timing and dose of MF/JH exposure, which are critically important. For example, ESA prior to 3rd zoea in *H. americanus* required a supernumerary molt and thereby delayed metamorphosis. But that after 3rd zoea accelerated molt and advanced metamorphosis (Table 7.5). Similarly, the exposure of 3rd instar larvae of *Artemia* to MF at dose of 1×10^{-7} delayed molting and hence metamorphosis. But that of 1×10^{-8} accelerated molt and metamorphosis (Ahl and Brown, 1990). Notably, MF had no effect on adult tadpole shrimp *Triops longicaudatus*, suggesting MF may not inhibit adult morphogenesis (Tsukimura et al., 2006).

MO-IH In decapod crustaceans, MO is under the negative control of MO-IH, a peptide originating from SG in eyestalks (Wainwright et al., 1996). The peptide-induced inhibition of MF was demonstrated *in vitro* in *H. americanus* (Waddy et al., 1995) and *O. senex senex* (Nagaraju et al., 2005). In *P. clarkii*, methyltransferase activity increased by 20-100 times eight-12 d after eyestalk ablation (Chaves, 2000), indicating the inhibitory role of MO-IH on MF synthesis. Purified MO-IH was shown to directly inhibit the secretion of MF *in vitro* (Wainwright et al., 1996). Bioassays measuring inhibition of MF synthesis in MO by MO-IH revealed the presence of MO-IH-1 and MO-IH-2 isoforms in *Cancer pagurus* (Borst et al., 2002). The amino acid sequences of MO-IH-1 of *C. pagurus* is 59% similar to that of its own MIH, 55% to MIH of *C. maenas* and 42% to VIH of *H. americanus* (see Nagaraju, 2007). Examining the structural organization of *MO-IH-1* and *MO-IH-2* genes coding the two isoforms of MO-IH-1 and MO-IH-2 peptides, Lu et al. (2000) reported that the genes were closely related to each other and resulted from the ancestral *CHH* gene with subsequent divergence producing the *MIH* and *MO-HIs*.

7.1.5 Vertebrate Type Hormones

In some crustaceans, many vertebrate type hormones and their corresponding receptors have been identified (Kashian and Dodson, 2004). Crustaceans have also the enzymatic capacity to synthesize vertebrate type of steroids (Table 7.6). For example, Baldwin and LeBlanc (1994) demonstrated that *Daphnia* is capable of hydroxylation of testosterone (T) at multiple sites by p 450 enzymes. The presence of 17β-estradiol (E_2) and progesterone and their positive relation to ovarian maturation were reported from commercially important crayfish, prawns and shrimps (Table 7.6). However, negative and no relationships were also reported in *Peneus japonicus* and *Scylla serrata*, respectively. In an attempt to induce spawning, a few authors administered human chorionic gonadotropin (hCG) by injection or through diet; the hCG stimulated ovarian growth and vitellogenesis. Du et al. (2015) reported the presence of the transcript *GnRH-like receptor (MnGnRHR)* stably expressing in the ovary of *Macrobrachium nipponense*. Malati et al. (2013) also found that the changes in levels of E_2 and 17β-hydroxyprogesterone were related to oocyte development in the narrow clawed crayfish *Astacus leptodactylus*. E_2 levels increase in hemolymph (from 58 pg/ml during post-spawning to 307 pg/ml during vitellogenesis) and ovary (from 545 pg/ml during post-spawning to 700 pg/ml during vitellogenesis and to 1,140 pg/ml during maturation) but decreased in hepatopancreas. Maximum levels of E_2 and 17β-hydroxyprogesterone were present in the ovary. Clearly, the reported evidences reveal that the vertebrate type steroids do have a role in ovarian vitellogenesis and maturation of crustaceans.

The presence of testosterone (T) is reported from mysid *Neomysis integer* and decapods *Neocaridina denticulata* and *M. rosenbergii* (Table 7.6). Interestingly, the presence of hCG-like substance is reported in the crab *Ovalipes oscillatus* (see Kashian and Dodson, 2000). In *N. integer*, T level is five fold higher in males than in females of *N. integer*. Through a series of publications, Verslycke et al. (2002, 2004, 2007, see also Mazurova et al., 2008) examined the effects of T, tributyltin chloride and methoprene on steroid metabolism of *N. integer*. Acute exposure to 2 µg/l T increased the level of 11β-hydroxytestosterone and induced synthesis of androstenedione. Exposure to 0.01-1.0 µg/l tributyltin chloride induced changes in hydroxylated T in whole body homogenate and increased excretion of sulfated T metabolites. Incidentally, sulfation is an important route of clearing active hormones through excretion (cf Thibaut and Porte, 2004). Exposure to high dose of 100 µg/l methoprene led to high levels of male specific sex hormones. In *Metapenaeus ensis*, Wu and Chu (2010) showed the indispensability of Glutathione Peroxidases (GPx) for normal gonadal development. Multiple sequence alignment of *MePGx* is evolutionarily conserved among invertebrates with common functionally important motifs. *In situ* hybridization revealed that *MeGPx* is highly expressed pre-vitellogenic and mid-vitellogenic oocytes while there is no expression in the late vitellogenic oocytes. *MeGPx* may play a pivotal role in preventing oocytes from oxidative damages.

TABLE 7.6

Occurrence and functions of vertebrate type hormones in crustaceans

Species and Reference	Reported Observations
Penaeus monodon Quinitio et al. (1994) *Mictyris brevidactylus* Shih (1997)	Positive relationship between 17β-estradiol and progesterone on one hand and Vg level on the other
Parapenaeopsis hardwickii Kulkurni et al. (1979)	Injection of progesterone induced ovarian development
Cherax abbidus Coccia et al. (2010)	17β-estradiol stimulated vitellogenesis in ovary
Austropotomobius pallipes Paolucci et al. (2002)	Progesterone and estradiol receptors detected in hepatopancreas and ovary
Crangon crangon Zukowska-Arendarczyk (1981)	Administration of hyperplasia gonadotropin induced ovarian maturation
P. indicus Jayaprakas and Shambu (1998)	Injection of human chorionic gonadotropin (hCG) stimulated spawnings
C.crangon, P. indicus, Idotea balthica Laufer and Landu (1991)	Dietary administration of hCG positively influenced vitellogenesis
P. japonicus Yano (1987)	Administration of 17β-hydroxyprogesterone stimulated ovarian growth and vitellogenesis
Oziotelphusa senex senex Reddy et al. (2006)	17α-hydroxyprogesterone increased ovarian vitellogenesis dose-dependently; it has a positive relation with the ovarian cycle
Macrobrachium nipponense Du et al. (2015)	26 transcripts representing genes associated vertebrate GnRH signaling induced the one encoding a *GnRH-like receptor (MnGnRHR)*, which expressed stably in ovaries at various stages
Scylla serrata Warrier et al. (2001)	Fluctuating levels of estradiol in ovary and hemolymph but different vitellogenic stages
P. japonicus Okumura and Sakiyama (2004)	Negative relationship between ovarian maturation and OSI on one hand and levels of sex steroids (estradiol, estriol, testosterone, 11-Ketotestosterone) in hemolymph on the other
Neocaridina denticulata Huang et al. (2004)	Testosterone (T) found in whole body homogenate
Neomysis integer Verslycke et al. (2002)	T is fivefold higher in males than in females
M. rosenbergii Martins et al. (2007)	T and 17α-hydroxytestosterone levels did not vary in females during ovarian cycle
Daphnia magna Kashian and Dodson (2004)	T inhibited fecundity without increasing mortality

7.2 Endocrine Disruption

Introduction

"More than 60% of the 100,000 man-made chemicals are in routine use worldwide since 1990s. Every year 200 to 1,000 new synthetic chemicals enter the market (Shane, 1994). Over 900 of these chemicals are identified as established or potential endocrine disruptors (Soffker and Tyler, 2012). These chemicals that either mimic or antagonize the actions of endogenous hormones, are known as endocrine disruptors (Himmatsu et al., 2006). After use in domestic (into sewage), agricultural (e.g. pesticides, fungicides) and industrial (e.g. 4-Nonyl phenol) sectors, hundreds of estrogens and their mimics are discharged into aquatic habitats" (Pandian, 2014, p 93). " A casual observation by Sweeting (1981) on 5% incidence of intersex roach *Rutilus rutilus* in the River Lee, Southeast England triggered the discovery of the occurrence of sexual disrupting estrogens in Sewage Treated Effluents (STE) water discharged into rivers" (Pandian, 2014, p 102). For detailed information on endocrine disruption, Pandian (2014) may be consulted. Available information from both field and experimental observations provide clear evidence that not only fishes but also crustaceans are vulnerable to diverse Endocrine Disrupting Chemicals (EDCs) in the environment. The occurrence of intersex harpacticoid copepods near an UK STE water discharge suggests a strong evidence of endocrine disruption in crustaceans too. "One of the most convincing evidences of endocrine disruption in crustaceans is the production of male offspring by *Daphnia magna* exposed to JH and JH-analogs such as methoprene, kinoprene, hydroprene, epofenonane and fenoxycarb" (Kusk and Wollenberger, 2007).

For crustaceans, a fairly large body of information is available on endocrine disruption. The ensuing account, however, is limited to reproduction and development. With a short life span, cladocerans and copepods have attracted relatively more attention. Besides exogenous natural hormones, a large number of insecticides, juvenile hormone analogs, bisphenol A, a raw material for plastic production, Nonyl Phenol (NP) and Octyl Phenol (OP) used as cleaners and detergents and tributyltin, used in anti-fouling paints, have been tested. The ECDs enter mostly through gills of crustaceans but some like *Nitocra spinipes* (Tables 7.7) ingest them as food particles.

Endosulfan is an insecticide. In water, its half life is 1-3 m, while that of endosulfan sulfate can be 2-6 y (Palma et al., 2009). Being a broad spectrum pesticide, endosulfan is widely used in agriculture fields of more than 70 countries. Immediately following its applications, its concentration is increased to 44-56 µg/l in Asian rice fields (Suvaparp et al., 2001). It is then rapidly bioconcentrated 2,682 times by freshwater algae (DeLorenzo et al., 2002). The river water concentrations of NP and OP are usually below 1 µg/l. Due to their lipophilicity, both NP and Op are capable of bioaccumulation (see Pandian, 2014, p 105). Mean levels of the organophosphorous compound methione is 9.4 – 49.4 µg/l in US waters (Toumi et al., 2015). Methioprene degrades rapidly in sun light but its breakdown products like methoprenic acid remains bioactive (Ghekiere et al., 2006). Toxaphene is an insecticide widely used in USA until it was banned in 1982. But it persists in the environment due to its lipophilic and volatile characteristics. In 1999, the International Maritime Organization resolved to avoid the use of tributyitin compound but it is still used in many Asian countries (see Pandian, 2014, p 105-106). Hence, studies on ensdocrine disruption in crustaceans continue to be relevant.

TABLE 7.7

Effects of natural hormones and exogenous hormone mimics on molting and development in Cladocera, Copepoda, Cirripedia and Amphipoda

Species and Reference	Reported Observations
Cladocera	
Daphnia magna	
Zou and Fingerman (1997a, b)	Exposure to endosulfan or diethylstilbestrol affected molting frequency
Mu and LeBlanc (2002)	Exposure to 5-40 µM testosterone (T) interfered with 20-hydroxy ecdysone (20-HE), prolonged the first molt and caused abnormal progeny development
Palma et al. (2009)	Exposure to 9.2-92 µg/l endosulfan sulfate reduced molt number from 8 to 7. The first molt was postponed from 7.0 to 7.7 d
Toumi et al. (2015)	Exposure to 90 µg/l melathion reduced molt number from 9.7 to 8.0
Copepoda	
Acartia tonsa	Exposure of eggs, nauplii, copepodites and adults to cypermethrin showed that nauplii
Barata et al. (2002)	were 28 times more sensitive than adults, and males were twice as sensitive as females
Andersen et al. (2001)	Exposure to EE_2 or octylphenol affected larval development
	Exposure to anti-estrogen tamoxifen or anti-androgen flutamide affected naupliar development
Nitocra spinipes	The ingestion of particles of polybromated diphenyl
Breitholz and Wollenberger (2003)	ethers, used to prevent fires, affected larval development
Cirripedia	
Elmiinius modestus	On pulse exposure to 10 µg nonyl phenyl/l on
Billinghurst et al. (2001)	5, 6 and 7th d, < 20% larvae alone developed into cyprids on 8th d, the day of settlement
Amphipoda	
Neomysis integer	Exposure to 10 µg/l methoprene extended intermolt
Ghekiere et al. (2006)	duration from 3.5 to 5.7 d. Consequently, growth was reduced to half, i.e. from 120 to 60 µm/d

The estrogens and estrogen-mimics bind to the estrogen receptor or nuclear receptors and interact with an Estrogen Responsive Element (ERE) or act via other pathways that alter estrogen signaling and elicit response (Zou and Fingerman, 1997a, b). Some crustaceans are able to metabolize vertebrate steroids (Baldwin et al., 1998), implying that there could be other routes of crustacean endocrine disruption than direct binding of them to steroid receptors (Breitholz and Bengtsson, 2001). Zou and Fingerman (1997a, b) reported that estrogens and their mimics inhibit chitobiase activity in *Uca pugilator*, which may explain the reduced molting frequency observed in many copepods and amphipods that were exposed to estrogens and their mimics.

7.2.1 Larvae and Maturation

Crustacean larval development involves a series of morphologically distinct molt stages culminating in metamorphosis. These events are regulated by an array of natural hormones. But the number and duration of larval molt cycle and the number of molts are influenced by many vertebrate hormones like Testosterone (T) and hormone mimics like NP, OP, endosulfan and methoprene (Table 7.7). In *Elminius modestus*, the exposure to NP during the labile period from 5th to 7th d is reported to reduce settling of cyprid larvae to 20% (Billinghurst et al., 2001).

Exposure to one or another estrogen mimics (e.g. endosulfan) and JH analogs postponed sexual maturity and reduced the number of broods and/or neonates produced by a female (Table 7.8). Similarly, fenarinol, a fungicide that inhibits ecdysone synthesis, decreased the mating ability of males and fertilization success. However, it is difficult to comprehend how the exposure of estrogen mimic bisphenol A enhances fertility in *Corophium volutator*. Incidentally, the exposure duration has to be synchronized with the labile period, during which an animal is sensitive to a hormone (see Pandian, 2013, p 184-187). For example, AG ablation or implantation regulates the sex differentiation process toward female or male in malacostracans, when surgery is made in sexually undifferentiated young ones alone (see p 178). Similarly, male differentiation is induced in parthenogenic daphnids at sub-lethal dose of MF or JH. At higher dose, it may postpone sexual maturity. Exposure to endosulfan up to 92 μg/l reduces the number of neonates per female and a higher dose inhibits reproduction *in toto*. Hence, the response of a crustacean to an exposure of a hormone may differ with exposure duration and hormone dose.

TABLE 7.8

Effects of natural hormones and exogenous hormones mimics on fertility, fecundity and larvae of Cladocera, Copepoda, Amphipoda and Decapoda

Species and Reference	Reported Observations
	Cladocera
Daphnia magna Palma et al. (2009)	Exposure to 9.2-92 µg/l endosulfan sulfate reduced the number of neonates/female from ~ 100 to 60-70. At 458.7 µg/l, it inhibited reproduction *in toto*
Toumi et al. (2015)	Exposure to 90 µg/l malathion reduced the brood number from 5.9 to 4.2 and number of neonates/female from 117 to 73
Miona macrocopa Chu et al. (1997)	Exposure to 0.1 mg/l methoprene postponed sexual maturity from 3 to 6 d
Ceriodaphnia dubia Brooks et al. (2003)	Exposure at environmentally relevant dose of 0.1 mg/l fluoxetine, a serotonin reuptake inhibitor, reduced fecundity
	Copepoda
Nitocra spinipes Breitholz and Bengtsson (2001)	Exposure to diethylstilbestrol (DES) but not E_2 or EE_2 increased the proportion of females producing nauplii. Neither development nor reproduction was disturbed at sub-lethal doses of E_2 (0.16 mg/l), EE_2 (0.05 mg/l) and DES (0.03 mg/l)
Acartia tonsa Andersen et al. (1999)	Exposure to 23 µg/l E_2 or 20 µg/l bisphenol A accelerated maturity and increased fecundity
Tisbe battagliai Hutchinson et al. (1999a)	Life time exposure to estrone EE_2 or E_2 reduced the number of nauplii/female
	Amphipoda
Gammarus fossarum Lacaze et al. (2011) Lewis and Ford (2012)	Exposure to genotoxicant methyl methanesulfonate (MMS) induced oxidative damage, breaks and cross links of DNA strand. > 20% sperm suffered DNA damage leading to defective fertilization, development and high offspring morbidity. Oocytes were less sensitive
Monoporeia affinis Jacobson and Sundelin (2006)	Fenarinol, decreased male mating ability and fertilization success
Melita mitida Borowsky et al. (1997)	57% females exposed to polluted estuarine sediment developed abnormal brood plate setae
Corophium volutator Brown et al. (1999)	Exposure to 50 µg/l nonylphenol (NP) enhanced fertility
	Decapoda
Procambarus clarkii Sarojini et al. (1997)	Exposure to methionine enkephalin dose dependently inhibited ovarian maturation
Macrobrachium rosenbergii Revathy and Munuswamy (2010)	Exposure of brooders to 1.56 tributyltin disorganized embryogenesis *in toto*. It disrupted elongation of body, formation of eyestalk and deposition of yolk

7.2.2 Induction of *Daphnia* Males

Many other investigations began to report that vertebrate hormone and especially insecticides induced production of male neonates in daphnids. For example, progesterone induced production of male neonates in the second clutch and also initiated meiosis in males of *D. pulex* (Table 7.9). Exposure to atrazine (Dodson et al., 1999a) or nonyl phenol (Baer and Owens, 1999) stimulated male production in *D. pulicaria* and *D. magna*, respectively. However, that of dieldrin reduced the male production in *D. galeata mendotae* (Dodson et al., 1999b). Realizing the role of toxicants on sex differentiation in many crustaceans, toxicologists increasingly used the induction of male production in daphnids as a highly specific bioassay end point and as an indicator of endocrine disruption.

TABLE 7.9

Induction of male sex differentiation by natural hormones and hormone mimics in Cladocera and Copepoda

Species and References	Reported Observations
	Cladocera
Daphnia pulex Peterson et al. (2001)	Exposure to environmentally relevant dose of 1-10 µg/l 20-OH ecdysone increased all-male broods but that at 10-100 µg/l decreased all male broods
D. pulex Olmstead and LeBlanc (2002)	Exposure to 400 n M MF for 12 h during ovarian oocyte maturation produced 100% male neonates. This level corresponds to those in males of 3 crab species (Borst et al., 1987).
Mu and LeBlanc (2004)	Exposure during the late ovarian development to MF caused the oocytes to develop into males. Not only MF, but also fenoxycarb and pyriproxyfen did the same
D. magna Palma et al. (2009)	Exposure to 229.3 µg/l endosulfan sulfate increased male ratio from 0.0 to 0.041 in 100% broods
D. magna Tatarazako et al. (2003)	On exposure of 24 h-old females to 21 d, 4 of 5 tested juvenoids increased male ratio but reduced fecundity
D. magna Oda et al. (2005)	On exposure of 24 h-old females to 21 d, the tested 6 juveniles produced male neonates. The more potent juvenoid caused greater reduction in fecundity
D. magna Matsumoto et al. (2008)	On exposure of 13 d-old females to 24 h, MF (50 µg/l), phenoxycarb (70 ng/l) or pyriphenoxyfen (100 ng/l) induced 100% male neonates without reducing fecundity
D. magna Ignace et al. (2011)	Exposure to 50 µg/l toxaphene, an insecticide, induced male sex differentiation in ovarian oocytes that were at initial cleavage stage, i.e. 0-12 h prior to their release into the brood chamber
	Copepoda
Tisbe battagliai Walts et al. (2002)	Exposure to EE_2 or E_2 did not affect sex ratio
Hutchinson et al. (1999b)	Exposure to 86.5 µg/l 20-hydroxyecdysone produced males

A series of relevant publications in this aspect shows that 1. Hitherto some 10 juvenoids were used to induce increasing male ratio in *D. magna, Moina macrocopa, M. microrura, Ceriodaphnia dubia* and *C. reticulata* (Fig. 7.3 A) 2. The potency of the juvenoids decreased in the order of pyriproxyfen > fenoxycarb > MF > epofenonane > JH II > methoprene > JH III i.e. (i) the doses required to induce 100% male production by fenoxycarb and pyriproxyfen were 500-700 fold lower than that of MF (Matsumoto et al., 2008) and (ii) JH II was about 450 times less potent than fenoxycarb (Oda et al., 2005). 3. At the peak of natural male production, the male ratio remains below 0.5 (Dosdon and Frey, 2009). However, the exposure of adult daphnids to these juvenoids, excepting that of methoprene, induced parthenogenic production of males up to 100%. This induction involved progeny mortality and reduced reproduction (Fig. 7.3B). It is not clear whether the dead progenies were genetic females. However, Oda et al. (2005) considered that the toxicity through endocrine disruption may not be the major cause for the death of the progenies 4. In fact, reducing the exposure duration from 21 d (see Tatarazako et al., 2003, Oda et al., 2005) to just 24 h in the 13 d-old females, i.e. synchronizing the exposure duration to the sensitive labile period, Matsumota et al. (2008) achieved the production of 100% male neonates but with no mortality and reduction in fecundity. This has an important implication to the issue whether the juvenoids determine the sex or they induce sex differentiation alone. That the proportion of naturally produced male remains below 50% and the juvenoids could induce 100% males with no progeny mortality clearly indicate that the juvenoids induce sex differentiation but not sex determination. Sex is probably determined during the transitory "synaptic concentration of chromosome" (Ojima, 1958). 5. Exposing *D. magna* females to toxaphene during the four selected stages, Ignace et al. (2011) showed that the sensitive labile period for sex differentiation is located in the ovarian oocytes that have not yet undergone the final cleavage, i.e. 1-12 h prior to their release into the brood chamber (Table 7.9). Incidentally, this finding confirms the earlier observation by Olmstead and LeBlanc (2002). Sagawa et al. (2005) found that germ cell lines are developed in chemically induced male embryos of *D. magna*. However, it is not clear whether the juvenoids induced morphological differentiation but not gonadal sex differentiation, i.e. the chemical induction is limited to the morphological differentiation alone but the gonads remain undifferentiated. For, Tatarazako et al. (2003) noted "the chemically induced males have not yet been found to be able to fertilize resting eggs". Similarly, Matsumota et al. (2008) also reported "diapausing eggs of females" fertilized by "chemical (ly) induced males have not been found to be able to hatch". Hence, the proposal of juvenoids and other toxicants determining sex (e.g. Olmstead and LeBlanc, 2002) is questionable. Experiments are urgently required to show that the chemically induced males and females do produce normal gametes resulting in successful fertilization and hatching, as those of natural ones.

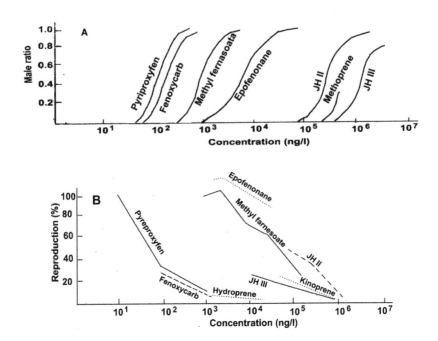

FIGURE 7.3
Free hand drawings to show A. Male ratio and B. Reproduction in *Daphnia magna* exposed to MF and juvanoids (compiled from Tatarazako et al., 2003, Oda et al., 2005, Matsumoto et al., 2008).

7.3 Parasitic Disruption

7.3.1 Parasites and Taxonomy

In Crustacea, parasitic taxons are prevalent in Copepoda, Rhizocephala (120 species, see Boxshall et al., 2005), Branchiura (175 species, see Boxshall et al., 2005) and Isopoda. Among the parasitic isopods, bopyrids, cryptoniscoids and cymothoids (including gnathids) are collectively called epicarideans. In the speciose Isopoda with 10,300 species, for example, 6.6% parasitize other crustacean hosts. Of 675 parasitic isopods on other crustaceans, the bulk of 76.0% belongs to the Bopyridae, followed by 12.5, 6.8 and 4.7% by Cryptoniscoidae, Dajidae and Entoniscoidae, respectively (Williams and Boyko, 2012). The epicardians cling firmly attached to the branchial chamber (Fig. 7.4 A) or abdomen (Fig. 7.4 B) and thereby feed on hemolymph and surrounding gill tissues or hepatopancreas using their piercing and sucking mouth parts. The cymothoids (~ 1,250 species) are ectoparasites on fishes feeding on their blood and gill tissues. However, some of them are endoparasites. For example, females and males of *Ichthyoxenus fushanensis* occur together within the parasitic sac and are detected by the presence of an orifice near the pectoral fin of *Varicorhinus bachatulus* (Tsai et al., 1999). Parasitic copepods infect a wide spectrum of hosts comprising crustaceans, molluscs, echinoderms and fishes (Boxshall et al., 2005). With no need for motility and search for food, the parasitic copepods and isopods, especially the adult females, undergo different levels of morphological aberrations (Fig. 7.5). Despite structural loss, these parasitic copepods and isopods have developed hooks, dactyl sockets to enable them to firmly cling to the hosts. Incidentally, the rhizocephalans are not known to infect hosts other than crabs; hence, the species numbers in rhizocephalan (120 species) as well as their host taxons are limited. The only rhizocephalan *Sylon hippolytes* parasitizes the shrimps *Pandalus platyceros* and *Spirontocaris lilljeborgi* (see Meyers, 1990).

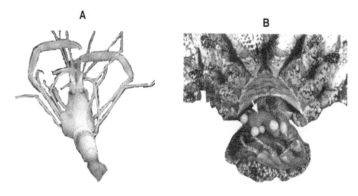

FIGURE 7.4
Location of parasitic *Argeiopsis inhacae* in (note the protruding) the left gill chamber of *Stenopus hispidus* (redrawn from Calado et al., 2008), B. Abdominal location of the parasitic *Liriopsis pygmaea* in the host *Paralomis granulosa*. Note the presence of hyperparasite *Briarosaccus callosus* (from Lovrich et al., 2004, permission by International Ecology Institute).

The protogynic isopod *Cyathura carinata* serves as a second intermediate host of microphallid trematodes. The presence of their cysts in *C. carinata* reduces the host's fecundity to a level of causing reproductive death of female function (Ferreira et al., 2005). The list of Lafferty and Kuris (2009) indicates that the copepod *Coelotropus* partially castrates nudibranchs. Infecting 58 to 63% of the freshwater fish *Cyphocharax gilbert*, the parasitic cymathoids like *Riggia pranensis* cause reproductive death of female function (Azevedo et al., 2006). Parasitized by another cymothoid *Anilocra apogonae*, females of the cardinalfish *Cheilodipterus quinquellineatus* produce fewer and smaller oocytes; occupying parts of oral cavity and branchial chamber *A. apogonae* also prevents mouthbrooding in males (Fogelman et al., 2009). Obviously, *A. apogonae* inhibits vitellogenesis and thereby causes reproductive death of female function. The cnidarians *Hydrichthys* is reported as a hypercastrator of copepods (Lafferty and Kuris, 2009).

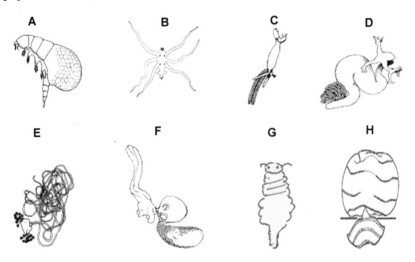

FIGURE 7.5
Range of gross morphology of parasitic A. Notodelphid B. Spanchnotrophid C. Lernaeopodid D. Pennellid E. Chitonophilid F. Herphyllobiid copepods (permission by University of New England, CSIRO Publisher, Victoria http://www.publish.csiro.au/pid/5045.htm. from Boxshall et al., 2005, Chapter 4, Crustacean parasites. pp 123-169. In: Marine Parasitology (ed) K. Rohde) and isopods G. *Hemioniscus balani* (hand drawn) and H. *Liriopsis pygmaea* (permission by International Ecology Institute).

7.3.2 Life Cycle and Biology

The life cycle of many parasitic crustaceans involves a single definitive host. For example, (i) the free-living nauplii released from egg sacs of the parasitic calicoid copepod females directly infect definitive host (e.g. *Lernaea chalkoensis* on *Catla catla*, Gnanamuthu cited by Ayyar and Ananthakrishnan, 2000), (ii) the free-living mancas released from marsupium of parasitic female directly infect the definitive host (e.g. *Ichthyoxenus fushanensis* on *Varicorhinus brachatulu*) and (iii) the non-feeding rhizocephalan female cyprid, on infecting the host, metamorphoses into kentogon and then to vermogon stages; the male cyprid, on settling over the external of female rhizocephala, metamorphoses into trichogen larva. Notably, the non-feeding adult and naupliar stages of the monstrilloid copepods are free-living but all their five copepodid stages are endoparasitic (Fig. 7.6 A) The life cycle of entoniscoid and most bopyrids involves pelagic calanoid copepod as an intermediate host prior to infecting the definitive hosts like the barnacle (e.g. *Balanus glandula*, Blower and Roughgarden, 1988) stenopodid shrimp (*Stenopus hispidus*, Calado et al., 2008) squat lobster (CMFRI). A generalized life cycle of these bopyrids pieced together from many sources is depicted in Fig 7.6 B.

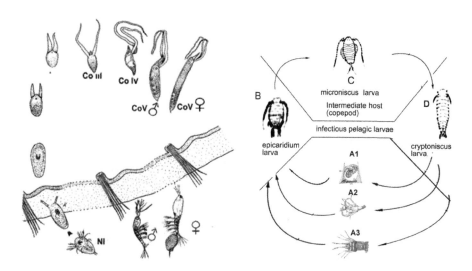

FIGURE 7.6

Left panel: Example for direct life cycle of a monstrilloid copepod. Note all the five stages of copepodid are endoparasitic, while adult and naupliar stages are non-feeding but free-living (permission by University of New England, CSIRO Publisher, Victoria http://www. publish.csiro.au/pid/5045.htm. from Boxshall et al., 2005, Chapter 4, Crustacean parasites. pp 123-169. In: *Marine* Parasitology (ed) K. Rohde). Right panel: An example for indirect life cycle of bopyrids involving B. epicardium and C. microniscus parasitizing a calanoid copepod D. cryptoniscus. Cryptoniscus infects the definitive hosts, which may be A1. barnacle or A2. shrimp (redrawn from R. Calado) or A3. squat lobster. This generalized life cycle of the bopyrid is pieced from many sources.

Hundreds of publications report on the occurrence and prevalence of crustacean parasites on edible crustaceans and fishes (Kinne, 1990). They are important to understand the recruitment dynamics and management of these fisheries. Microbial parasites are very important, as they inflict mass mortality within a short period of time (Brock and Lightner, 1990). However, the number and diversity of parasitic crustaceans provide a unique opportunity to analyze the disruption of sex differentiation process. The enigmatic protozoan parasite *Ellobiopsis* sp inhibits the ovarian development of *Calanus helgolandicus* (Albaina and Irigoien, 2006) and the parasitic nematode *Contracaecum aduncum* inhibits egg production in *Pandalus borealis* (Margolis and Butler, 1954), apparently by depriving nutrients and hormones for the ovary and oocyte development in respective hosts. Being small in size and perhaps as poor energy drainers, parasitic copepods are not known to interfere with the sex differentiation process of their relatively large crustaceans. With size compatibility, at least one parasitic copepod castrates its ostracod host. *Sphaeronellopsis monothrix* is a parasite on myodocopid ostracod *Pasterope pollex*. Sucking hemolymph of the host, *S. monothrix* inhibits ovulation in sexually maturing hosts. Known for its mimicry, the parasite deposits her eggs in the brood chamber of the virgin *P. pollex* female. The female ostracod broods parasite's eggs, as if they are her own eggs (Bowman and Kornicker, 1967), a situation similar to that of *Petrolithes cabrilloi* (see p 204). More publications on parasitic castration by size-compatible copepods are awaited. Those interested in non-crustacean parasites may consult Meyers (1990).

TABLE 7.10

Fecundity of bopyrids releasing manca or epicardium as a function of body size

Parameter/Host/Reference	Body Size (mm)/ Fecundity (No.)	r Value*/Egg Size (mm)	Host Size (mm)
Ichthyoxenus fushanensis on *Varicorhinus bachalutus* Tsai et al (1999)	11.6-25.9 107-820 Manca	0.935	71-127 BL
Bopyrinella thorii on *Thor floridanus* Romero-Rodiguez and Roman Contreras (2008)	1.0-3.5 83-1.036 Eggs	0.769 0.11-0.18 mm	1.2-3.8 CL
Stegias clibanarius on *Clibanarius tricolor* McDermott (2002)	191-667 Embryos	0.378† 0.157 mm	2.4-5.0 BL
Bopyrinella ocellata Bourdon (1968)	1.5-3.1 330-1,410 Eggs	–	
Bopyrissa wolffi on *C. tricolor* McDermott (2002)	Mean = 1.91 Mean	–	
Anuropodione carolinensis on *Munida iris iris* Wenner and Windsor (1979)	5-16 9,500-28,300 Epicardium	0.8 ~1.45 µm	10-25 BL
Argeia pugattensis on *Crangon franciscorum* Jay (1989)	4.2-9.0 1,600-33,300 Epicardium	0.81	25-60 BL
Aporobopyrus curtatus on *Peterolithes armatus* Miranda and Mantelatta (2010)	0.5-3.5	–	6-12 BL

*correlation coefficient between body size and fecundity.

Rhizocephala Following the injection of germinal cells by kentogon larva into a host, the rapidly growing rootlets completely occupy the host's viscera so as to soak up nutrients and hormones from the host's fluid. The mass of interna weighs as much as 14.1 g in the crabs *Paralithodes platypus* and 21.6 g in *Lithodes aequispinei* (Hawkes et al., 1986). The length of externa ranges from 12 mm in a newly extruded virginal one to 77 mm in a matured parasite *Briarosaccus callosus*. Depending on externa size, 3.1 to 3.9 × 10⁵ nauplii are produced (Hawkes et al., 1985). Hence, the unique publication by Hawkes et al. provides a scope to assess fecundity of a rhizocephalan parasite as a function of externa size (see Fig. 7.7). Following a number of molts, the rhizocephalan nauplii attain the infective cyprid stage within 25 d at 7°C. The non-feeding cyprids may survive another 16 d prior to infecting a suitable host.

Even after settling on a host, the kentogons have to struggle a lot to complete the process of infection by penetration and injection of the germ cells. Among the crabs, the porcellanids are excellent groomers (Boxshall et al., 2005) and make every effort to escape from parasitic infection. However, crabs with damaged or impaired grooming appendages run the risk of infection. Experimental exposure of an unparasitized porcelain crab *Petrolithes cabrilloi* to several hundreds of infective larvae of *Lernaeodiscus porcellanae*, caused physical exhaustion from frequent grooming behavior and eventual death of the larvae from stress (Ritchi and Hoeg, 1981). Hence, grooming ability of the host may play an important role in preventing infection.

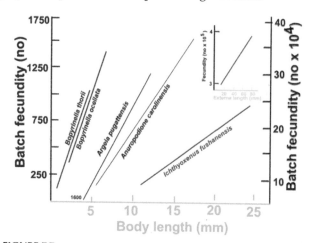

FIGURE 7.7
Fecundity as a function of body/externa length of parasitic bopyrids and the rhizocephala *Braiarosaccus callosus* (in window, data from Hawkes et al., 1985). Note fecundity increases with increasing body length at maturity of bopyrids. Fecundity (epicardium no) of *Anuropodione carolinensis* (Wenner and Windsor, 1979) and *Argeia pugattensis* (Jay, 1989) are shown in black color. Fecundity of *Ichthyoxenus fushanensis* releasing manga (Tsai et al., 1999) is shown in orange to distinguish from other low-fecund, epicardium releasing *Bopyrinella ocellata* (Bourdon, 1968) and *B. thorii*, Romero-Rodriquez and Roman-Contreras, 2008).

Injection of aqueous extracts of rootlets of the sacculinid *Loxothylacus panopei* into unparasitized *Rhithropanopeus harrissii* and *Panopeus herbstii* males resulted in mesodermal degeneration, pyknosis in the testis, degenerative changes in the brain and thoracic ganglion, hypersecretion of Sinus Gland (SGs) and signs of degeneration of Androgenic Glands (AGs) (Rubiliani, 1985). Similar experiments are yet to be undertaken in female crabs. *Heterosaccus californicus* induces precocious maturity in the majid crab *Pugettia product* (O'Brien, 1984), presumably facilitating free flow of MF by destroying SGs, the source of GIH. The rhizocephalans go to the extreme of destroying AGs, through hormonal or chemical means and thereby castrate the host. Consequently, the morphology, physiology and behavior of the host is so altered that the host simply remains an automaton serving to sustain and promote reproduction of the parasite alone.

In the fields, the parasites typically infect immature crabs of both sexes but with preference for females (e.g. *Pagurus pubescens* infected by *Peltogaster paguri*, Reinhard, 1942) and retard growth by preventing molt and thereby produce dwarf or button crabs (e.g. *Callinectes sapidus*, Overstreet, 1978). Rhizocephalans inflict different levels of negative effects. For example, the colonial parasitic rhizocephalan *Thompsoni* sp did not inhibit molting and alter any structure of *Secondary Sexual Characteristics* (SSCs) and reproductive behavior of *Portunus pelagicus* (Phang, 1975). *Triangulus munidae* too had no effect on the SSCs of *Munida sarsi* (Attrill, 1989). *Peltogaster paguri* also caused no structural change in SSCs and no negative effect on the testis (Reinhard, 1942). But *Lernaediscus porcellanae* altered the behavior of *Petrolithes cabrilloi* male to an extent that it acted like an ovigerous female by actively grooming and ventilating the externae, as if they were its own berried eggs (Ritche and Hoeg, 1981). However, *P. paguri* inflicted the degeneration of ovaries completely but no negative effect on the testis and SSCs of the hermit crab *Pagurus pubescens* (Reinhard, 1942). *Petrolithes cabrilloi* was castrated by *Peltogaster paguri* regardless of sex (Ritchi and Hoeg, 1981). *Sacculina carcini* inhibited spermatogenesis and arrested vitellogenesis and thereby caused reproductive death in both males and females of *Carcinus maenus* (Rubiliani et al., 1980).

Bopyrids Prevalence of bopyrid infection ranges from 0.4% for *Eophryxus lysmatae* on *Lysmata seticauda* (Calado et al., 2005) to 3.4% for *Argeia pugattensis* on *Crangon francisorum* (Jay, 1989) and to as high as 58% for *Ichthyoxenus fuschanensis* on the fish *Varicorhinus bachatulus* (Tsai et al., 1999). A wide spectrum of decapod hosts like shrimps, squat lobsters, anomurans and brachyurans are infected by bopyrids (Romero-Rodriguez and Roman-Contreras, 2008). Bopyrids and rhizocephalans seem to mutually exclude each other. In general, prevalence of a parasite may be determined by many factors like population density and grooming ability of the host as well as the natality and risks involved in reaching the definitive host directly by a manca (e.g. *I. fuschanensis* on *V. bachatulus*) or indirectly by an epicardium larva (e.g. *L. seticauda* and *A. pugattensis*), which has to pass through a pelagic intermediate host prior to infecting the definite host.

Many bopyrids have synchronized their molt cycle with the respective hosts, as the parasite cannot molt, when the host's carapace is hard (Wenner and Windson, 1979). Hence, an increase in parasite size along with that of host size indicates the synchronization of molt cycle by the parasite with that of the host (e.g. *Aporobopyrus curtatus* on *Petrolithes armatus*, Miranda and Mantellato, 2010).

Known for their direct life cycle and production of fully developed manca, the isopods spawn large and smaller number of eggs (e.g. 53 eggs in *Ligia oceanica*, Pandian, 1972). However, the parasitic isopods produce hundreds and thousands of small eggs (Table 7.10). Being ectoparasites of relatively large fishes, the cymothoids are fairly large and produce a few hundred mancas (e.g. Host: *V. bachatulus*, body length [BL]: 71-127 mm; parasite: *I. fushanensis*, BL 11.6-25.9 mm; Table 7.10). With indirect life cycle involving an intermediary host, the cyprinidscoids and endoniscoids are prodigious, producing thousands of eggs. Being the smallest parasite *Bopyrinella thorii* (1.0-3.5 mm) infecting the smallest host *Thor floridanus* (1.2-3.8 mm Carapace Length), *B. thorii* produces a maximum of 1,036 eggs. The largest parasite is *Anuropodione carolinensis* (5-16 mm) but the largest host is *Crangon franciscorum* (25-60 mm). Both *A. carolinensis* and *Argeia pugattensis* produce a maximum of 28,000 and 38,300 eggs. In fact, Beck (1980) indicated that some bopyrids produce up to 41,000 eggs. Secondly, abdominal parasites produce more embryos than branchial parasites. For example, infecting *Pagurus longicarpus*, the abdominal parasites *Stegophryxus hypticus* brood a maximum of 3,437 embryos, as compared to the branchial parasite *Stegias clibanarius* brooding the maximum of 667 embryos only (see McDermott, 2002).

The following may be inferred from Fig. 7.7 showing the relationship between body size and batch fecundity of bopyrids, for which information is available: 1. The body size-fecundity relationships of all these bopyrids are positive and linear. 2. Body length, at which the bopyrids begin to produce epicardium, determines the level of batch fecundity, resembling that described for egg shedding penaeids (Fig. 3.5 A). 3. Despite being the largest, the cymothoids with direct life cycle are not as fecund as those of bopyrids characterized by indirect life cycle involving three larval stages and an intermediate host. 4. Unlike the free-living peracarids (see Fig. 3.8 B, C), the parasitic bopyrids do not display any sign of reproductive senescence.

7.3.3 Energy Drainers

The parasitic bopyrids weigh about 3-4% of their respective hosts (e.g. 3.1% in *Pseudione tuberculata* Romero-Rodriguez and Roman-Contreras, 2008, 4% *Parabopyrella* sp on *Lysmata ambionensis*, Calado et al., 2006). They consume from 8% (Andersen, 1977) to 25% (Walker, 1977) hemolymph of the host every day. *Pagurus* crabs parasitized by *Peltogaster* have much lower fat content in the hepatopancreas and in the whole body than normal crabs (Reinhard and von Brand, 1944). Thus, depriving nutrients and hormones required for molt, growth and reproduction, the parasites reduce molting frequency and/or molt increment in *Palaemonetes paludosus* infected by *Probopyrus pandalicola* (Beck, 1980). Consequent to parasitization, the hosts incur weight loss and fecundity. The weight loss ranges from about 10% in *Crangon francisorum* infected by *Argeia pugattensis* (Jay, 1989) to 10-30% in squat lobsters *Cervimunida johni* and *Pleuroncodes monodon* parasitized by *Pseudione humboldtensis* (Gonzalez and Acuna, 2004). Since molting is a pre-reqisite for egg fertilization and oviposition, reduced molting frequency caused by the parasite may reduce fertility or induce infertility (Thomas et al., 1996b). For example, parasitic reduction in fecundity ranges from 18% in the sponge-dwelling snapping shrimp *Synalpheus yano* (Hernaez et al., 2010) to 50% in *Palaemonetes* infected by *Probopyrus* and hippolytid carideans infected by *Bopyrina abbreviata* (Romero-Rodriguez and Roman-Contreras, 2008). In general, the parasitic infection occurs, when the crustacean host has already been irrevocably differentiated into a male or female. Hence, the parasitic castration/sterilization of the gonochoric crustaceans is limited to reproductive death of one sex but the host is not capable of automatic revival of developing the female sex. Consequently, it is to be expected that the response of gonochores and hermaphrodites to parasitic castration/sterilization differs. The bopyrids attached to the abdomen usually castrate their host (Romero-Rodriguez and Roman-Contreras, 2008).

7.3.4 Reproductive Death

The host crustaceans invest > 5-15% of their body mass on the structure of the reproductive system including copulatory/ovipositing organ, SSCs, ornamentation and bright color as well as on packing materials like cyst shell, spermatophore. Substantial energy is allocated on their functioning inclusive of mate searching (e.g. *Eulimnadia texana*, Zucker et al., 2000), competition and brood care (see Fig. 3.7). Parasitic castration is defined as an infectious strategy that eventually eliminates reproduction from the hosts, as primary means of sustained acquisition of resources and energy but without affecting longevity of the host (Lafferty and Kuris, 2009). Thus parasitic castrators cause reproductive death of the hosts. Castration is an elegant and sophisticated strategy employed by highly evolved castrators, who can 'read and understand' the host's neuroendocrine system and synchronize their molting with that of the host. For example, molting of *Bopyrus fougerouxi* is synchronized with its host *Palaemon serratus* (see Pandian, 1994, p. 83). In the Monaco shrimp *Lysmata seticauda*, the mean intermolt duration is 10 and 11 d in unparasitized and parasitized shrimps, respectively. The bopyrid *Eophryxus lysmatae* has also synchronized its molt cycle with that of its host. Consequently, the bopyrid is never lost from its host (Calado et al., 2005). When intermolt duration of the parasite is not synchronized with that of the host (Calado et al., 2006), the parasite runs the risk of being eliminated, leaving a scar, as in *Carcinus maenas* left by the externae of *S. carcini* (Kristensen et al., 2012). In general, the parasites, which have synchronized their molt cycle with their respective hosts, are highly host specific. Others, who have not synchronized the cycle, may opt for more than one host species. For example, the bopyroid *Hermiarthrus abdominalis* is recorded from as many as 22 host species (see Pandian, 1994, see p 83).

In zoological research, parasitic castration is a classical discovery. Not surprisingly, a volume of literature on this aspect is available. Pandian (1994, p 82-83) has enumerated interesting anecdotes that followed the classical discovery. Some of which are listed: 1. The incidence of castration ranges from 25 to 92% (e.g. *Hemigrapsus oregonensis*) infected by *Portunion californicus*. 2. A number of instances of parasitic infection within the same crustacean order is known, which may be of interest to molecular phylogeneticists (e.g. *Balanus improvisus* infected by *Boschmaella balani*, *Clepeoniscus hanseni* by *Iodotea pelagicus*). There is also an example of infection within the same family of Bopyridae: (e.g. *Gyge branchialis* by *Pseudione euxinica*) 3. Hypercastrators are not uncommon among crustaceans. In the event of multiple or hyperparasitism, the crustacean castrator adopts one of the following strategies to bring the host under the 'command' by a single parasite: (i) masculinization of the second infecting larva (e.g. *Ione thoracica*) or feminization of the existing parasite (e.g. *Ichthyoxenus fushanensis*), (ii) elimination of the first castrator e.g. *Cryptoniscus pagurus*, a hypercastrator on *Clibanarius erythropus* becomes the primary castrator by eliminating the first one *Septosaccus rodriquezi*, (iii) mutual exclusion e.g. the bopyroid excludes sacculinid and the sacculinid the bopyrid from the host crab *Petrolithes boschi*. Incidentally, the ovarian atrophy of the first parasite for the second one seems to be a common means of elimination of the first parasite. For example, the lithodid false king crab *Paralomis granulosa* is first infected by *Briarosaccus callosus*, whose externae is protected within the abdomen, normally occupied by the egg mass of the host (Fig. 7.4 B). The cryptoniscid *Liriopsis pygmaea*, a hyperparasite on the king crab, is found inside the mantle cavity or attached to the outer surface of the externae of *B. callosus*. In the presence of *L. pygmaea*, only 19% of the externae were ovigerous but in its absence as much as 86% externae were ovigerous (Lovrich et al., 2004).

In those involving castration of the crab *Carcinus maenas* by *Sacculina carcini*, broadening of the abdomen is induced, but not as broad as a normal female, or as that in the sacculinized *Portunus pelagicus* (Table 7.11). With complete protection by the well broadened abdomen, the loss of externae from *P. pelagicus* is < 1% (Weng, 1987). But the loss is as high as 24% in the relatively less broadened abdomen of the sacculinized *C. maenas*. Notably, the loss is just 15% from the normal female. Secondly, of 249 sacculinized green crabs, " not even a single crab develop any trace of the setose(d) pleopods 2-5, characteristic of female" (Kristensen et al., 2012). Clearly, the sacculinids are able to induce different levels of feminization.

TABLE 7.11

Summary of parasitic castration/sterilization in gonochoric and hermaphroditic crustaceans and others

Host/Reference	Parasite	Remarks
Energy drained Castration and Reproductive Death in Gonochores		
Palaemon squilla Callan (1940)	*Bopyrus*	Castrated prawn with pleopods not setosed and hence not berried. Reproductive death of male and female.
Gammarus sp Le Roux (1931a,b)	*Polymorphus minutus* larva	Castrated isopod with no oostegite setose and vitellogenesis. Reproductive death of male and female.
Portunus pelagicus Weng (1987)	Sacculinid	Castrated crab resegmented and broadened abdomen, as in female
Carcinus maenas Kristensen et al. (2012)	*Sacculina carcini*	Castrated crab resegmented but not broadened abdomen, as in female. Pleopods not setosed.
P. pelagicus Shields and Woods (1993)	*S. granifera*	Male and female are partially castrated/sterilized. Reduced GSI. Fewer clutch of eggs but not berried
Cyphochara gilbert Azevedo et al. (2006)	*Riggia pranensis*	Oogenesis inhibited
Cheilodipterus quinquellineatus Fogelman et al. (2009)	*Anilocra apogynae*	Vitellogenesis inhibited. Physical presence of parasite inhibits mouth brooding in males
Energy drained Sterilization of Ovary in Hermaphrodites		
Balanus glandula Blower and Roughgarden (1988)	*Hemioniscus balani*	Ovary destroyed. Testes function normally
Lysmata seticauda Calado et al. (2005)	*Eophryxux lysmatae*	Ovary partially sterilized; restored at the loss of parasite. Testes function normally.
L. amboinensis Calado et al. (2006)	*Parabopyrella* sp	Ovary completely sterilized. Testes function normally.
Physical Presence inducing Sex Change		
Ione thoracica Reverberi and Pittoti (1942)	*Callianassa laticauda*	First *I. thoracica* female chemically induces the second to become male
Varicorhinus bachatulus Tsai et al. (1999)	*Ichthyoxenus fushcanensis*	Arrival of second male induces existing male to change sex into female
Physical Presence Halting Spawning		
Crepidula cachimilla Ocampo et al. (2014)	*Calyptraeotheres garthi*	Presence obstructs spawning & brooding. Pest removal restores reproduction
Chemical Induction of Castration and Feminization/Sex Change		
Gammarus duebeni Bulnhein (1975)	*Octosporia effeminans*	AG destroyed by excreted toxicant. All female progeny
Copepods Shields (1994)	Dinoflagellate *Blastodinium* spp	Released toxicant castrated and feminized copepodites.
Armadillidium vulgare Bouchon et al. (2008), Cordaux et al. (2011)	*Wolbachia*	Production of ZZ phenotypic female leading to complete elimination of ZW female and ZZ male

Reporting partial castration of *P. pelagicus* by *Sacculinia granifera*, Shields and Wood (1993) brought to light that the sacculinid only partially castrated the males and females. The sacculinid affected individual hosts through retardation of chelar growth, inhibition of molt cycle and castration/sterilization. The OSI and TSI decreased from three to one and five to one in relatively more infected males and females, respectively. Hence, the infected hosts were capable of mating and in a few cases even produced clutches of eggs. However, Shields and Wood did not indicate whether the spawned few eggs were brooded on the setosed appendages.

7.3.5 Secondary Sexual Characteristics (SSCs)

With reference to manifestation of SSCs in the parasitized crustaceans, information available is limited and remains piecemeal and contradictory. Visibly impressed by chela, many have described their differences between sexes, and parasitized and unparasitized ones. Tucker (1930) noted that the chelar length and breadth are around 50% smaller in females and parasitized males, as well. In decapods, the chelae are used for defense, food acquisition and sexual display. However, spermatophore transferring appendages and appendixes masculina in males, and setosed pleopods in females are more important indicators of gonadal status in the castrated/sterilized decapods. It must be noted that the male and female specific SSCs are developed in the presence of Androgenic Glands (AGs) and Androgenic Hormone (AH) in the male and the ovary in the female (see p 175). Attrill (1989) reported that the rhizocephalan *Triangular munida* had no effect on the pleopod structure of both the sexes in the galatheid *Munida sarsi*. There are many reports on the 'feminization' of appendages of parasitized male decapods. However, the critically important characteristic is the setose. 'Castrated' *Carcinus maenas* by sacculinid failed to develop setose on pleopods (Kristensen et al., 2012). Even with the presence of smaller ovary in *Portunus pelagicus*, the eggs are not berried, indicating that the appendages were not setosed (Shields and Wood, 1993). Callan (1940) found that the castrated prawns *Palaemon squilla* and *P. xiphias*, were neither setosed nor berried. Among the female specific SSCs, the 'feminine' characteristics of appendages may be revived in partially castrated decapods but not the setose. Once the setose are lost, they are not revived any more. Le Roux (1931a, b) reported that the formation of oostegite setose and vitellogenesis in *Gammarus* coincides in time that the presence of oostigite setose can be used as a morphological indicator of vitellogenesis in oocytes. Parasitized by the acanthocephalan larva *Polymorphus minutus*, *Gammarus* female underwent puberty molt but failed to develop oostegite setose. In males too, the coupling hooks cincinnuli (Fig. 3.17B) are lost during the natural addition of female sex in protrandric simultaneous hermaphrodites, but they are not restored, despite the continued presence of AG. It is likely that the setose on the appendages of decapods female as well as cincinnuli are not developed solely by the AG. And the gene (s) responsible for these structures are expressed only once during the life of decapods.

7.3.6 Single Sexed Hermaphrodites

Bopyrids and cryptoniscids infecting protandric simultaneous hermaphroditic shrimps and hermaphroditic barnacles induce reproductive death of female function but not male function. For example, the epicarid isopod *Hemioniscus balani* parasitically sterilizes adults of three barnacle species *Chthmalus dalli, C. fissus* and *Balanus glandula*. The hosts are simultaneous hermaphrodites and the parasite is a protandric. *H. balani* passes through epicardium-microniscus-cryptoniscus stages prior to infecting its definitive host and metamorphosing into a female. The female attaches to the surface of the barnacle's ovariolar follicles (Fig. 3.2B) and sucks out the ovarian fluid and thereby prevents oocyte development without impairing sperm production in the testicular follicle located at different analagen (Blower and Roughgarden, 1988).

In the more interesting group of protandric simultaneous hermaphrodic, shrimp *Lysmata seticauda* (infected by *Eophryxus lysmatae*) was unable to produce embryos, due to total elimination of the female function. But both parasitized and unparasitized male (phase) shrimps were able to fertilize unparasitized female (phase) shrimps and produced almost equal number of (346 and 344) embryos (Calado et al., 2005). In another protandric simultaneous hermaphrodite *L. amboinensis,* the infection by *Parabopyrella* sp induced partial elimination of female function. Hence, parasitized hermaphrodites during the female phase carried 363 embryos, in comparison to the unparasitized female (phase) 1,409 embryos. Notably, the molt cycle of *Parabopyrella* sp was not synchronized with the host. Consequently, the parasite was lost during host molting. The formerly parasitized hermaphrodite during its female phase could also pair with the male (phase) and produce as many as 1,362 embryos. Hence, the partial elimination of female function in *L. ambionensis* is reversible, implying that the adult ovotestis has retained the PGCs/Oogonial Stem Cells (OSCs). It may be interesting to know whether *L. seticauda,* in which, the parasite totally eliminates the female function, is also able to reverse functional female phase.

The separate anlage of the ovary and testis in these simultaneous hermaphrodites renders the testicular function. In the barnacle too, the ovarian and testicular follicles are located at the different analage (Fig. 7.8A). In the protandric simultaneous hermaphroditic shrimps too, the ovary and testis are located at the anterior and posterior anlage of the ovotestis and are separated by a connective tissue (Fig. 7.8 B). In the parasitized shrimps, the testes retain their Primordial Germ Cells (PGCs) and or Spermatogonial Stem Cells (SSCs), which facilitates the continued normal function of the testis. With the fair chances of retaining PGCs and/or Oogonial Stem Cells

(OSCs), *L. ambionensis* is able to restore completely the female function after *Parabopyrella* sp has been eliminated. Similar chances of recovery of female or male reproductive function exist in crabs with H-shaped ovary (Fig. 7.8C), in which the posterior end may still retain a vestigial testis. However, the chance for similar restoration of ovarian function in peracarids is very limited, as they have ovary (Fig. 7.8 D) and testis rather than an ovotestis.

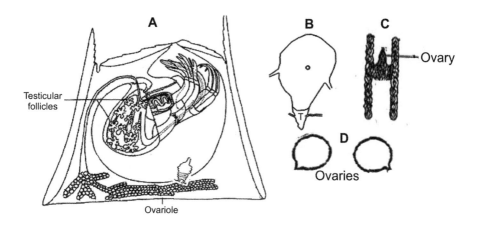

FIGURE 7.8

A. Location-wise separation of ovariolar and testicular follicles in a balanoid. Note an isopod parasite sucking hemolymph from the ovarioles. B. Ovotestis in a protandric simultaneous hermaphrodite. Note the ovary (o) and testis (T) are clearly separated by a connective tissue. C. The H-shaped ovary of a crab (see Meyers, 1990). The vestigial testis is presumably situated at the posterior end of the ovary. D. The ovaries of a typical peracarid (A, B, C and D are free hand drawings).

7.3.7 Induction of Sex Change

In many bopyrids, the first cryptoniscus that settles on the definitive host differentiates into a female (protogyny) and the second into a neotonic/dwarf male, presumable bypassing the female phase. Should many cryptoniscus larvae simultaneously enter the hosts, then all of them become females but only one of them sexually matures (Meyers, 1990). In the 'protogynic' *Ione thoracica* parasitic on the green crab *Callianassa laticauda*, the first one infecting the crab differentiates into a female. The female induces the second undifferentiated young one settling on its body to differentiate into a dwarf male (Reverberi and Pittoti, 1942, see also Anderson, 1977). On transfer of *Ione thoracica* male to an uninfected crab, the differentiated adult male is transdifferentiated into a female (Reverberi and Pittoti, 1942). In another bopyrid *Stegophryxus hyptius*, a similar sex differentiation process has been described (Reinhard, 1949). Clearly, PCGs/OSCs with bisexual potency are retained in this 'protogynic' bopyrid (see Pandian, 2011, p 146-159, 2012, p 96-102, 2014, p 82-86). On the other hand, the cymathoids seem to be 'protandrics'. The manca of the protandric cymothoid *Ichthyoxenus fushanensis* infecting the freshwater fish *Variocorhinus baclatulus* differentiates into a male. In *I. fuscanensis*, the female (19.1 mm) of the protandric cymothooid *I. fushanensis* infecting the freshwater fish *V. baclatulus* is larger than a male (9.1 mm). Mostly (91%), the female and male occur together as pair; some % occur as solitary males and the remaining are too small for identification of sex. Notably, the female never occurs alone. Being endoparasitic, the parasitic male and female inhabit within the parasitic sac (Tsai et al., 1999). The large size and presence of vestigial penis and appendixes masculina in the female clearly indicate that the newly arriving manca male induces sex change in the already existing male into female. Hence, the arrival of second male on the host induces the first male to undertake protandric sex change into a female. Obviously, the first male as an adult that existed in the host, has retained PCGs/OSCs with bisexual potency and enabled its sex change.

7.3.8 Obstruction of Spawning

Among brachyurans, the pinnothorid pea crabs (303 species) are an interesting group. They are known for the miniaturistic feature. The smallest pinnothorid *Nannotheres moorei* with a carapace width of 1.5 mm (Manning and Felder, 1996), measures just 3.3% of the largest crab *Macrocheirus kampferi* (CW: 46 cm). The miniaturization in the pinnothorids may be of interest to cytologists, endocrinologists and evolutionary biologists (cf Pandian, 2010, p 2). As they molt, the molt increment for every molt may be minimal to miniaturize the pinnothorids. However, experimental evidences are needed. McDermont (2009) provided a long list of pinnothorids that are commensals inside bivalve molluscs and in the tubes and burrows of polychaetes. The bopyroid *Rhophalione atrinicolae* is a parasite on *Pinnothorids atrinicola*. The infected females have poorly developed ovaries and the males display female characteristics including the female like pleopods. Some pea crab species infecting bivalves and gastropods are reported to partially or completely halt reproduction of the hosts (Chapparro et al., 2001). The pea crab *Calyptraeotheres garthi* inhabits the incubating egg chamber of the slipper limpet *Crepidula cachimilla*. Of course, the crab steals some food from the host but the host's overall health remains unaffected. On attaining hard-shell stage, the pea crab leaves the host in search of a mating partner. But at the post-molt, it returns to the same or another host (see also Hernandeuz et al., 2012). A 90 d experiment has shown that in the presence of the pea crab (soft stage II-IV) or its paraffin mimic completely halts spawning and brooding in the limpet. Within 7-10 d after the removal of pea crab, the limpet quickly resumes normal spawning and brooding (Ocampo et al., 2014).

7.3.9 Chemical Induction

Whereas crustaceans and other parasites cause reproductive death in crustacean hosts by consuming hemolymph and/or tissues, or physically obstructing spawning, some microbes, dinoflagellates and protozoans employ chemical induction to reverse sex and/or cease reproduction. The transovarian transmission of the sporozoan parasite *Octosphorea effemionaus* produces all-female progenies (thelygeny) in *Gammarus duebeni* (Bulnhein, 1975). Evidently, the sporozoan parasite inhibits the development of AG through its excreted toxicant (Pandian, 1994, p 84). For an annotated list of protozoan parasites and hyperparasites on decapods crustacean, Sprague and Couch (1971) may be consulted. Most interestingly, the parthenogenic *Daphnia magna* has developed counter-adaptation against infection by the bacterium *Pasteuria ramosa*. When infected, *D. magna* grows faster and enhances fecundity, prior to the negative effects that are inflicted by *P. ramosa* (Ebert et al., 2004). Of 2,000 recognized species of dinoflagellates, 140 are parasites on other organisms. Dinoflagellates belonging to the orders Blastodinida and Syndinida infect copepods, amphipods, mysids, euphausiids and decapods. Some of them destroy host eggs (e.g. chytriodinids), while others castrate and feminize the host, the degree of feminization is related to the stage at which parasitization occurs. Infection of copepods by *Blastodinium* spp results in cessation of molt or feminization of the host (Shields, 1994). Evidently, the parasitic dinoflagellate too releases unknown chemical (s) that inhibit molt and/or castrate the host. Employing cytoplasmic strategy, *Wolbachia*, an endosymbiont, transforms ZZ genetic males into functional phenotypic females in the female heterogametic amphipod *Armadillidium vulgarae* (Bouchon et al., 2008, Cordaux et al., 2011).

7.3.10 Validation of Hypotheses

At this junction, it is appropriate to consider the less-explored 'energy drain' and 'steric interference' hypothesis. The first one considers that castration is caused by the consumption of host derived resource by the parasite and thereby deprives the host of the resources required for its own reproduction. The second one proposes that the parasite physically inhibits host's reproduction. Employing different strategies, the parasites castrate the hosts. Some of them are: 1. The biomass of many castrators is ~ 5% of its host. For example, *Parapopyrella* sp is just 4% of *Lysmata ambionensis* (Calado et al., 2006). But they consume 8 to 25% of hemolymph of the host/day (see Gonzalez and Acuna, 2004). *H. balani* sucks the entire ovarian fluid in the barnacles (Blower and Roughgarden, 1988). This sort of 'energy drain' at such high levels deprives the host the resources required for growth and hormones required to initiate and sustain reproduction. 2. Others specifically inhibit vitellogenesis and thereby cause reproductive death of female function. *Anilocra apogynae* inhibits vitellogenesis to produce fewer and smaller oocytes in the host and in the presence of the parasite in the gills, the cardinalfish male is unable to orally incubate even the fewer fertilized eggs (Fogelman et al., 2009). However, microbes and algae do not employ the 'energy drain' strategy to achieve castration/sterilization. They employ cytological or chemical strategies. Sporozoans are known to excrete toxicans that destroy AG (Bulnhein, 1975). Many dinoflagellates also employ the same chemical strategy. *Wolbachia* adopts cytogenetic transformation to produce males alone. Hence, the 'energy drain' hypothesis alone cannot explain the different strategies employed by castrators. The steric interference hypothesis fits in with transient and reversible sterilization by the physical presence of *Calyptraeotheres garthi* in *Crepidula cachimilla* (Ocampo et al., 2014).

7.4 Food Disruption

To avoid predation, many algae have developed specific chemicals to interfere with larval development (e.g. Carotenuto et al., 2006) and alter sex differentiation. For example, the diatom *Thalassiosira rotula* arrests the embryonic development of the copepod *Calanus helgolandicus* (Poulet et al., 1994). Despite flexibility, the sex differentiation process of crustaceans is sensitive to quantity and quality of algal feed during larval development and natural sex change in adults. From limited publications, the following are described.

7.4.1 Sex Change in Copepods

In a land-mark review, Gusamao and McKinnon (2009) proposed that the undifferentiated C_4 presumptive calanoid male may undergo protandric sex change to phenotypic female, when food supply is abundant. This proposal is important in view of the species richness (18,000 species) of the calanoid and their role in trophic dynamics as primary consumers. Firstly, two important observations reported from both laboratory and field must be noted: (i) Female-biased sex ratios are common among copepods (Peterson, 1986, Irigoien et al., 2000); for example, the ratio of adult paracalanoids is 0.8 ♀ : 0.2 ♂ (Kiorboe, 2006) and (ii) High incidence of intersexuality is also common in copepods (e.g. 2.5 – 23% in *Paracalanus*, Liang and Uye, 1996, 41% in *Calanus finmarchichus*, Svensen and Tande, 1999). Intersex individual copepods originating from both field and culture can readily be distinguished from the males by the presence and size of the fifth appendage in C_5 and from female, in which the fifth appendage is reduced to the basopodite. For more information on intersexuals in copepods, the long list of Gusamao and McKinnon (2009) may be consulted. Intersexes have the potential to have a strong effect on sex ratio.

The number of molts in the life cycle of copepods is defined and once maturity is attained, the copepods do not molt further. As SSCs are not developed up to copepodid$_4$ (C$_4$), the copepod remains morphologically undifferentiated. Hence the molt from C$_4$ to C$_5$ stage is the crucial time point, at which sex is differentiated (see also Fig. 7.6). Fleminger (1985) was the first to analyze the seasonal variation of sex ratios accounting for intersexes and to suggest an interaction between food supply on one hand and sex differentiation and intersexuality on the other. Irigoien et al. (2000) reported higher proportions of *Calanus helgolandicus* males, when the copepods were reared from the egg to adult stage with abundant food supply. From their experimental culture of *Acrocalanus gracilis*, Gusamao and McKinnon (2012) resolved that the proportions of intersex (neofemale) individuals are equal to the increase in female ratio above 0.5 (see also Irigoien et al., 2000, Miller et al., 2005). They also found that the limited food supply induce some presumptive males to transdifferentiate into females. Apparently, suppression of AG and its hormone AH by the diminishing food supply induces altered sex differentiation to produce neofemales in copepods. Following sex change, the development of the fifth appendage of presumptive male copepods ceases or is reduced. Hence the phenotypic neo-female can be identified from the normal female by the number and male-like aesthetacs (e.g. *Simodiaptomus indicus*, Dharani and Altaff, 2002). Despite the antennular differences, spermatophores are attached to the phenotypic females of *Pseudocalanus elongatus*, indicating that these neofemales are as attractive as normal females (Cattley, 1948). In fact, the carriage of egg sacs by phenotypic calanoids suggests the successful fertilization of their eggs in the field (Dharani and Altaff, 2002). The phenotypic females produce more eggs and thrice the number of clutches, as normal females (Gusamao and McKinnon, 2009).

Contrastingly, the diminishing food supply induces occasional males and sexual females through MO and its hormone MF in cyclic parthenogentic cladocerans (p 96). Apparently, the suppression of AG and its hormone AH by the diminishing food supply induces altered sex differentiation to produce neofemales in copepods.

7.4.2 Induction to Females

The hippolytids include species that are gonochoric (e.g. *Hippolyte obliquimanus*) and protandric (e.g. *H. inermis*) Espinoza-Fuenzalida et al. (2008) considered 33 hippolytid species and suggested that *H. inermis*, *H. karaussiana*, *H. longirostris*, *H. micholsoni*, *H. varians* and *H. vetricusa* are to be protandric hermaphrodites. However, there are adequate field and experimental evidences for the existence of protandric hermaphroditism only in *H. inermis*. In *H. inermis*, differentiation of male sex is completed by the 49th d after hatching. After 12 m of existence as males, the sex changing process is initiated in them. Le Roux (1963) is perhaps the first to locate the sensitive labile period of sex differentiation to lie between larval stage IV and the last zoeal stage of the protandric digynic hermaphrodite *H. inermis.* From his field study, Zupo (1994) found that the labile period in *H. inermis* was terminated by the 30th d of post-hatching (dph), when the shrimp attains a body length of 5 mm. From a well designed field study of *H. inermis* and cluster analysis of food items, Zupo (2001) showed that the abundantly available diatom *Coconeis neothumensis* on the leaf canopy of *Posidonia oceanica* during spring, on being consumed, induced direct development to primary females bypassing a male phase. The food composition of these primary females was significantly different from that of others. To confirm the induction of primary female by *C. neothumensis,* Zupo (2000) reared the shrimp larvae on diet supplemented with and without *C. neothumensis.* Most larvae underwent sex differentiation between 35 and 45 dph; the fact that primary females became abundant, even when 35% of larvae were still in juvenile stage indicated that the juveniles did not pass through a short male phase but directly differentiated into primary females. Thus Zupo demonstrated that *C. neothumensis* indeed induced direct production of 61% primary females, against the control production of 5% primary females.

Using the TUNEL technique, Zupo and Messina (2007) detected rapid destruction of Androgenic Glands (AG), testes and vas deferens by apoptosis during a period of 7 d-all within a single molt cycle of *H. inermis* zoea. *H. inermis* is characterized by the absence of an ovotestis, as in other decapods (Cobos et al., 2005). As a result, female gonad is developed from undifferentiated cells, after the complete disruption of testis. Zupo and Messina (2007) also reported that *C. neothumensis* ingested by *H. inermis* contains a chemical that induces apoptosis to selectively destroy cells in AG, testes and vas deferens. Earlier Zupo (2000) also demonstrated that other congeneric diatoms *C. scutellum parva* and *Diploneis* sp induced apoptosis of AG. The research is in progress to identify the chemical in these diatoms that induces apoptosis. However, a more important question to be answered is: from where do these undifferentiated cells arise? Are they oogonial stem cells?

8

Highlights and Directions

Being arthropods, crustaceans molt to grow and/or reproduce. As a hallmark strategy, a vast majority (> 96% i.e. except 1,800 calanoids and ~ 200 penaeids and euphausiids) of crustaceans brood their eggs on their body. Consequently, they share the available energy among intense competing processes (i) growth including molting, (ii) breeding and (iii) brooding (Fig. 3.7). Hence, crustaceans stand unique among invertebrates and render the study of reproduction and development a fascinating field of research. In this context, this book represents perhaps the first attempt to comprehensively elucidate almost all aspects of reproduction and development covering taxons from anostracan *Artemia* to xanthid crabs. This holistic approach has led to highlight many new findings and indicates the future direction of research.

Sexuality   Of > 52,000 crustacean species, ~ 92% are gonochores. The non-gonochorics employ mutually exclusive parthenogenesis and hermaphroditism, which are more prevalent among entomostracans. However, parthenogenesis (e.g. *Procambarus fallax, Orconectes limosus*) and sequential hermaphroditism also occur in malacostracans. Briefly, there may be more but at present 12 ontogenetic pathways are known, through which crustaceans are sexualized (Fig. 8.1).

Sex Determination and Differentiation   In crustaceans, cytogenetic and breeding investigations have revealed the existence of the basic male and female heterogametic sex determining mechanisms. Uniquely, ostracods have explored the possibility of employing numerically more X (1-11) and W (1, 2) chromosomes to determine the female sex (Table 6.3). Others like the anostracan *Branchipus schaefferi* employ B-chromosomes carrying the higher load of male sex determining genes (Beladjal et al., 2002). Besides, the presence of sex linked markers (e.g. *Tisbe reticulata*, Battaglia, 1961; *Paracerceis sculpta*, Shuster et al., 2014) and sex associated/linked molecular markers (Table 6.2) evidence the existence of genetic sex determining mechanism in crustaceans. Yet, for 52,000 species evidences to show the existence of heterochromatism (Table 6.3) in about 0.1% (66 species) of crustaceans are too small, and imply the need for more research in this area as well as caution that it may be premature to make any generalization. Nevertheless, the existence of heterogametism across crustacean taxa from Anostraca to Decapoda (Table 6.3) emboldens to state that sex in crustaceans is mostly determined by gene (s) harbored on male or female heterogametic chromosomes.

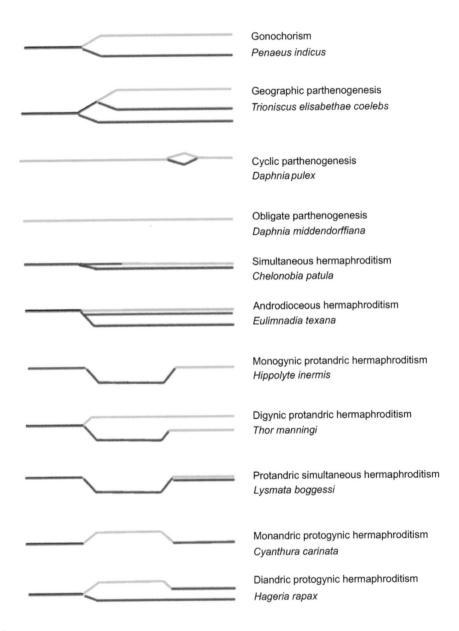

FIGURE 8.1
Known ontogenetic pathways of sex differentiation in Crustacea. Red lines indicate females, green lines males and black lines undifferentiated neuter.

TABLE 8.1

Genetic sex determination and sex differentiation during embryonic/larval stages of crustaceans

---

Anostraca

B-chromosomes transmit a higher load of male sex determining genes in *Branchipus schaefferi* (Beladjal et al., 2002)

Cladocera

In daphnids, male and female are parthenogenically differentiated during ovarian oocyte maturation 12 h prior to the release of oocytes into the brood chamber (Olmstead and LeBlanc, 2002, Ignace et al., 2011)

Spinicaudata

In *Eulimnadia texana*, male and hermaphroditic sexes are determined by female heterogametic ZW system (Weeks et al., 2010)

Copepoda

In calanoids, female sex is irrevocable differentiated earlier. Transdifferntiation of presumptive males into phetotypic females is completed between C_4 and C_5 copepodid stages (Gusamao, and McKinnon, 2009). In parasitic monstrilloid copepods too, sex is differentiated between C_4 and C_5 copepodid stages (Fig. 7.6).

Maxillopoda

(i)   Existence of sex determination system in *Conopea* (*Balanus*) *galeatus* (Gomez, 1975)
(ii) Not more than 50% cyprids settle as males in *Scalpellum scalpellum* (Svana, 1986)
(iii) Sexually dimorphic larvae occur in *Ulophysema oersundense* (Melander, 1950)
(iv) In rhizocephalans, female cyprid larvae pass through kentogen and vermogon stages but male cyprids through trichogen stage

Isopoda

AG is present at birth in male manga (Johnson et al., 2001)

Decapoda

(i)   In field populations of protandric simultaneous hermaphroditic *Thor manningi*, 50% shrimps are primary males and remaining 50% pass through female phase prior to differentiation into simultaneous hermaphrodite
(ii) In *Fenneropenaeus chinensis*, the maternally inherited PGCs are recognized at the limb bud stage. Germ Stem Cells become histologically detectible at post larvae but male-specific *Transformer-2C* gene commences expression at mysis stage itself.
(iii) In decapods, the parasitic destruction of AG results in castration, but not in revival of functional female sex in adult host (see Section 7.3.4)
(iv) In *Litopenaeus vannamei*, appearance of first pleopods in 50 d-old post larva marks irrevocable sex differentiation. Subsequently, the differentiation is not amenable to change by any environmental cue, as differentiation has reached a point of no return (Campos-Ramos et al., 2006).
(v)   In *Macrobrachium rosenbergii*, sex is irrevocably differentiated in post larva between 25 and 100 d (at < 1 g body weight) (Aflalo et al., 2006)

Unlike vertebrates, endocrine and spermatogenic functions are clearly separated into androgenic glands and testes, respectively. Hence, sex is more or less genetically determined at fertilization in crustaceans. The following sex differentiation process is also completed on or before the penultimate larval stage. In isopods characterized by direct life cycle, the male manca at birth possesses masculanizing androgenic glands (Table 8.1). In many decapods, sex differentiation is irrevocably completed at post larva or even earlier at mysis stage. In copepods too, female sex is irrevocably differentiated at C_4 stage but the presumptive males may change sex to females at copepodid stages between C_4 and C_5 alone but not after (Gusamao and McKinnon, 2009). In parasitic monstrilloid copepods too, sex is differentiated between C_4 and C_5 stages (see Fig. 7.6, Boxshall et al., 2005). In the Maxillopoda, cyprid larvae are already differentiated and are sexually dimorphic. More interestingly, not only larvae are sexually dimorphic but also the rhizocephalan larvae pass through sex specific different stages (Table 8.1). The already sex differentiated scallpellid male cyprids settle always on the mantle rim and develop into dwarf males, while the hermaphroditic cyprids settle on the substratum and develop into hermaphrodites (Spremberg et al., 2012). On exposure to 17α-methyltestosterone from egg to adult stage, the crayfish *Procambarus alleni* is not amenable to masculinization (Vogt, 2007), undoubtedly indicating that testosterone cannot induce the development of androgenic gland. Clearly, sex in crustaceans is irrevocably differentiated prior to metamorphosis or prior to birth. Unlike in fishes, sex differentiation in crustaceans is neither a flexible nor a protracted process. The labile period, during which sex remains manipulable, is terminated at the embryonic and larval stages in some fishes like *Poecilia reticulata, Salmo fontinalis, Betta splendens* but it is a long protracted process terminated only at puberty in many fishes like *Mugil cephalus, Anguilla anguilla* (Pandian, 2014, Chapter 7.3). In comparison to fishes, crustaceans complete sex differentiation much faster and earlier. Hence, they may not possess the Primordial Germ Cells (PGCs) or their derivatives Oogonial Stem Cells (OSCs)/Spermatogonial Stem Cells (SSCs) with bisexual potential. This implies that when sexual maturity is completed, the ovaries in female are left with oogonial cells alone and the testes in males with spermatogonial cells alone.

Understandably, PGCs or their derivatives OSCs/SSCs with bisexual potency are retained in simultaneous and sequential hermaphroditic crustaceans. To date, as many as 75 protrandric and 14 protogynic crustaceans are known to exist (Table 3.8). Of them, 36 are carideans, which possess a pair of ovotestes (Fig. 3.3). In them, it is possible that the undeveloped/vestigial ovaries harboring PGCs/OSCs develop functional ovaries, when female sex is added to the existing male sex. In amphipods and isopods, females and males possess ovaries and testes, respectively (Fig. 3.3). It is difficult to comprehend, from where these protandric amphipods and isopods bring PGCs/OSCs to develop ovaries, when testes are destroyed. The same applies to protogynic peracarids. Incidentally, the protandric shrimp *Hippolyte innermis* is reported to possess only testis, unlike the other decapods. During the food-induced sex change from male to female, the androgenic glands, testes and vas deferens are selectively apoptosized. Zupo and Messina (2007) reported that the ovaries are developed from some undifferentiated cells.

Research in this area is required to know whether these undifferentiated cells are PSCs/OSCs? If so, where were they harbored earlier? Incidentally, sequential fishes constitute up 0.5% of fish species (see Pandian, 2012, p 104, 105) but those of crustaceans 0.1% only. Surgical removal of ovarian zone of the ovotestis in the protandric black porgy *Acanthopagurus schlegeli* induced precocious sex change from female to male but that of ovary in the protogynous *Thalassoma bifasciatum* resulted in the death of reproductive function (Pandian, 2014, p 173-174).

It may be recalled that unlike vertebrates, endocrine and gametogenic functions in crustacean males are clearly separated into distinct organs, the androgenic glands (AGs) and the testis, respectively. Hence, it is possible for a few protandric simultaneous hermaphroditic carideans (19 species, Table 3.8) and a barnacle (e.g. *Chelonobia patula*) to add female sex and retain the function of male sex. Arguably, the genetic cascades underlying male and female sexual differentiation must have largely been independent in the gonochoric ancestors of Crustacea. In fishes, exogenous estrogens are reported to induce almost functional sex reversal from male to female (e.g. *Cyprinus carpio*) (see also Pandian, 2014, Chapter 4). In gonochoric crustaceans, exposure to Juvenile Hormone (JH) or its analogs could only alter the duration of larval stage or at the best add a supernumerary stage (Table 7.5). In parthenogenic cladocerans alone, the juvenoids induce production of parthenogenic males and females; however, the functional ability of chemically induced male to fertilize the eggs remains questionable (see p 197). Functional sex reversal is also not yet reported in gonochoric crustaceans exposed to one or other pollutant or steroid (e.g. Vogt, 2007). Similarly, many animal parasites do castrate their gonochoric crustacean hosts, but are unable to revive automatic feminization in mature males. Understandably, the cascades of male and female sex differentiation genes largely remain separated in gonochoric crustaceans. Yet, it must be noted that the prokaryotic parasites like the bacterium *Wolbachia*, the dinoflagellates like *Blastodinium* spp and protozoans like *Octosporea effemionans* simultaneously castrate and feminize their hosts. Clearly, these parasites are able to chemically induce the simultaneous deregulation of both cascades of male and female sex differentiation, despite the large level of separation between these two cascades. More information on the biochemical pathways of altered sex differentiation process by microbial/protozoan parasites is desired.

Hermaphroditism and Parthenogenesis Isopods are the only crustacean taxon, in which representatives of both protandrics and protogynics are present. Parasitic bopyrids successfully undergo protrandric (cymothoids) or protogynic (cryptoniscids and entoniscids) sex change once in their life time in a single direction. In cymothoids, the newly arriving manca induces the existing male to change sex to female. Expected of protogynics, the first cryptoniscus larva of cryptoniscids and entoniscids that settles on the definitive host always differentiates into a female. But the second one to settle on the same host directly differentiates into a male, bypassing the female phase. Sex change and bypassing one or another sexual phase are not uncommon among sequential hermaphrodites (cf Fig. 3.16). Notably, *Ione thoracica* must be recognized as a serial hermaphrodite capable of undertaking sex change more than once in either direction. For the first time, this book has clearly recognized and designated the bopyrids as sequential hermaphrodites. Hence, these bopyrids have necessarily retained PGCs or OSCs/SSCs with bisexual potency, even after sexual maturity and reproduction. Secondly, another 2,000 bopyrids (i.e. 1,250 cymothoids + 765 crustacean species parasitizing other crustaceaens) may have to be added to those listed in Table 3.8. With this addition of ~ 2,000 species, i.e. ~ 4% crustaceans (2,000 of 52,000) may prove to be sequential hermaphrodites.

In crustaceans, the coexistence of parthenogenesis and hermaphroditism is not so far reported. Hence, parthenogenics and hermaphrodites may mutually exclude each other. Parthenogenics are common among 720 species of cladocerans. They occur sporadically in Anostraca, Copepods and Isopoda, and rarely in Decapoda. There are no estimates on the number of ostrocodan species, which employ parthenogenic mode of reproduction (cf Cohen and Morin, 1990). Neither are there estimates on the number of species, in which hermaphroditism occurs in Notostraca, Spinicaudata and Thecostraca. However, it is likely parthenogens (720 cladoceran species) and hermaphrodites (~ 2,000 species in sequential hermaphroditic bopyrids alone) occur in 1 : 3 ratio, i.e. ~ 1,000 parthenogenic species: 3,000 (i.e 2,000 bopyrids + other entomostracan hermaphrodites like spinicaudates and so on) hermaphroditic species of 52,000 crustaceans. This would imply that gonochorism is limited to < 92% in crustaceans in comparison to > 98% (Pandian, 2011, p 8) in relatively more motile fishes. With the large number of sessile, sedentary and low motile species, it may not be a surprise that gonochorism is limited to < 92% in crustaceans.

Investment and Fecundity　In general, Generation Time (GT) and Life Span (LS) of tropical crustaceans are shorter than those in temperate and Arctic counterparts (Table 2.2). On a average, the investment on GT as a fraction of LS is 28, 48 and 85% in tropical, temperate and Arctic crustaceans. Considering parthenogenic cladocerans as an example, tropical cladocerans are about 10 times more fecund than their temperate counterparts (Table 2.2). Parthenogenic crustaceans and *Artemia* allocate 69 and 28% of the available energy on egg production (Table 2.7). With shortest GT investment, the largest fraction of available energy allocated for reproduction and 10-times more fecundity, the tropical cladocerans, *Artemia* and perhaps copepods may prove to be the best fodder live organism. Therefore, the development of mass culture techniques for these fodder crustaceans may greatly benefit tropical aquaculturists.

Aging and Reproductive Senescence　For the first time, it has been demonstrated that brooding crustaceans do undergo aging related reproductive senescence.

Oogonial Stem Cells undergo asymmetric mitotic divisions to produce OSCs and oogonia. The decrease in fecundity in older/larger females may therefore indicates the diminishing rate of oogonial production. Evidence for decreasing fecundity in older/larger females was adduced from almost all taxons across Crustacea; the evidences include also for the crustaceans characterized by gonochoric, parthenogenic and hermaphroditic modes of reproduction as well as free living and sessile modes of life. Not only laboratory reared crustaceans but also those collected from the field were shown to undergo reproductive senescence. Incidentally, the parthenogenic crayfish *Procambarus fallax* has also been shown to undergo reproductive senescence (Fig. 3.14 C). Possibly, brooding stress imposes reproductive senescence.

With regard to aging and reproductive senescence, the following have to be reconsidered: The fecundity-body size relationships are positive and linear in egg shedding calanoids, euphasiids and penaeids. Clearly, their OSCs continue to produce oogonia in females of all sizes (Fig. 8.2). However, in the egg shedding calanoids *Parvocalanus crassrostris*, the fecundity remains flat at low density or decreased at high density, as a function of culture age (Fig. 3.8E). Hence, the accumulated excretory wastes reduce the rate at which OSCs produce oogonia, i.e. external factors alter the rate of oogonia production in egg shedders. On the other hand, the fecundity (epicardium/manca, Table 7.10) and body size relationships are positive and linear in all egg shedding parasitic bopyrids and rhizocephala, unlike in other brooders, which are free-living. There are adequate indications that these bopyrids produce more eggs that exceed the capacity of the female's marsupium and host's branchiostegites to contain the egg mass, which results in loss of eggs, as in other brooders (e.g. Romero-Rodriguez and Roman-Conteras, 2008). Clearly, the OSCs in these brooding bopyrids continued to actively produce more than adequate oogonia in females of all size. Hence, brooding alone may not impose reproductive senescence. It is the interaction between stress induced by brooding and external factors (e.g. excretory wastes, predators, parasites) that ultimately determines to impose or not impose reproductive senescence in crustaceans.

Figure 8.2 shows fecundity as a function of age/size of females in copepods, euphasiids and penaeids/palaemonids; these are taxonomic groups, in which some species are egg shedders, while others are brooders. From field collected data, it is shown that the fecundity of 29 species of calanoid copepods shed almost 7.5 times more eggs than that of 16 species of egg brooding copepods (Kiorboe and Sabatini, 1995) and three times more eggs by egg shedding euphausiids than their counterpart brooders (Pandian 1994, p 94). Penaeids shed > 100 times more eggs than their counterpart palaemonids. Notably, the size of shedded egg is smaller than that of brooded eggs. For example, it is three times smaller in egg shedding calanoids than that of brooders. The presence of eggs on pleopods postpones the ensuing spawning. But the continuous removal of brooded embryos increases fecundity one and a half times in *Macrobrachium nobilii* (Table 2.4). Egg brooding includes egg carriage, ventilation and grooming. For example, the fanning frequency of pleopods increases from 2,676 time/h during the initial incubation to 8,760 time/h during the terminal period in *M. nobilii* (Pandian and Balasundaram, 1980). Among aquatic invertebrates, crustaceans are the only taxon that does not secrete mucus to clean themselves chemically. Physical cleaning by grooming the body and brooded eggs requires the development of setae, combs and the like. For example, the porcellanid crabs are excellent groomers and remove all parasites before they settle (isopods)/penetrate into the body (rhizocephalans) (Boxshall et al., 2005). Hence, the high investment on egg brooding seems to have induced reproductive senescence. All egg shedding calanoids, euphausiids and penaeids seem to have escaped from it.

Not only these shedders but also the epicardium releasing bopyrids have also eliminated the reproductive senescence. In support of this new finding, available evidence has been listed in Table 7.10 and depicted in Fig 7.7. Interestingly, all the parasitic crustaceans like Rhizocephala and Bopyridae including the manga producing cymothoids show linear and positive relationships between fecundity and body size. The elimination of reproductive senescence is a strategic adaptation, as the cryptoniscids and entoniscids have to pass through a series of epicardium-microniscus-cryptoniscus larval stages involving a pelagic calanoid copepod as an intermediate host. However, the free-living (but within the spongocoel of *Leucetta longelensis*) isopod *Paracerceis sculpta* display a trend characterized by reproductive senescence (Fig. 8.2 D). Briefly, parasitism in bopyrids has retained sexual plasticity until the cryptoniscus larvae settle on the definitive host and also eliminated the reproductive senescence.

Asexualism and Regeneration   In colonial parasitic rhizocephalans, asexual reproduction involves just a few tissue types (epidermis, nerve cells, muscle cells and connective tissues) derived from ectoderm and mesoderm. The cluster of cells, probably the Oligopotent Stem Cells (OlSCs), injected into the host by vermogon is exhausted after asexual reproduction of a few externae. Consequently, the loss of externae results in death of interna and the parasite, as well. The lost part of the chelate appendage or chelar appendage itself is completely regenerated in the presence of OlSCs among the muscle tissues in the chela. The regeneration also involves just a few tissue types (epidermis, nerve cells, muscle cells and connective tissues) derived from ectoderm and mesoderm. Hence, asexual reproduction in the parasitic colonial rhizocephalans and regeneration of the chelipede involve the same tissue types derived from ectoderm and mesoderm. Therefore, asexual reproduction of the externae in colonial parasitic rhizocephalans simulates more of regeneration rather than agametic cloning of a whole animal with many tissue types derived from not only ectoderm and mesoderm but also endoderm.

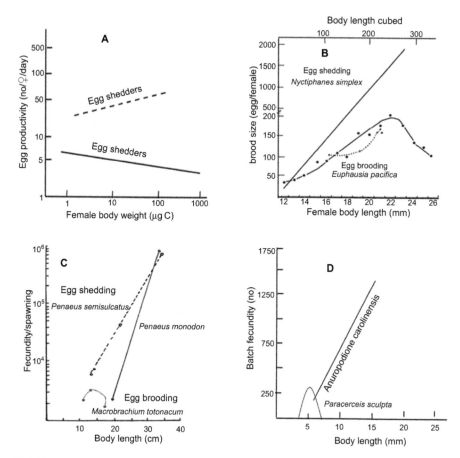

FIGURE 8.2

A. Egg productivity (no/♀/d)-body weight relationships reported for egg shedding and brooding copepods (from Kiorboe and Sabatini, 1995, permission by International Ecology Institute), B. Brood size as a function of female length in egg shedding *Nyctiphanes simplex* (redrawn from Lavaniegos, 1995) and egg brooding *Euphausia pacifica* (dotted line for 5 populations and continuous line for 3 populations) (drawn using data of Gomez-Gutierrez et al., 2006) C. Fecundity as a function of body length in selected egg shedding Indian penaeids (drawn using data of CMFRI, 2013) and egg brooding *Macrobrachium totonacum* (drawn using data of Mejia-Oritz et al., 2010), D. Fecundity as function of body size in parasitic bopyrid *Anuropodione carolinensis* (drawn from data of Wenner and Windsor, 1979) and free-living isopod *Paracerceis sculpta* (drawn from data of Shuster, 1991).

Secondly, no one has so far described the effect of natural or experimental loss and regeneration of spermatophore-transferring thoracic appendages in males or embryo-brooding pleopods in females. Experiments are also required to know whether the spermatophore-transferring thoracic appendage can be regenerated on a stump of eyestalk. Experimental investigations in this area may be rewarding.

Egg Banks and Cysts   Cyst production requires larger quantum of yolk deposition and is to be followed by shell gland activity to envelope a fertilized/unfertilized oocyte with a hard shell and in many species attractive ornanamention. Consequently, a female has to decide prior to oogenesis whether to produce subitaneous eggs or cysts. Multiple environmental factors are known to trigger cysts hatching. However, not all the cysts in an egg bank hatch simultaneously. Hatching in batches is a crustacean trait. In brooded eggs, it is initiated by the mother with the release of a chemical substance (Crisp and Spencer, 1958). But pelagic spawning and cyst producing mothers do not have direct control over hatching in batches. Obviously the presence of an inherent genetic/chemical factor in the cysts that induces earlier hatching in some cysts at an appropriate environmental cue, while others remain unhatched. Investigations are required to know (i) the differences in the cyst size from a single spawn by a female, and (ii) whether some batches of cysts within single spawn by a female hatch in response to temperature, while others to osmolarity, still others to oxygen level and yet others to photoperiod.

Secondly, the crustacean cysts are of two types namely the 'dry' and 'wet' ones. In the dry cysts like those of *Artemia*, the proteins contained in a single tissue namely the syncytium with 4,000 nuclei withstand dehydration up to 0.7 µg residual water/g cysts. Prior to hatching, the cysts are hydrated by one million times to 0.7 ml/g nauplii (Clegg et al., 2009). As in endosperm of dry seeds, *Artemia* cysts also have a single tissue type of a syncytium. Hence, dehydration and hydration of this single tissue may not pose a serious problem. The human body has 200 and odd tissue types, some like bones have > 10% water and others like the blood has < 90% water. Space travel requiring long years may resort to 'cystification' of astronauts. However, hydration and dehydration of an astronaut may prove to be an onerous task.

Conversely, the 'wet cysts' rather diapausing eggs of daphnids and perhaps copepods contain 70% water (Clark, 1978). But the diapausing eggs of cladocerans too contain a single type of tissue. Fortunately, the wet diapausing cyclopoid copepodites and ostracodan larvae (7th instar) with many more tissue types do also pass through a wet diapausing phase, during which they can withstand freezing or drought for an extended period of time. In these aerobes, oxygen uptake is somehow depressed to a level that it is scarcely measurable. Unfortunately, metabolic studies like those undertaken by Dr. J. S. Clegg for *Artemia* cysts have not yet been undertaken on the wet diapausing copepodites and ostracod larvae. Research in this area of wet diapausing copepodites and ostracod larvae may lead to generation of a wet diapausing astronaut, who could withstand long durations (days/years) of passage to a desired planet and return home to the mother Earth safely.

9

References

Aarestrup, K., F. Økland, M. M. Hansen et al. 2009. Oceanic spawning migration of the European eel (*Anguilla anguilla*). Science, 325: 1660-1660.

Abdu, U., A. Barki, I. Karplus et al. 2001. Physiological effects of methyl farnesoate and pyriproxyfen on wintering female crayfish *Cherax quadricarinatus*. Aquaculture, 202: 163-175.

Abe, M. and H. Fukuhara. 1996. Protogynous hermaphroditism in the brackish and freshwater isopod, *Gnorimosphaeroma naktongense* (Crustacea: Isopoda, Sphaeromatidae). Zool Sci, 13: 325-329.

Acharya. K., J. D. Jack and P. A. Bukaveckas. 2005. Dietary effects on life history traits of riverine *Bosmina*. Freshwater Biol, 50: 965-975.

Adiyodi, K. G. 1985. Reproduction and its control. In: *The Biology of Crustacea*. (eds) D. E. Liss and L. H. Mantel. Academic Press, New York. 9: 146-215.

Adiyodi, K. G. and R. G. Adiyodi. 1989. *Reproductive Biology of Invertebrates*. Oxford and IBH Publishing, New Delhi, Vol 4, Part A. p 463.

Adiyodi, K. G. and R. G. Adiyodi. 1990. *Reproductive Biology of Invertebrates*. Oxford and IBH Publishing, New Delhi, Vol 4, Part B. p 527.

Adiyodi, K. G. and R. G. Adiyodi. 1993. *Reproductive Biology of Invertebrates*. Oxford and IBH Publishers, New Delhi, Vol 6 Part A. p 410.

Adiyodi, K. G. and R. G. Adiyodi. 1994. *Reproductive Biology of Invertebrates*. Oxford and IBH Publishers, New Delhi, Vol 6 Part B. p 432.

Adiyodi, R. G. and T. Subramoniam. 1993. Oogenesis, oviposition, oosorption. In: *Reproductive Biology of Invertebrates*: Arthropoda, Crustacea. (eds) K. G. Adiyodi and R. G. Adiyodi. John Wiley, New York. Vol. 1: 443-495.

Adolfsson, S., Y. Michalakis, D. Paczeniak et al. 2009. Evaluation of elevated ploidy and asexual reproduction as alternative explanations for geographic parthenogenesis in *Eucypris virens* ostracods. Evolution, 64: 986-997.

Aflalo, E. D., T. T. T. Hoang, V. H. Nguyen et al. 2006. A novel two-step procedure for mass production of all-male populations of the giant freshwater prawn *Macrobrachium rosenbergii*. Aquaculture, 256: 468-478.

Ahl, J. S. B. and J. J. Brown. 1990. Salt-dependent effects of juvenile hormone and related compounds in larvae of the brine shrimp *Artemia*. Comp Biochem Physiol, 95 A: 491-496.

Ahl, J. S. B. and H. Laufer. 1996. The pubertal molt in Crustacea revisited. Invert Reprod Dev, 30: 177-180.

Alajmi, F. and C. Zeng. 2014. The effects of stocking density on key biological parameters influencing culture productivity of the calanoid copepod, *Parvocalanus crassirostris*. Aquaculture, 434: 201-207.

Albaina, A. and X. Irigoien. 2006. Fecundity limitation of *Calanus helgolandicus* by the parasite *Ellobiopsis* sp. J Plank Res, 28: 413-418.

Alfaro-Montoya, J. and L. A. Vega. 2011. The effect of environmental cues and neurotransmitters on male sexuality of the eastern Pacific *Penaeus* species. Aquaculture, 316: 60-67.

Alfaro-Montoya, J. and L. Hernández. 2012. The histological structure of the androgenic gland and cellular cord of the male reproductive system of adult *Litopenaeus* and *Rimapenaeus byrdi.* J Crust Biol, 32: 351-357.

Altaff, K., and M. R. Chandran. 1994. Oviducal gland of planktonic copepod, *Heliodiaptomus viduus* Gurney-A new report. Curr Sci, 66: 81-83.

Alvarino, A. 1994. Chaetognatha. In: *Reproductive Biology of Invertebrates.* (eds) K. G. Adiyodi and R. G. Adiyodi. Oxford and IBH Publishers, New Delhi, Vol 6 Part B. pp 329-338.

Ambler, J. W. 1985. Seasonal factors affecting egg production and viability of eggs of *Acartia tonsa* Dana from East Lagoon, Galveston, Texas. Estu Coast Shelf Sci, 20: 743-760.

Anastasiadou, C. and I. D. Leonardos. 2010. Karyological analysis of the freshwater shrimp *Atyaephyra desmarestii* (Decapoda: Caridea: Atyidae). J Crust Biol, 30: 332-335.

Anders, F. 1957. Über die Geschlechts bei Einflussende Wirkung von Farbellen bei *Gammarus pulex spp subterranus* (Schneider). Z Indukt Abstanm Vererbungsl, 88: 291-332.

Andersen, H. R., B. Halling-Sorensen and K. O. Kusk. 1999. A parameter for detecting estrogenic exposure in the copepod *Acartia tonsa*. Ecotoxicol Environ Saf, 44: 56-61.

Andersen, H. R., L. Wollenberger, B. Halling-Sorensen and K. O. Kusk. 2001. Development of copepod nauplii to copepodites — A parameter for chronic toxicity including endocrine disruption. Environ Toxicol Chem, 20: 2821-2829.

Anderson, G. 1977. The effects of parasitism on energy flow through laboratory shrimp populations. Mar Biol, 42: 239-251.

Anderson, G. 1990. Post-infection mortality of *Palaemonetes* spp. (Decapoda: Palaemonidae) following experimental exposure to the bopyrid isopod *Probopyrus pandalicola* (Packard) (Isopoda: Epicaridea). J Crust Biol, 10: 284-292.

Andersson, M. B. 1994. *Sexual Selection.* Princeton University Press, Princeton.

Anger, K. 2013. Neotropical *Macrobrachium* (Caridea: Palaemonidae): On the biology, origin, and radiation of freshwater-invading shrimp. J Crust Biol, 33: 151-183.

Anker, A., J. A. Baeza and S. De Grave. 2009. A new species of *Lysmata* (Crustacea, Decapoda, Hippolytidae) from the Pacific Coast of Panama, with observations of its reproductive biology. Zool Stud, 48: 682-692.

Anon. 1993. Natural history of the Monica Lake brine shrimp. J Nat His Monica Lake brine shrimp. Appendix J. 1-8.

Aravind, N. P., P. Sheeba, K. K. C. Nair and C. T. Achuthankutty. 2007. Life history and population dynamics of an estuarine amphipod, *Eriopisa chilkensis* Chilton (Gammaridae). Estu Coast Shelf Sci, 74: 87-95.

Arcos, G. F., A. M. Ibarra, C. Vazquez-Boucard et al. 2003. Haemolymph metabolic variables in relation to eyestalk ablation and gonad development of Pacific white shrimp *Litopenaeus vannamei* Boone. Aquac Res, 34: 749-755.

Arcos, G. F., I. S. Racotta and A. M. Ibarra. 2004. Genetic parameter estimates for reproductive traits and egg composition in Pacific white shrimp *Penaeus (Litopenaeus) vannamei*. Aquaculture, 236: 151-165.

Ariani, A. P. and K. J. Wittmann. 2010. Feeding, reproduction, and development of the subterranean peracarid shrimp *Spelaeomysis bottazzii* (Lepidomysidae) from a brackish well in Apulia (southeastern Italy). J Crust Biol, 30: 384-392.

Atashbar, B., N. Agh, L. Beladjal et al. 2012. Effects of temperature on survival, growth, reproductive and life span characteristics of *Branchinecta orientalis* GO Sars, 1901 (Branchiopoda, Anostraca) from Iran. Crustaceana, 85: 1099-1114.

Atienza, D., E. Saiz and A. Skovgaard. 2008. Life history and population dynamics of the marine cladoceran *Penilia avirostris* (Branchiopoda: Cladocera) in the Catalan Sea (NW Mediterranean). J Plank Res, 30: 345-357.

Attrill, M. J. 1989. A rhizocephalan (Crustacea; Cirripedia) infestation of the deep-sea galatheid *Munida sarsi* (Crustacea; Decapoda), the effects on the host and the influence of depth upon the host-parasite relationship. J Zool, 217: 663-682.

Avarre, J. C., E. Lubzens and P. J. Babin. 2007. Apolipocrustacein, formerly vitellogenin, is the major egg yolk precursor protein in decapod crustaceans and is homologous to insect apolipophorin II/I and vertebrate apolipoprotein. BMC Evol Biol, 7: 3. Doi: 10.1186/471-2148.7.3.

Ayyar, M. E. and T. Ananthakrishnan. 2000. *A Manual of Zoology.* Viswanathan Publishers, Chennai, Vol. 2 Part II pp 479-991.

Azevedo, J. D. S., L. G. D. Silva, C. Bizerri et al. 2006. Infestation pattern and parasitic castration of the crustacean *Riggia paranensis* (Crustacea: Cymothoidea) on the fresh water fish *Cyphocharax gilbert* (Teleostei: Curimatidae). Neotropical Ichthyol, 4: 363-369.

Baer, K. N. and K. D. Owens. 1999. Evaluation of selected endocrine disrupting compounds on sex determination in *Daphnia magna* using reduced photoperiod and different feeding rates. Bull Environ Contam Toxicol, 62: 214-221.

Baeza, J. A. 2006. Testing three models on the adaptive significance of protandric simultaneous hermaphroditism in a marine shrimp. Evolution, 60: 1840-1850.

Baeza, J. A. 2007a. Male mating opportunities affect sex allocation in a protandric-simultaneous hermaphroditic shrimp. Behav Ecol Sociobiol, 61: 365-370.

Baeza, J. A. 2007b. Sex allocation in a simultaneously hermaphroditic marine shrimp. Evolution, 61: 2360-2373.

Baeza, J. A. 2009. Protandric simultaneous hermaphroditism is a conserved trait in *Lysmata* (Caridea: Lysmatidae): Implications for the evolution of hermaphroditism in the genus. Smith Contrib Mar Sci, 38: 95-110.

Baeza, J. A. and R. T. Bauer. 2004. Experimental test of socially mediated sex change in a protandric simultaneous hermaphrodite, the marine shrimp *Lysmata wurdemanni* (Caridea: Hippolytidae). Behav Ecol Sociobiol, 55: 544-550.

Baeza, J. A. and A. Anker. 2008. *Lysmata* Hochi N. Sp., a new hermaphroditic shrimp from the Southwestern Caribbean Sea (Caridea: Hippolytidae). J Crust Biol, 28: 148-155.

Baeza, J. A., J. M. Reitz and R. Collin. 2007. Protandric simultaneous hermaphroditism and sex ratio in *Lysmata nayaritensis* Wicksten, 2000 (Decapoda: Caridea). J Nat Hist, 41: 2843-2850.

Baeza, J. A., A. A. Braga, L. S. López-Greco et al. 2010. Population dynamics, sex ratio and size at sex change in a protandric simultaneous hermaphrodite, the spiny shrimp *Exhippolysmata oplophoroides*. Mar Biol, 157: 2643-2653.

Baeza, J. A., M. Furlan, A. C. Almeida et al. 2013. Population dynamics and reproductive traits of the ornamental crab *Porcellana sayana*: implications for fishery management and aquaculture. Sex Early Dev Aquat Org, 1: 1-12.

Baeza, J. A., D. C. Behringer, R. J. Hart et al. 2014. Reproductive biology of the marine ornamental shrimp *Lysmata boggessi* in the south-eastern Gulf of Mexico. J Mar Biol Ass UK, 94: 141-149.

Baguna, J. A. U., M. E. E. Saló and C. Auladell. 1989. Regeneration and pattern formation in planarians. 3. That neoblasts are totipotent stem cells and the cells. Development, 107: 77-86.

Bailie, D. A., S. Fitzpatrick, M. Connolly et al. 2014. Genetic assessment of parentage in the caridean rock shrimp *Rhynchocinetes typus* based on microsatellite markers. J Crust Biol, 34: 658-662.

Balasundaram, C. 1980. Ecophysiological studies in prawn culture (*Macrobrachium nobilii*). Ph.D. Thesis, Madurai Kamaraj University, Madurai.

Balasundaram, C. and T. J. Pandian. 1980. Contribution to the reproductive biology and aquaculture of *Macrobrachium nobilii*. Proc Symp Invert Reprod, Madras University, Madras. 1L: 183-193.

Balasundaram, C. and T. J. Pandian. 1981. *In vitro* culture of *Macrobrachium* eggs. Hydrobiologia, 77: 203-208.

Balasundaram, C. and T. J. Pandian. 1982. Egg loss during incubation in *Macrobrachium nobilii* (Henderson & Mathai). J Exp Mar Biol Ecol, 59: 289-299.

Baldwin, A. P. and R. T. Bauer. 2003. Growth, survivorship, life-span, and sex change in the hermaphroditic shrimp *Lysmata wurdemanni* (Decapoda: Caridea: Hippolytidae). Mar Biol, 143: 157-166.

Baldwin, W. S. and G. A. LeBlanc. 1994. Identification of multiple steroid hydroxylases in *Daphnia magna* and their modulation by xenobiotics. Environ Toxicol Chem, 13: 1013-1021.

Baldwin, W. S., S. E. Graham, D. Shea and G. A. LeBlanc. 1998. Altered metabolic elimination of testosterone and associated toxicity following exposure of *Daphnia magnato* nonylphenol polyethoxylate. Ecotoxicol Environ Saf, 39: 104-111.

Bama, A. and K. Atlaff. 2004. Male reproductive system of the calanoid copepod *Allodiaptomus (Redudiaptomus) raoi* Kiefer. 1936. J Aquat Biol, 19: 37-41.

Banta, A. M. 1939. Studies on the physiology, genetics and evolution of some Cladocera. Carnegie Institution Carnegie, Washington, D. C: 182-200.

Banta, A. M. and L. A. Brown. 1929a. Control of sex in Cladocera. I. Crowding the mothers as a means of controlling male production. Physiol Zool, 2: 80-92.

Banta, A. M. and L. A. Brown. 1929b. Control of sex in Cladocera. II. The unstable nature of the excretory products involved in male production. Physiol Zool, 2: 93-98.

Banta, A. M. and L. A. Brown. 1929c. Control of sex in Cladocera. III. Localization of the critical period for control of sex. Proc Natl Acad Sci USA, 15: 71.

Banzai, K., S. Izumi and T. Ohira. 2102. Molecular cloning and expression analysis of cDNAs encoding an insulin-like androgenic gland factor from three palaemonid species *Macrobrachium lar, Palaemon paucidens* and *P. pacificus*. JARQ, 46: 105-114.

Barata, C., M. Medina, T. Telfer and D. J. Baird. 2002. Determining demographic effects of cypermethrin in the marine copepod *Acartia tonsa*: stage-specific short tests versus life-table tests. Archiv Environ Contam Toxicol, 43: 0373-0378.

Barki, A., I. Karplus, I. Khalaila et al. 2003. Male-like behavioral patterns and physiological alterations induced by androgenic gland implantation in female crayfish. J Exp Biol, 206: 1791-1797.

Barnes, H. and M. Barnes. 1968. Egg numbers, metabolic efficiency of egg production and fecundity; local and regional variations in a number of common cirripedes. J Exp Mar Biol Ecol, 2: 135-153.

Barnes, M. 1989. Egg production in cirripedes. Oceanogr Mar Biol Annu Rev, 27: 91-166.

Barría, E. M. and M. I. González. 2008. Effect of autotomy and regeneration of the chelipeds on growth and development in *Petrolisthes laevigatus* (Guérin, 1835) (Decapoda, Anomura, Porcellanidae), Crustaceana, 81: 641-652.

Bas, C. C. and E. D. Spivak. 2000. Effect of salinity on embryos of two southwestern Atlantic estuarine grapsid crab species cultured *in vitro*. J Crust Biol, 20: 647-656.

Battaglia, B. 1961. Rapporti tra geni per la pigmentazione e la sessualita in *Tisbe reticulata*. Atti Accad Gioenia Sci Nat Catina, 6: 439-447.

Battaglia, B. and B. Volkmann-Rocco. 1973. Geographic and reproductive isolation in the marine harpacticoid copepod *Tisbe*. Mar Biol, 19: 156-160.

Baudoin, M. 1975. Host castration as a parasitic strategy. Evolution, 29: 335-352.

Bauer, H. 1940. Über die Chromosomen der Bisexuellen und der Parthenogenetischen Rasse des Ostracoden *Heterocypris incongruens* Ramd. Chromosoma, 1: 620-637.

Bauer, R. T. 1979. Antifouling adaptations of marine shrimp (Decapoda: Caridea): gill cleaning mechanisms and grooming of brooded embryos. Zool J Linn Soc, 65: 281-303.

Bauer, R. T. 1986. Sex change and life history pattern in the shrimp *Thor manningi* (Decapoda: Caridea): A novel case of partial protandric hermaphroditism. Biol Bull, 170: 11-31.

Bauer, R. T. 2000. Simultaneous hermaphroditism in caridean shrimps: A unique and puzzling sexual system in the decapoda. J Crust Biol, 20: 116-128.

Bauer, R. T. and W. A. Newman. 2004. Protrandric simultaneous hermaphroditism in the marine shrimp *Lysmata californica* (Caridae: Hipplytidae). J Crust Biol, 24: 131-139.

Beck, J. T. 1980. The effects of an isopod castrator, *Probopyrus pandalicola* on the sex characters of one of its caridean shrimp hosts, *Palaemonetes paludosus*. Biol Bull, 158: 1-15.

Béguer, M., J. Bergé, M. Girardin and P. Boët. 2010. Reproductive biology of *Palaemon longirostris* (Decapoda: Palaemonidae) from Gironde estuary (France), with a comparison with other European populations. J Crust Biol, 30: 175-185.

Begum, F., I. Nakatani, S. Tamotsu and T. Goto. 2010. Reproductive characteristics of the albino morph of the crayfish, *Procambarus clarkii* (Girard, 1852) (Decapoda, Cambaridae). Crustaceana, 83: 169-178.

Beladjal, L., T. T. M. Vandekerckhove, B. Muyssen et al. 2002. B-chromosomes and male-biased sex ratio with paternal inheritance in the fairy shrimp *Branchipus schaefferi* (Crustacea, Anostraca). Heredity, 88: 356-360.

Beladjal, L., N. Peiren, T. T. Vandekerckhove and J. Mertens. 2003a. Different life histories of the co-occurring fairy shrimps *Branchipus schaefferi* and *Streptocephalus torvicornis* (Anostraca). J Crust Biol, 23: 300-307.

Beladjal, L., E. M. Khattabi and J. Mertens. 2003b. Life history of *Tanymastigites perrieri* (Anostraca) in relation to temperature. Crustaceana, 76: 135-147.

Belanger, R. M. and P. A. Moore. 2013. A comparative analysis of setae on the pereiopods of reproductive male and female *Orconectes rusticus* (Decapoda: Astacidae). J Crust Biol, 33: 309-316.

Belanger, R., X. Ren, K. McDowell et al. 2008. Sensory setae on the major chelae of male crayfish, *Orconectes rusticus* (Decapoda, Astacidae): impact of reproductive state on function and distribution. J Crust Biol, 28: 27-36.

Bell, G. 1982. *The Masterpiece of Nature: The Evolution and Genetics of Sexuality*. University of California Press, Berkeley, CA.

Benazzi, M. and G. Benazzi-Lentati. 1993. Platyhelminthes-Turbellaria. In: *Reproduction Biology of Invertebrates*. (eds) K. G. Adiyodi and R. G. Adiyodi. Oxford and IBH Publishers, New Delhi, Vol 6, Part A. pp 107-141.

Benvenuto, C., A. Calabrese, S. K. Reed et al. 2009. Multiple hatching events in clam shrimp: Implications for mate guarding behaviour and community ecology. Curr Sci, 96: 130-136.

Benzie, J. A. H., M. Kenway, E. Ballment et al. 1995. Interspecific hybridization of the tiger prawns *Penaeus monodon* and *Penaeus esculentus*. Aquaculture, 133: 103-111.

Bernice, R. 1972. Ecological studies on *Streptocephalus dichotomus* Baird. Hydrobiologia, 39: 217-240.

Billinghurst, Z., A. S. Clare and M. H. Depledge. 2001. Effects of 4-n-nonylphenol and 17β-oestradiol on early development of the barnacle *Elminius modestus*. J Exp Mar Biol Ecol, 257: 255-268.

Biswas, A., R. K. Saha, A. Sengupta and H. Saha. 2014. Life cycle of a new bosminid Cladocera: *Bosmina (Bosmina) tripurae* (Korinek, Saha, and Bhattacharya, 1999). Proc Natl Acad Sci, Ind, 84B: 953-960.

Blower, S. M. and J. Roughgarden. 1988. Parasitic castration: host species preferences, size-selectivity and spatial heterogeneity. Oecologia, 75: 512-515.

Blueweiss, L., H. Fox, V. Kudzma et al. 1978. Relationships between body size and some life history parameters. Oecologia, 37: 257-272.

Boddeke, R., J. R. Bosschiter and P. G. Goudswaard. 1991. Sex change, mating and sperm transfer in *Crangon crangon* (L.). In: *Crustacean Sexual Biology*. (eds) R. T. Bauer and J. W. Martin. Columbia University Press, New York. pp 164-182.

Bode, S. N. S., S. Adolfsson, D. K. Lamatsh et al. 2010. Exceptional cryptic diversity and multiple origins of parthenogenesis in a freshwater ostracod. Mol Phylogenet Evol, 54: 542-552.

Boore, J. L. and W. M. Brown. 1995. Complete DNA sequence of the mitochondrial genome of the annelid worm, *Lumbricus terrestris*. Genetics, 141: 305-319.

Borok, Z., C. Li, J. Liebler et al. 2006. Developmental pathways and specification of intrapulmonary stem cells. Pediat Res, 59: 84R-93R.

Borowsky, B. 1983. Placement of eggs in their brood pouches by females of the amphipod Crustacea *Gammarus palustris* and *Gammarus mucronatus*. Mar Freshwat Behav Physiol, 9: 319-325.

Borowsky, B., P. Aitken-Ander and J. T. Tanacredi. 1997. Changes in reproductive morphology and physiology observed in the amphipod crustacean, *Melita nitida* Smith, maintained in the laboratory on polluted estuarine sediments. J Exp Mar Biol Ecol, 214: 85-95.

Borst, D. W. and B. Tsukimura. 1992. Methyl farnesoate levels in crustaceans. In: *Insect Juvenile Hormone Research*. (eds) B. Mauchamp, F. Couillaud and J. C. Baehr. INRA Editions, Paris. pp 27-35.

Borst, D. W., H. Laufer, M. Landau et al. 1987. Methyl farnesoate and its role in crustacean reproduction and development. Insect Biochem, 17: 1123-1127.

Borst, D. W., J. Ogan, B. Tsukimura et al. 2001. Regulation of the crustacean mandibular organ. Am Zool, 41: 430-441.

Borst, D. W., G. Wainwright and H. H. Rees. 2002. *In vivo* regulation of the mandibular organ in the edible crab, *Cancer pagurus*. Proc R Soc Lond, 269B: 483-490.

Bosch, T. C. G. and C. N. David. 1987. Stem cells of *Hydra magnipapillata* can differentiate into somatic and germ line cells. Dev Biol, 121: 182-191.

Bottrell, H. H. 1975. Generation time, length of life, instar duration and frequency of moulting, and their relationship to temperature in eight species of Cladocera from the River Thames, Reading. Oecologia, 19: 129-140.

Bouchon, D., R. Cordaux and P. Greve. 2008. Feminizing *Wolbachia* and the evolution of sex determination in the isopods. In: *Insect Symbiosis*. (eds) K. Bourtzis and T. Miller. Taylor and Francis, CRC Press, Boca Raton, pp 273-294.

Bourdon, R. 1968. Les Bopyridae des mers européennes. Mem Mus Natl Hist Nat, Paris, 50A: 387-405.

Bowman, T. E. and L. S. Kornicker. 1967. Two new crustaceans. The parasitic copepod *Sphaeropsis monothrix* (Choniostomaditae) and its myodocopid ostracod host *Parasterope pollex* (Cylindroleberidae) from the southern New England coast. Proc US Natl Mus, 123: 1-28.

Boxshall, G. A. and R. Huys. 1989. New tantulocarid, *Stygotantulus stocki*, parasitic on harpacticoid copepods, with an analysis of the phylogenetic relationships within the Maxillopoda. J Crust Biol, 9: 126-140.

Boxshall, G. A., R. J. Lester, G. Grygier et al. 2005. Crustacean parasites. In: *Marine Parasitology*. (ed) K. Rohde. University of New England, CSIRO Publishers, Victoria. pp 123-169.

Brantner, J. S., D. W. Ott, R. J. Duff et al. 2013a. Androdioecy and hermaphroditism in five species of clam shrimps (Crustacea: Branchiopoda: Spinicaudata) from India and Thailand. Invert Biol, 132: 27-37.

Brantner, J. S., D. W. Ott, R. J. Duff et al. 2013b. Evidence of selfing hermaphroditism in the clam shrimp *Cyzicus gynecia* (Branchiopoda: Spinicaudata). J Crust Biol, 33: 184-190.

Breitholtz, M. and B. E. Bengtsson. 2001. Oestrogens have no hormonal effect on the development and reproduction of the harpacticoid copepod *Nitocra spinipes*. Mar Pollut Bull, 42: 879-886.

Breitholtz, M. and L. Wollenberger. 2003. Effects of three PBDEs on development, reproduction and population growth rate of the harpacticoid copepod *Nitocra spinipes*. Aquat Toxicol, 64: 85-96.

Brendonck, L. 1996. Diapause, quiescence, hatching requirements: what we can learn from large freshwater branchiopods (Crustacea: Branchiopoda: Anostraca, Notostraca, Conchostraca). Hydrobiologia, 320: 85-97.

Brendonck, L., M. D. Centeno and G. Persoone. 1996. The influence of processing and temperature conditions on hatching of resting eggs of *Streptocephalus proboscideus* (Crustacea: Branchiopoda: Anostraca). Hydrobiologia, 320: 99-105.

Brock, J. A. and D. V. Lightner. 1990. Diseases of crustacea. Diseases caused by microorganisms. In: *Diseases of Marine Animals*. (ed) O. Kinne. Biologische Anstalt Helgoland, Hamburg, 3: 245-349.

Brook, H. J., T. A. Rawlings and R. W. Davies. 1994. Protogynous sex change in the intertidal isopod *Gnorimosphaeroma oregonense* (Crustacea: Isopoda). Biol Bull, 187: 99-111.

Brooks, B. W., P. K. Turner and J. K. Stanley. 2003. Water bourne and sediment toxicity of fluoxetine to select organisms. Chemosphere, 52: 135-142.

Brown, R. J., M. Conradi and M. H. Depledge. 1999. Long-term exposure to 4–nonylphenol affects sexual differentiation and growth of the amphipod *Corophium volutator* (Pallas, 1766). Sci Total Environ, 233: 77-88.

Browne, R. A. 1980. Reproductive pattern and mode in the brine shrimp. Ecology, 61: 466-470.

Browne, R. A. 1982. The costs of reproduction in the brine shrimp. Ecology, 63: 43-47.

Browne, R. A. 1992. Population genetics and ecology of *Artemia*: insights into parthenogenetic reproduction. Trends Ecol Evol, 7: 232-237.

Browne, R. A. and G. Wanigasekara. 2000. Combined effects of salinity and temperature on survival and reproduction of five species of *Artemia*. J Exp Mar Biol Ecol, 244: 29-44.

Browne, R. A., S. E. Sallee, D. S. Grosch et al. 1984. Partitioning genetic and environmental components of reproduction and lifespan in *Artemia*. Ecology, 949-960.

Browne, R. A., L. E. Davies and S. E. Sallee. 1988. Effects of temperature and relative fitness of sexual and asexual brine shrimp *Artemia*. J Exp Mar Biol Ecol, 124: 1-20.

Buhl-Mortensen, L. and J. T. Hoeg. 2013. Reproductive strategy of two deep-sea scalpellid barnacles (Crustacea: Cirripedia: Thoracica) associated with decapods and pyconogonids and the first description of a penis in scalpellids dwarf males. Org Divers Evol, 13: 545-557. Doi 10.1007/s13127-013-0137-3.

Buikema Jr, A. L. 1975. Some effects of light on the energetics of *Daphnia pulex* and implications for the significance of vertical migration. Hydrobiologia, 47: 43-58.

Bulnheim, H. P. 1975. Microsporidian infections of amphipods with special reference to host-parasite relationships: a review. Mar. Fish. Rev, 37: 39-45.

Buřič, M., M. Hulák, A. Kouba et al. 2011. A successful crayfish invader is capable of facultative parthenogenesis: a novel reproductive mode in decapod crustaceans. PLoS One, 6: e20281, doi: 10.1371/journal.pone.0020281.

Burnett, A. L., L. E. Davis and R. E. Ruffing. 1966. A histological and ultra-structural study of germinal differentiation of interstitial cells using from gland cells in *Hydra viridis*. J Morphol, 120: 1-9.

Butler, T. H. 1964. Growth, reproduction, and distribution of pandalid shrimps in British Columbia. J Fisher Res Board Can, 21: 1403-1452.

Butlin, R. 2002. Evolution of sex: the cost and benefits of new insights from old sexual lineage. Nat Rev Genet, 3: 311-317.

Cáceres, C. E. 1998. Interspecific variation in the abundance, production and emergence of *Daphnia* diapausing eggs. Ecology, 79: 1699-1710.

Cáceres, C. E and D. A. Soluk. 2002. Blowing in the wind: a field test of overland dispersal and colonization by aquatic invertebrates. Oecologia, 131: 402-408.

Cáceres, C. E and A. J. Tessier. 2003. How long to rest: the ecology of optimal dormancy and environmental constraint. Ecology, 84: 1189-1198.

Cáceres, C. E and A. J. Tessier. 2004a. Incidence of diapause varies among populations of *Daphnia pulicaria*. Oecologia, 141: 425-431.

Cáceres, C. E. and A. J. Tessier. 2004b. To sink or swim: variable diapause strategies among *Daphnia* species. Limnol Oceanogr, 49: 1333-1340.

Caceres, C. E., A. N. Christoff and W. J. Boeing. 2007. Variation in ephippial buoyancy in *Daphnia pulicaria*. Freshwater Biol, 52: 313-318.

Cáceres, C. E., C. J. Knight and S. R. Hall. 2009. Predator-spreaders: predation can enhance parasite success in a planktonic host-parasite system. Ecology, 90: 2850-2858.

Calado, R., C. Bartilotti and L. Narciso. 2005. Short report on the effect of a parasitic isopod on the reproductive performance of a shrimp. J Exp Mar Biol Ecol, 321: 13-18.

Calado, R., A. Vitorino and M. T. Dinis. 2006. Bopyrid isopods do not castrate the simultaneously hermaphroditic shrimp *Lysmata amboinensis* (Decapoda: Hippolytidae). Dis Aquat Org, 73: 73-76.

Calado, R., C. Bartilotti, J. Goy and M. T. Dinis. 2008. Parasitic castration of the stenopodid shrimp *Stenopus hispidus* (Decapoda: Stenopodidae) induced by the bopyrid isopod *Argeiopsis inhacae* (Isopoda: Bopyridae). J Mar Biol Asso UK, 88: 307-309.

Callaghan, T. R., B. M. Degnan and M. J. Sellars. 2010. Expression of sex and reproduction-related genes in *Marsupenaeus japonicus*. Mar Biotechnol, 12: 664-677.

Callan, H. G. 1940. The effects of castration by parasites and X-rays on the secondary sex characters of prawns (*Leander* spp.). J Exp Biol, 17: 168-179.

Campos, E. O., D. Vilhena and L. Caldwell. 2012. Pleopod rowing is used to achieve high forward swimming speeds during the escape response of *Odontodactylus havanensis* (Stomatopoda). J Crust Biol, 32: 171-179.

Campos-Ramos, R. 1997. Chromosome studies on the marine shrimps *Penaeus vannamei* and *P. californiensis* (Decapoda). J Crust Biol, 17: 666-673.

Campos-Ramos, R., R. Garza-Torres, D. A. Guerrero-Tortolero et al. 2006. Environmental sex determination, external sex differentiation and structure of the androgenic gland in the Pacific white shrimp *Litopenaeus vannamei* (Boone). Aquac Res, 37: 1583-1593.

Campos-Ramos, R., H. Obregon-Barboza and A. M. Maeda-Martinez. 2009. Species representation and gender proportion from mixed *Artemia franciscana* and *A. parthenogenetica* (Anostraca) commercial cysts over a wide range of temperatures. Curr Sci, 96: 111-113.

Carbonell, A., A. Grau, V. Lauronce and C. Gómez. 2006. Ovary development of the red shrimp, *Aristeus antennatus* (Risso, 1816) from the northwestern Mediterranean Sea. Crustaceana, 79: 727-743.

Carius, H. J., T. J. Little and D. Ebert. 2001. Genetic variation in a host-parasite association: potential for coevolution and frequency-dependent selection. Evolution, 55: 1136-1145.

Carlisle, D. B. and F. G. W. Knowles. 1959. *Endocrine Control in Crustaceans*. Cambridge University Press, Cambridge, Vol 10.

Carotenuto, Y., A. Ianora, M. Di Pinto et al. 2006. Annual cycle of early developmental stage survival and recruitment in the copepods *Temora stylifera* and *Centropages typicus*. Mar Ecol Prog Ser, 314: 227-238.

Carr, S. D., A. T. Richard, J. L. Hench et al. 2004. Movement patterns and trajectories of ovigerous blue crab *Callinectes sapidus* during the spawning migration. Estu Coast Shelf Sci, 60: 567-570.

Cartes, J. E. and F. Sardà. 1992. Abundance and diversity of decapod crustaceans in the deep-Catalan Sea (Western Mediterranean). J Natl Hist, 26: 1305-1323.

Cattley, J. G. 1948. Sex reversal in copepods. Nature, 161: 937.

Chakraborthy, C. and G. Agoramoorthy. 2012. Stem cells in the light of evolution. Ind J Med Res, 135: 813-819.

Chamberlain, G. W. and A. L. Lawrence. 1981. Effect of light intensity and male and female eyestalk ablation on reproduction of *Penaeus stylirostris* and *P. vannamei*. J World Maricult Soc, 12: 357-372.

Chan, S. M., X. G. Chen and P. L. Gu. 1998. PCR cloning and expression of the molt-inhibiting hormone gene for the crab (*Charybdis feriatus*). Gene, 224: 23-33.

Chang, E. S. 1985. Hormonal control of molting in decapod crustacea. Integ Comp Biol, 25: 179-185.

Chaparro, O. R., C. L. Saldivia and K. A. Paschke 2001. Regulatory aspects of the brood capacity of *Crepidula fecunda*, Gallardo 1979 (Gastropoda: Calyptraeidae). J Exp Mar Biol Ecol, 266: 97-108.

Chaplin, J. A., J. E. Havel and P. D. Herbert. 1994. Sex and ostracods. Trend Ecol Evol, 9: 435-439.

Chapman, A. D. 2009. Numbers of living species in Australia and the world. Report for the Australian Biological Resource Study, Department of Environment, Government of Australia, p 80.

Charmantier, G. and J. P. Trilles. 1979. La degenerescence de l'organe Y chez *Sphaeroma serratum* (Fabricius, 1787) (Isopoda, Flabellifera): etude ultrastructurale. Crustaceana, 36: 29-38.

Charmantier, G., M. Charmantier-Daures and D. E. Aiken. 1988. Larval development and metamorphosis of the American lobster *Homarus americanus* (Crustacea, Decapoda): effect of eyestalk ablation and juvenile hormone injection. Gen Comp Endocrinol, 70: 319-333.

Charniaux-Cotton, H. 1953. Etude du determinisme des caracteres sexuels secondaires par castration chirurgicale et implantation dovaire chez un crustace amphipode (*Orchestia-gammarella*). C R Hebd Scan Acad Sci, 236: 141-143.

Charniaux-Cotton, H. 1954. Decouverte chez un crustace amphipode (*Orchestia-gammarella*) dune glande endocrine responsable de la differenciation de caracteres sexuels primaires et secondaires males. C R Hebd Scan Acad Sci, 239: 780-782.

Charniaux-Cotton, H. 1958. The androgenic gland of some decapodal crustaceans and particularly *Lysmata seticauda,* a species with functional protandrous hermaphroditism. C R Hebd Scan Acad Sci, 246: 2814-2817.

Charniaux-Cotton, H. 1959. Masculinisation des femelles de la crevette a hermaphrodisme proterandrique *Lysmata seticauda,* par greffe de glandes androgenes-interpretation de lhermaphrodisme chez les decapodes. C R Hebd Scan Acad Sci, 249: 1580-1582.

Charniaux-Cotton, H. 1960. Sex determination. In: *The Physiology of Crustacea. 1. Metabolism and Growth.* (ed) H. W. Talbot. Academic Press, New York. pp 441-447.

Charniaux-Cotton, H. 1961. Physiologie de Inversion sexuelle chez la (Crevette a hermaphrodisme proterandrique) chez les individus normauset les femelles masculinisees. C R Hebd Scan Acad Sci, 252: 199-201.

Charniaux-Cotton, H. 1965. Hormonal control of sex differentiation in invertebrates. In: *Organogenesis.* (eds) R. De Hann and H. Urspring. Holt, New York. pp 701-740.

Charniaux-Cotton, H. 1970. Hermaphroditism and gynandromorphism in mala-costracan Crustacea. In: *Intersexuality in the Animal Kingdom.* (ed) R. Reinboth. Springer Verlag, Berlin. pp 91-105.

Charnov, E. L. 1982. *The Theory of Sex Allocation.* Princeton University Press, Princeton.

Charnov, E. L., J. M. Smith and J. J. Bull. 1976. Why be an hermaphrodite? Nature, 263: 125-126.

Chaves, A. R. 2000. Effect of x-organ sinus gland extract on (35S) methionine incorporation to the ovary of the red swamp crayfish, *Procambarus clarkii.* Comp Biochem Physiol A Mol Integr Physiol 126: 407-413.

Chaves, A. R. 2001. Effects of sinus gland extract on mandibular organ size and methyl farnesoate synthesis in the crayfish. Comp Biochem Physiol, 128: 327-333.

Chen, T., T. S. Villeneuve, K. A. Garant et al. 2007. Functional characterization of artemin, a ferritin homolog synthesized in *Artemia* embryos during encystment and diapause. FEBS J, 274: 1093-1101.

Chittleborough, R. G. 1976. Breeding of *Panulirus longipes cygnus* George under natural and controlled conditions. Mar Freshwat Res, 27: 499-516.

Choy, S. C. 1987. Growth and reproduction of eyestalk ablated *Penaeus canaliculatus* (Olivier, 1811) (Crustacea: Penaeidae). J Exp Mar Biol Ecol, 112: 93-107.

Christy, J. H. 2011. Timing of hatching and release of larvae by brachyuran crabs: patterns, adaptive significance and control. Integ Comp Biol, 51: 62-72.

Chu, K. H., C. K. Wong and K. C. Chiu. 1997. Effects of the insect growth regulator (S)-methoprene on survival and reproduction of the freshwater cladoceran *Moina macrocopa*. Environ Pollut, 96: 173-178.

Clark, M. S., N. Y. Denekamp, M. A. Thorne et al. 2012. Long-term survival of hydrated resting eggs from *Brachionus plicatilis*. PLoS One, 7(1): e29365.

Clegg, J. S. 1962. Free glycerol in dormant cysts of the brine shrimp *Artemia salina*, and its disappearance during development. Biol Bull, 123: 295-301.

Clegg, J. S. 1965. The origin of threhalose and its significance during the formation of encysted dormant embryos of *Artmia salina*. Comp Biochem Physiol, 14: 135-143.

Clegg, J. S. 1974. Biochemical adaptations associated with the embryonic dormancy of *Artemia salina*. Trans Am Microsc Soc, 93: 481-490.

Clegg, J. S. 1978. Hydration-dependent metabolic transitions and the state of cellular water in *Artemia* cysts. In: *Dry Biological Systems*. (eds) J. Crowe and J. S. Clegg. Academic Press, New York. pp 117-153.

Clegg, J. S. 1997. Embryos of *Artemia franciscana* survive four years of continuous anoxia: the case for complete metabolic rate depression. J Exp Biol, 200: 467-475.

Clegg, J. S. and S. A. Jackson. 1997. Significance of cyst fragments of *Artemia* sp. recovered from a 27, 000 year old core taken under the Great Salt Lake, Utah, USA. Int J Salt Lake Res, 6: 207-216.

Clegg, J. S., J. K. Willsie and S. A. Jackson. 1999. Adaptive significance of a small heat shock/α-crystallin protein (p 26) in encysted embryos of the brine shrimp, *Artemia franciscana*. Am Zool, 39: 836-847.

CMFRI. 2013. *Handbook of Marine Prawns of India*. Central Marine Fisheries Research Institute, Kochi. p 414.

Cobos, V., V. Diaz, J. E. Garcia-Raso and M. E. Manjòn-Cabeza. 2005. Insights on the female reproductive system in *Hippolyte inermis* (Decapoda, Caridea): is this species really hermaphroditic? Invert Biol, 124: 310-320.

Coccia, E., E. De Lisa, C. Di Cristo et al. 2010. Effects of estradiol and progesterone on the reproduction of the freshwater crayfish *Cherax albidus*. Biol Bull, 218: 36-47.

Cohen, A. C. and J. G. Morin. 1990. Patterns of reproduction in ostracodes: a review. J Crust Biol, 10: 184-211.

Cohen, F. P., B. F. Takano, R. M. Shimizu and S. L. Bueno. 2011. Life cycle and population structure of *Aegla paulensis* (Decapoda: Anomura: Aeglidae). J Crust Biol, 31: 389-395.

Colbourne, J. K., T. J. Crease, L. J. Weider et al. 1998. Phylogenetics and evolution of a circumartic species complex (Cladocera: *Daphnia pulex*). J Biol Linn Soc, 65: 347-365.

Colbourne, J. K., V. Singan and D. Gilbert. 2005. wFleaBase: the *Daphnia* genome database. BMC, Bioinformatics, 6: 45.

Colbourne, J. K., C. C. Wilson and P. D. N. Hebert. 2006. The systematics of Australian *Daphnia* and *Daphniopsis* (Crustacea: Cladocera): a shared phylogenetic history transformed by habitat-specific rates of evolution. Biol J Linn Soc, 89: 469-488.

Colbourne, J. K., M. E. Pfrender, D. Gilbert et al. 2011. The ecoresponsive genome of *Daphnia pulex*. Science, 331: 555-561.

Coman, F. E., M. J. Sellars, B. J. Norris et al. 2008. The effects of triploidy on *Penaeus (Marsupenaeus) japonicus* (Bate) survival, growth and gender when compared to diploid siblings. Aquaculture, 276: 50-59.

Conklin, E. G. 1906. Does half an ascidian egg gives rise to a whole larva? Arch Entwicklungsmech, 2: 727-753.

Cordaux, R., D. Bouchon and P. Greve. 2011. The impact of endosymbionts on the evolution of host sex-determination mechanism. Trends Genet 27: 332-341.

Corley, L. S., M. A. White and M. R. Strand. 2005. Both endogenous and environmental factors affect embryo proliferation in the polyembryonic wasp *Copidosoma floridanism*. Evol Dev, 3: 432-442.

Correa, C. and M. Thiel. 2003. Mating systems in caridean shrimp (Decapoda: Caridea) and their evolutionary consequences for sexual dimorphism and reproductive biology. Revista Chile Hist Nat, 76: 187-203.

Costa-Souza, A. C., S. S. da Rocha, L. E. A. Bezerra and A. O. Almeida. 2014. Breeding and heterosexual pairing in the snapping shrimp *Alpheus estuariensis* (Caridea: Alpheidae) in a tropical bay in northeastern Brazil. J Crust Biol, 34: 593-603.

Costlow, J. D. 1963. The effect of eyestalk extirpation on metamorphosis of megalopa of the blue crab, *Callinects sapidus* Rathbun. Gen Comp Endocrinol, 3: 120-130.

Costlow, J. D. 1966. The effect of eyestalk extirpation on larval development of the crab, *Sesarma reticulatum* Say. In: *Some Contemporary Studies in Marine Science.* (ed) H. Barnes. Appleton-Century-Crofts, New York. pp 209-224.

Costlow, J. D. and C. G. Bookhout. 1960. A method for developing brachyuran eggs *in vitro*. Limnol Oceanogr, 5: 212-215.

Crisp, D. J. 1983. *Chelonobia patula* (Ranzani), a pointer to the evolution of complete male. Mar Biol Lett 4: 281-294.

Crisp, D. J. and C. P. Spencer. 1958. The control of the hatching process in barnacles. Proc R Soc Lond, 149B: 278-299.

Crocos, P. J. and J. D. Kerr. 1983. Maturation and spawning of the banana prawn *Penaeus merguiensis* de Man (Crustacea: Penaeidae) in the Gulf of Carpentaria, Australia. J Exp Mar Biol Ecol, 69: 37-59.

Cronin, L. E. 1947. Anatomy and histology of the male reproductive system of *Callinectes sapidus* Rathbun. J Morph, 81: 209-239.

Crowe, J. H. 1971. Anhydrobiosis: an unsolved problem. Am Nat, 105: 563-573.

Cui, Z., H. Liu, T. S. Lo and K. H. Chu. 2005. Inhibitory effects of the androgenic gland on ovarian development in the mud crab *Scylla paramamosain*. Comp Biochem Physiol Part A: Mol Integ Physiol, 140: 343-348.

D'Agaro, E., E. A. Ferrero and P. G. Giulianini. 2010. Induction of ovarian maturation by means of dietary hormonal treatment in *Austropotamobius pallipes*. Ital J Ani Sci, 4: 583-585.

Damare, S., C. Raghukumar and S. Raghukumar. 2006. Fungi in deep-sea sediments of the Central Indian Basin. Deep Sea Res, 53: 14-27.

Damrongphol, P. and P. Jaroensastraraks. 2001. Morphology and regional distribution of the primordial germ cells in the giant freshwater prawn *Macrobrachium rosenbergii*. Sci Asia, 27: 15-19.

Damrongphol, P., N. Eangchuan and B. Poolsanguan. 1990. Simple *in vitro* culture of embryos of the giant freshwater prawn (*Macrobrachium rosenbergii*). J Sci Soc Thailand, 16: 17-24.

Dan, S., T. Kaneko, S. Takeshima et al. 2014. Eyestalk ablation affects larval morphogenesis in the swimming crab *Portunus trituberculatus* during metamorphosis into megalopae. Sex Early Dev Aquat Org, 1: 57-73.

Darwin, C. 1859. *On the Origins of Species by Means of Natural Selection.* Murray, London. p 247.

David, C. J. and T. J. Pandian. 2006. Cadaveric sperm induces intergeneric androgenesis in the fish, *Hemigrammus caudovittatus*. Theriogenology, 65: 1048-1070.

Dawidowicz, P. and Z. M. Gliwicz. 1983. Food of brook charr in extreme oligotrophic conditions of an alpine lake. Environ Biol Fish, 8: 55-60.

Decaestecker, E., L. De Meester and J. Mergeay. 2009. Cyclical parthenogenesis in *Daphnia*: sexual versus asexual reproduction. In: *Lost Sex*. (ed) I. Schon. Springer Verlag, Berlin. pp 295-315.

Deiana, A. M., E. Coluccia, A. Milia and S. Salvadori. 1996. Supernumerary chromosomes in *Nephrops norvegicus* L. (Crustacea, Decapoda). Heredity, 76: 92-99.

DeLorenzo, M. E., L. A. Taylor, S. A. Lund et al. 2002. Toxicity and bioconcentration potential of the agricultural pesticide endosulfan in phytoplankton and zooplankton. Arch Environ Contam, 42: 173-181.

Delorme, L. D. 2009. Ostracoda. In: *Ecology and Classification of North American Freshwater*. (eds) J. H. Thorp and A. P. Covich. Academic Press, San Diego, 2nd edn. pp 811-848.

Demeusy, N. 1962. Role de la glande de mue dans revolution ovarrienne du crabe *Carcinus maenas* Linne. Can Biol Mar, 3: 37-56.

Desrosiers, R. R., A. Gérard, J. M. Peignon et al. 1993. A novel method to produce triploids in bivalve molluscs by the use of 6-dimethylaminopurine. J Exp Mar Biol Ecol, 170: 29-43.

Devlin, R. H. and Y. Nagahama. 2002. Sex determination and sex differentiation in fish: an overview of genetic, physiological, and environmental influences. Aquaculture, 208: 191-364.

DeWalsche, C., N. Munuswamy and H. J. Dumont. 1991. Structural differences between the cyst walls of *Streptocephalus dichotomus* (Baird), *S. torvicornis* (Waga), and *Thamnocephalus platyurus* (Packard) (Crustacea: Anostraca), and a comparison with other genera and species. In: *Studies on Large Branchiopod Biology and Aquaculture*. Springer Verlag, Netherlands. pp 195-202.

Dharani, G. and K. Altaff. 2002. Facultative sex reversal in the freshwater plankton *Simodiaptomus* (*Rhinediaptomus*) *indicus* (Calanoida: Copepoda). Curr Sci, 82: 794-795.

Dharani, G. and K. Altaff. 2004a. Life cycle of a copepod. In: *A Manual of Zooplankton*. (ed) K. Altaff. Department of Zoology, The New College, Chennai. pp 66-75.

Dharani, G. and K. Altaff. 2004b. Ultrastructure of subitaneous and diapausing eggs of planktonic copepod *Simodiaptomus* (*Rhinediaptomus*) *indicus*. Curr Sci, 87: 109-112.

Djamali, M., P. Ponel, T. Delille et al. 2010. A 200,000-year record of the brine shrimp *Artemia* (Crustacea: Anostraca) remains in Lake Urmia, NW Iran. J Aquat Sci, 1: 14-18.

Dodson, S. I. and D. G. Frey. 2009. Cladocera and other Branchiopoda. In: *Ecology and Classification of North American Freshwater*. (eds) J. H. Thorp and A. P. Covich. Academic Press, San Diego, 2nd edn. pp 849-913.

Dodson, S. I., C. M. Merritt, J. P. Shannahan and C. M. Schults, 1999a. Low doses of atrazine increase male production in *Daphnia pulicaria*. Environ Toxicol Chem, 18: 1568–1573.

Dodson, S. I., C. Merritt, L. Torrentera et al. 1999b. Dieldrin reduces male production and sex ratio in *Daphnia galeata mendotae*. Toxicol Ind Health, 15: 192-199.

Domes, K., R. A. Norton, M. Maraun and S. Scheu. 2007. Re-evolution of sexuality breaks Dollo's law. Proc Natl Acad Sci USA, 104: 7139-7144.

Donnell, D. M., L. S. Corloy, G. Chan and M. R. Strand. 2004. Caste determination in a polyembryonic wasp involves inheritance of germ cells. Proc Natl Acad Sci USA, 101: 10095-10100.

Downer, D. F. and D. H. Steele. 1979. Some aspects of the biology of *Amphiprotic lawrenciana* Shoemaker (Crustacea: Amphipoda) in Newfoundland waters. Can J Zool, 75: 257-263.

Du, Y. X., K. Y. Ma and G. F. Qiu. 2015. Discovery of the genes in putative GnRH signaling pathway with focus on characterization of GnRH-like receptor transcripts in the brain and ovary of the oriental river prawn *Macrobrachium nipponense*. Aquaculture, 442: 1-11.

Duffy, J. E. 1996. Eusociality in a coral-reef shrimp. Nature, 381: 512-514.

Duffy, J. E. 2007. The ecology and evolution of eusociality in sponge-dwelling shrimp. In: *Evolution of Social and Sexual Systems-Crustaceans as Model Systems*. (eds) J. E. Duffy and M. Thiel. Oxford University Press, Oxford. pp 386-408.

Dufresne, F. and P. D. Hebert. 1994. Hybridization and origins of polyploidy. Proc R Soc Lond, 258 B: 141-146.

Dufresne, F. and P. D. N. Hebert. 1997. Pleistocene glaciations and polyphyletic origins of polyploidy in an arctic cladoceran. Proc R Soc Lond, 264 B: 201-206.

Dufresne, F., S. Markova, R. Vergilliro et al. 2011. Diversity in the reproductive modes of European *Daphnia pulicaria* deviates from the geographical parthenogenesis. PLoS ONE 6(5): e 20049. Doi: 10.1371/journal.pone.0020049.

Dumont, H. J. and N. Munuswamy. 1997. The potential of freshwater Anostraca for technical applications. Hydrobiologia, 358: 193-197.

Dumont, H. J. and S. Negrea. 2002. *Introduction to the Class Branchiopoda. Guides to the Identification of the Microinvertebrates of the Continental Waters of the World*. Backhuys Publishers, Leiden. 19: 1-397.

Dumont, H. J., P. Casier, N. Munuswamy and C. De Walsche. 1992. Cyst hatching in Anostraca accelerated by retinoic acid, amplified by calcium ionophore A23187 and inhibited by calcium-channel blockers. Hydrobiologia, 230: 1-7.

Dumont, H. J., S. Nandini and S. S. S. Sarma. 2002. Cyst ornamentation in aquatic invertebrates: a defence against egg-predation. Hydrobiologia, 486: 161-167.

Dupré, E. M. and C. Barros. 2011. *In vitro* fertilization of the rock shrimp, *Rhynchocinetes typus* (Decapoda, Caridea): a review. Biol Res, 44: 125-133.

Eads, B. D., J. Andrews and J. K. Colbourne. 2008. Ecological genomics in *Daphnia*: stress response to environmental sex determination. Heredity, 100: 184-190.

Ebert, D. 1994. Virulence and local adaptation of a horizontally transmitted parasite. Science, 265: 1084-1084.

Ebert, D. 2005. *Introduction to the Ecology, Epidemiology, and Evolution of Parasitism in Daphnia*. Natural Library of Medicine (USA), National Centre for Biotechnology Information, Bethesda.

Ebert, D., M. Lipsitch and K. L. Mangin. 2000. The effect of parasites on host population density and extinction: experimental epidemiology with *Daphnia* and six microparasites. Am Nat, 156: 459-477.

Ebert, D., H. J. Carius, T. Little and E. Decaestecker. 2004. The evolution of virulence when parasites cause host castration and gigantism. Am Nat, 164: S19-S32.

Edomi, P., E. Azzoni, R. Mettulio et al. 2002. Gonad inhibiting hormone of the Norway lobster (*Nephrops norvegicus*): cDNA cloning, expression, recombinant protein production, and immunolocalization. Gene, 284: 93-102.

Eppley, S. M. and L. K. Jesson. 2008. Moving to mate: the evolution of separate and combine sexes in multicellular organisms. J Evol Biol, 21: 727-736.

Espinoza-Fuenzalida, N. L., M. Thiel, E. Dupre and J. A. Baeza. 2008. Is *Hippolyte williamsi* gonochoric or hermaphroditic? A multi-approach study and a review of sexual systems in hippolyte shrimps. Mar Biol, 155: 623-635.

Farazmand, A., K. Inanloo and N. Agh. 2010. Expression of *Dmrt* family genes during gonadal differentiation in two species of *Artemia* (Branchiopoda, Anostraca) from Urmia Lake (Iran). Crustaceana, 83: 1153-1165.

Farmer, A. S. 1972. A bilateral gynandromorph of *Nephrops norvegicus* (Decapoda: Nephropidae). Mar Biol, 15: 344-349.

Fenchel, T. 1965. On the ciliate fauna associated with the marine species of the amphipod genus *Gammarus* JG Fabricius. Ophelia, 2: 281-303.

Feng, Z. F., Z. F. Zhang, M. Y. Shao and W. Zhu. 2011. Developmental expression pattern of the Fc-vasa-like gene, gonadogenesis and development of germ cell in Chinese shrimp, *Fenneropenaeus chinensis*. Aquaculture, 314: 202-209.

Ferreira, S. M., K. T. Jensen, P. A. Martins et al. 2005. Impact of microphallid trematodes on the survivorship, growth, and reproduction of an isopod (*Cyathura carinata*). J Exp Mar Biol Ecol, 318: 191-199.

Fingerman, M. 1987. The endocrine mechanisms of crustaceans. J Crust Biol, 7: 1-24.

Fingerman, M. 1997a. Crustacean endocrinology: a retrospective, prospective and introspective analysis. Physiol Zool, 70: 257-269.

Fingerman, M. 1997b. Roles of neurotransmitters in regulating reproductive hormone release and gonadal maturation in decapod crustaceans. Invert Reprod Dev, 31: 47-54.

Fingerman, M., R. Nagabhushanam, R. Sarojini and P. S. Reddy. 1994. Biogenic amines in crustaceans: identification, localization, and roles. J Crust Biol, 14: 413-437.

Fitzer, S. C., J. D. Bishop, G. S. Caldwell et al. 2012. Visualisation of the copepod female reproductive system using confocal laser scanning microscopy and two-photon microscopy. J Crust Biol, 32: 685-692.

Fitzsimmons, J. M. and D. J. Innes. 2006. Inter-genotype variation in reproductive response to crowding among *Daphnia pulex*. Hydrobiologia, 568: 187-205.

Fleminger, A. 1985. Dimorphism and possible sex change in copepods of the family Calanidae. Mar Biol, 88: 273-294.

Flossner, D. 1972. Kiemen and Blattfussen Branchiopoda. Fischlause Branchiura. Tierwelt Deutschland. pp 60.

Fogelman, R. M., A. M. Kuris and A. S. Grutter. 2009. Parasitic castration of a vertebrate: effect of the cymothoid isopod, *Anilocra apogonae*, on the five-lined cardinalfish, *Cheilodipterus quinquelineatus*. Int J Parasitol, 39: 577-583.

Foote, A. R., G. C. Mair, A. T. Wood and M. J. Sellars. 2012. Tetraploid inductions of *Penaeus monodon* using cold shock. Aquac Internatl, 20: 1003-1007.

Ford, A. T. 2012. Intersexuality in crustacean: an environmental issue? Aquat Toxicol, 108: 125-129.

Ford, A. T. and T. F. Fernandes. 2005. Notes on the occurrence of intersex in amphipods. Hydrobiologia, 540: 313-318.

Ford, A. T., T. P. Rodgers-Gray, I. M. Davies et al. 2005. Abnormal gonadal morphology in intersex, *Echinogammarus marinus* (Amphipoda): a possible cause of reduced fecundity?. Mar Biol, 147: 913-918.

Forró, L., N. M. Korovchinsky, A. A. Kotov and A. Petrusek. 2008. Global diversity of cladocerans (Cladocera; Crustacea) in freshwater. Hydrobiologia, 595: 177-184.

Foster, B. A. 1983. Complemental males in the barnacle *Bathylasma alearum* (Cirripedia: Pachylasmidae). Austr Mus Mem, 18: 133-139.

Fowler, R. J. and B. V. Leonard. 1999. The structure and function of the androgenic gland in *Cherax destructor* (Decapoda: Parastacidae). Aquaculture, 171: 135-148.

Frank, U., G. Pliokert and W. A. Muller. 2009. Cnidarian interstitial cells: the dawn of stem cell research. In: *Stem Cells in Marine Organisms*. (eds) B. Rinkevich and V. Matranga. Springer Verlag, Dordrecht. pp 33-59.

Franquinet, R. 1976. Etude comparative del'evolution des cellus de la planaaried'e and once *Polycelis tenuis* (Iyima) doms des fragments dissocies en culture *in vitro* aspects ultra structuraux incorporation de leucine et d'uridinetritice. J Emb Exp Morph, 36: 41-54.

Fraser, A. J., J. R. Sargent, J. C. Gamble and D. D. Seaton. 1989. Formation and transfer of fatty acids in an enclosed marine food chain comprising phytoplankton, zooplankton and herring (*Clupea harengus* L.) larvae. Mar Chem, 27: 1-18.

Frechette, J. G., G. W. Corrrivult and R. Courture. 1970. Hermaphroditisme proterandrieque chez une crevette de la famille deo crangonides *Argis dentata* Rathbun. Naturaliste Can, 97: 805-822.

Freeman, J. A. and J. D. Costlow. 1980. The molt cycle and its hormonal control in *Rhithropanopeus harrisii* larvae. Dev Biol, 74: 479-485.

Frey, D. G. 1965. Gynandromorphism in the chydorid Cladocera. Limnol Oceanogr, 10: R103-R114.

Frisch, A. 2004. Sex-change and gonadal steroids in sequentially-hermaphroditic teleost fish. Rev Fish Biol Fisher, 14: 481-499.

Fryer, G. 1997. A defence of arthropod polyphyly. In: *Arthropod Relationships Systematics Association*. (eds) R. A. Fortey and R. H. Thomas. Chapman and Hall, London. pp 23-33.

Fu, C., X. Huang, J. Gong et al. 2014. Crustacean hyperglycaemic hormone gene from the mud crab, *Scylla paramamosain*: cloning, distribution and expression profiles during the moulting cycle and ovarian development. Aquac Res, 1-12 doi:10.1111/are.12671.

Fyhn, U. E. and J. D. Costlow. 1977. Histology and histochemistry of the ovary and oogenesis in *Balanus amphitrite* L. and *B. eburneus* Gould (Cirripedia, Crustacea). Biol Bull, 152: 351-359.

Gadagkar, R. 1997. *Survival Strategy*. University Press, Hyderabad. p 196.

Gadgil, M. 1972. Male dimorphism as a consequence of sexual selection. Am Nat, 106: 574-580.

Gajardo, G. M. and J. A. Beardmore. 1989. Ability to switch reproductive mode in *Artemia* is related to maternal heterozygosity. Mar Ecol Prog Ser, 55: 191-195.

García-Velazco, H., H. Obregón-Barboza, C. Rodríguez-Jaramillo and A. M. Maeda-Martínez. 2009. Reproduction of the tadpole shrimp *Triops* (Notostraca) in Mexican waters. Curr Sci, 96: 91-97.

Garnica-Rivera, C., J. L. Arredondo-Figueroa and I. de los Angeles Barriga-Sosa. 2004. Optimization of triploidy induction in the Pacific white shrimp, *Litopenaeus vannamei*. J Appl Aquac, 16: 85-94.

Gaudy, R. 1974. Feeding four species of pelagic copepods under experimental conditions. Mar Biol, 25: 125-141.

Gavio, M. A., J. M. Orensanz and D. Amstrong. 2006. Evaluation of alternative life history hypotheses for the sand shrimp *Crangon franciscorum* (Decapoda: Caridea). J Crust Biol, 26: 295-307.

Geneviere, A. M., A. Aze, Y. Evan et al. 2009. Cell dynamics in early embryogenesis and pluripotent embryonic cell lines: from sea urchin to mammals. In: *Stem Cells in Marine Organisms*. (eds) B. Rinkevich and V. Matranga. Springer Verlag, Dordrecht. pp 245-266.

George, T. and T. J. Pandian. 1995. Production of ZZ females in the female-heterogametic black molly, *Poecilia sphenops*, by endocrine sex reversal and progeny testing. Aquaculture, 136: 81-90.

Ghanawi, J and I. P. Saoud. 2012. Molting, reproductive biology, and hatchery management of redclaw crayfish *Cherax quadricarinatus* (von Martens 1868). Aquaculture, 358: 183-195.

Ghekiere, A., T. Verslycke, N. Fockedey and C. R. Janssen. 2006. Non-target effects of the insecticide methoprene on molting in the estuarine crustacean *Neomysis integer* (Crustacea: Mysidacea). J Exp Mar Biol Ecol, 332: 226-234.

Ghiselin, M. T. 1969. The evolution of hermaphroditism among animals. Q Rev Biol, 44: 189-208.

Ghiselin, M. T. 1974. *The Economy of Nature and The Evolution of Sex*. University of California Press, Berkeley.

Ghomari, M. S., G. S. Selselet, F. Hontoria and F. Amat. 2011. *Artemia* biodiversity in Algerian sebkhas. Crustaceana, 84: 1025-1039.

Gierer, A., H. Berking, C. Bode et al. 1972. Regeneration of hydra from reaggregated cells. Nature New Biol, 239: 98-101.

Gimenez, L. and K. Anger. 2001. Relationships among salinity, egg size, embryonic development, and larval biomass in the estuarine crab *Chasmagnathus granulata* Dana, 1851. J Exp Mar Biol Ecol, 260: 241-257.

Ginburger-Vogel, T. and H. Charniaux Cotton. 1982. Sex determination. In: *The Biology of Crustacea*. (ed) L. G. Abele. Academic Press, New York. pp 257-289.

Giovagnoli, A., R. B. Ituarte and E. D. Spivak. 2014. Effects of removal from the mother and salinity on embryonic development of *Palaemonetes argentinus* (Decapoda: Caridea: Palaemonidae). J Crust Biol, 34: 174-181.

Girish, B. P., C. H. Swetha and P. S. Reddy. 2015a. Induction of ecdysteroidogenesis, methyl farnesoate synthesis and expression of ecdysteroid receptor and retinoid X receptor in the hepatopancreas and ovary of the giant mud crab, *Scylla serrata* by melatonin. Gen Comp Endocrinol, 217-218: 37-42.

Girish, B. P., C. H. Swetha and P. S. Reddy. 2015b. Expression of RXR, EcR, E75 and VtG mRNA levels in the hepatopancreas and ovary of the freshwater edible crab, *Oziothelphusa senex senex* (Fabricius, 1798) during different vitellogenic stages. Sci Nat, 102: 1-10.

Glazier, D. S., T. L. Brown and A. T. Ford. 2012. Similar offspring production by normal and intersex females in two populations of *Gammarus minus* (Malacostraca, Amphipoda) with high levels of intersexuality. Crustaceana, 85: 801-815.

Glenner, H., J. Lützen and T. Takahashi. 2003. Molecular and morphological evidence for a monophyletic clade of asexually reproducing rhizocephala: *Polyascus*, new genus (Cirripedia). J Crust Biol, 23: 548-557.

Gliwicz, Z. M. 1990. Food thresholds and body size in cladocerans. Nature, 343: 638-640.

Gliwicz, Z. M. 2003. *Between Hazards of Starvation and Risk of Predation: The Ecology of Offshore Animals*. International Ecology Institute, Nordbunte, Germany. p 379.

Gliwicz, Z. M. and H. Stibor. 1993. Egg predation by copepods in *Daphnia* brood cavities. Oecologia, 95: 295-298.

Gliwicz, Z. M. and P. Dawidowicz. 2001. Roach habitat shifts and foraging modified by alarm substance - 1. Field evidence, Arch Hydrobiol, 150: 357-376.

Glossener, R. R. and D. Tilman. 1978. Sexuality and the components of environmental uncertainity: clues from geographic parthenogenesis in terrestrial animals. Am Nat, 112: 659-673.

Glynn, P. W. 1968. Ecological studies on the associations of chitons in Puerto Rico, with special reference to sphaeromid isopods. Bull Mar Sci, 18: 572-626.

Gnanamuthu, C. P. 1954. *Choniosphaera indica*, a copepod parasitic on the crab *Neptunus* sp. Parasitology, 44: 371-378.

Golubev, A. P., N. N. Khmeleva, A. V. Alekhnovich et al. 2001. Influence of reproduction on variability of life history parameters in *Artemia salina* (Crustacea, Anostraca). Entomol Rev, 81: S 96-S 107.

Gomes, L. A. and J. H. Primavera. 1993. Reproductive quality of male *Penaeus monodon*. Aquaculture, 112: 157-164.

Gomez, E. D. 1975. Sex determination in *Balanus* (*Conopea*) *galeatus* (L.) (Cirripedia Thoracica). Crustaceana, 28: 105-107.

Gomez, E. D., D. J. Faulkner, W. A. Newman and C. Ireland. 1973. Juvenile hormone mimics: effect on cirriped crustacean metamorphosis. Science, 179: 813-814.

Gomez, R. and K. K. Nayar. 1965. Certain endocrine influences in the reproduction of the crab *Paratelphusa hydrodromous*. Zool Jahrb, Allgem Zool Physiol Tiere, 71: 694-701.

Gómez-Gutiérrez, J., L. R. Feinberg, T. Shaw and W. T. Peterson. 2006. Variability in brood size and female length of *Euphausia pacifica* among three populations in the North Pacific. Mar Ecol Prog Ser, 323: 185-194.

González, M. T. and E. Acuna. 2004. Infestation by *Pseudione humboldtensis* (Bopyridae) in the squat lobsters *Cervimunida johni* and *Pleuroncodes monodon* (Galatheidae) off northern Chile. J Crust Biol, 24: 618-624.

González-Tizón, A. M., V. Rojo, E. Menini et al. 2013. Karyological analysis of the shrimp *Palaemon serratus* (Decapoda: Palaemonidae). J Crust Biol, 33: 843-848.

Graham, D. J., H. Perry, P. Biesiot and R. Fulford. 2012. Fecundity and egg diameter of primiparous and multiparous blue crab *Callinectes sapidus* (Brachyura: Portunidae) in Mississippi waters. J Crust Biol, 32: 49-56.

Gregory, T. R., P. D. Hebert and J. Kolasa. 2000. Evolutionary implications of the relationship between genome size and body size in flatworms and copepods. Heredity, 84: 201-208.

Greve, P., O. Sorokine, T. Berges et al. 1999. Isolation and amino acid sequence of a peptide with vitellogenesis inhibiting activity from the terrestrial isopod *Armadillidium vulgare* (Crustacea). Gen Comp Endocrinol, 115: 406-414.

Grigarick, A. A., W. H. Lange and D. C. Finfrock. 1961. Control of the tadpole shrimp, *Triops longicaudatus* in California rice fields. J Econ Entomol, 54: 36-40.

Grygier, M. J. 1982. *Gorgonolaureus muzikae* sp. nov. (Crustacean: Ascothoracida) parasitic on a Hawaiian gorgonian, with special reference to its protandric hermaphroditism. J Nat Hist, 15: 1019-1045.

Guan, G., T. Kobayashi and Y. Nagahama. 2000. Sexually dimorphic expression of two types of DM (Doublesex/Mab-3)-domain genes in a teleost fish, the Tilapia (*Oreochromis niloticus*). Biochem Biophy Res Comm, 272: 662-666.

Gusamao, L. F. M. and A. D. McKinnon. 2009. Sex ratios, intersexuality and sex change in copepods. J Plank Res, 31: 1101-1117.

Gusamao, L. F. M. and A. D. McKinnon. 2012. *Acrocalanus gracilis* (Copepoda: Calanoida) development and production in the Timor Sea. J Plank Res, 31: 1089-1100.

Gyllström, M. and L. A. Hansson. 2004. Dormancy in freshwater zooplankton: induction, termination and the importance of benthic-pelagic coupling. Aquat Sci, 66: 274-295.

Haag, C. R., S. J. McTaggart, A. Didier et al. 2009. Nucleotide polymorphism and within-gene recombination in *Daphnia magna* and *D. pulex,* two cyclic parthenogens. Genetics, 182: 313-323.

Hairston Jr, N. G. 1987. Diapause as a predator-avoidance adaptation. In: *Predation: Direct and Indirect Impacts on Aquatic Communities.* (eds) W. C. Kerfoot and A. Sih. University Press of New England, London. pp 281-290.

Hairston, N. G. 1996. Zooplankton egg banks as biotic reservoirs in changing environments. Limnol Oceanogr, 41: 1087-1092.

Hairston Jr, N. G., R. A. Van Brunt, C. M. Kearns and D. R. Engstrom. 1995. Age and survivorship of diapausing eggs in a sediment egg bank. Ecology, 76: 1706-1711.

Hairston, N. G. Jr, A. M. Hansen and W. R. Schaffner. 2000. The effect of diapause emergence on the seasonal dynamics of a zooplankton assemblage. Freshwat Biol, 45: 133-145.

Hairston Jr, N. G., C. M. Kearns, L. P. Demma and S. W. Effler. 2005. Species-specific *Daphnia* phenotypes: a history of industrial pollution and pelagic ecosystem response. Ecology, 86: 1669-1678.

Hall, M. M., R. R. Mastro and G. G. Prestwich. 1999. Hormonal modulation of spawner quality in *Penaeus monodon*. In: *World Aquaculture 1999: Bridging the gap.* Ann Int Conf, World Aquacult Soc, Australia. p 308.

Hamr, P. 2002. *Orconectes*. In: *Biology of Freshwater Crayfish.* (ed) D. M. Holdich. Blackwell Science, London. pp 585-608.

Hand, S. C. 1995. Heat flow is measurable from *Artemia franciscana* embryos under anoxia. J Exp Zool, 273: 445-449.

Hanström, B. 1933. Neue Untersuchungen über Sinnesorgane und Nervensystem der Crustaceen. I. Zoo J, Apt Zool Physiol Tiere, 56: 387-520.

Haq, S. M. 1973. Factors affecting production of dimorphic males of *Euterpina acutifrons*. Mar Biol, 19: 23-26.

Hardewig, I., T. J. Anchordoguy, D. L. Crawford and S. C. Hand. 1996. Profiles of nuclear and mitochondrial encoded mRNAs in developing and quiescent embryos of *Artemia franciscana*. Mol Cell Biochem, 158: 139-147.

Harris, R. P. 1973. Feeding, growth, reproduction and nitrogen utilization by the harpacticoid copepod, *Tigriopus brevicornis*. J Mar Biol Asso U K, 53: 785-800.

Harrison, K. E. 1990. The role of nutrition in maturation, reproduction and embryonic development of decapod crustaceans: a review. J Shellfish Res, 9: 1-28.

Hartnoll, R. G. 2009. Sexual maturity and reproductive strategy of the rock crab *Grapsus adscensionis* (Osbeck, 1765) (Brachyura, Grapsidae) on Ascension Island. Crustaceana, 82: 275-291.

Harvey, A. W. 1990. Sexual differences in contemporary selection acting on size in the hermit crab *Clibanarius digueeti*. Am Nat, 136: 292-304.

Hathaway, S. A., D. P. Sheehan and M. A. Simovich. 1996. Vulnerability of branchiopod cysts to crushing. J Crust Biol, 16: 448-452.

Hawkes, C. R., T. R. Meyers and T. C. Shirley. 1985. Larval biology of *Briarosaccus callosus* Boschma (Cirripedia: Rhizocephala). Proc Biol Soc Washington, 98: 935-944.

Hawkes, C. R., T. R. Meyers, T. C. Shirley and T. M. Koeneman. 1986. Prevalence of the parasitic barnacle *Briarosaccus callosus* on king crabs of southeastern Alaska. Trans Am Fisher Soc, 115: 252-257.

Hedgecock, D., M. L. Tracey and K. Nelson. 1982. Genetics. In: *The Biology of Crustacea, Vol. 2: Embryology, Morphology and Genetics*. (ed) L. G. Abele. Academic Press, New York. pp 283-403.

Heeley, W. 1941. Observations on the life-histories of some terrestrial isopods. In: Proc Zool Soc. Blackwell Publishing, London. 111: 79-149.

Heinle, D. R., R. P. Harris, J. F. Ustach and D. A. Flemer. 1977. Detritus as food for estuarine copepods. Mar Biol, 40: 341-353.

Hemamalini, A. K. and N. Munuswamy. 1994. Variations in the activity of some metabolic enzymes during development of *Artemia parthenogenetica* (Crustacea: Anostraca). Arch Physiol Biochem, 102: 107-110.

Hentig, R. 1971. Einfluss von Salzgehalt und Temperatur auf Entwicklung. Wachstum, Fortpflanzung von *Artemia salina*. Mar Biol, 45: 255-260.

Herbert, P. D. N. 1978. The population biology of *Daphnia* (Crustacea, Daphnidae). Biol Rev, 53: 387-426.

Herbert, P. D. N. 1987. Evolution of cyclic parthenogenesis. In: *Evolution of Sex and Consequences*. (ed) S. Stearns. Birkhauser Verlag, Basel. pp 175-195.

Herbert, P. D. N. and R. D. Ward. 1972. Inheritance during parthenogenesis in *Daphnia magna*. Genetics 72: 639-642.

Herbert, P. D. N. and T. Crease. 1983. Clonal diversity in populations of *Daphnia pulex* reproducting by obligate parthenogenesis. Heredity, 51: 353-369.

Herbert, P. D. N. and K. C. Wilson. 1994. Provincialism in plankton-endemism and allopatric speciation in Australian *Daphnia*. Evolution, 48: 1333-1349.

Herbert, P. D. N. and T. L. Finston. 2001. Macrogeographic patterns of breeding system diversity in *Daphnia pulex* group from the United States and Mexico. Heredity, 87: 153-161.

Herbert, P. D. N., S. S. Schwartz, S. S. and T. L. W. Finston. 1993. Macrogeographic patterns of breeding system diversity in *Daphnia pulex* group. 1. Breeding system of Canadian populations. Heredity, 70: 148-161.

Herbert, P. D. N. and C. C. Wilson. 2000. Diversity of the genus *Daphniopsis* in the saline waters of Australia. Can J Zool, 78: 794-808.

Herbert, P. D. N., C. I. Rowe and S. J. Adarnowicz. 2007. Life at low temperatures. A novel breeding system adjustment in a polar cladoceram. Limnol Oceanogr, 52: 2507-2518.

Hernáez, P., B. Martínez-Guerrero, A. Anker and I. S. Wehrtmann. 2010. Fecundity and effects of bopyrid infestation on egg production in the Caribbean sponge-dwelling snapping shrimp *Synalpheus yano* (Decapoda: Alpheidae). J Mar Biol Ass U K, 90: 691-698.

Hinegardener, R. T. 1975. Morphology and genetics of sea urchin development. Am Zool, 15: 679-689.

Hinsch, G. W. 1972. Some factors controlling reproduction in the spider crab, *Libinia emarginata*. Biol Bull, 143: 358-366.

Hinsch, G. W. 1980. Effect of mandibular organ implants upon spider crab ovary. Trans Am Microsc Soc, 99: 317-322.

Hinsch, G. W. and M. H. Walker. 1974. The vas deferens of the spider crab, *Libinia emarginata*. J Morphol, 143: 1-1.

Hiramatsu, N., T. Matsubara, T. Fujita et al. 2006. Multiple piscine vitellogenins: biomarkers of fish exposure to estrogenic endocrine disruptors in aquatic environments. Mar Biol, 149: 35-47.

Hodgkin, J. 1990. Sex determination compared in *Drosophila* and *Caenorhabditis*. Nature, 344: 721-728.

Høeg, J. T. and J. Lützen. 1995. Life cycle and reproduction in the Cirripedia Rhizocephala. Oceanogr Mar Biol Annu Rev, 427-485.

Høeg, J. T., M. Pérez-Losada, H. Glenner et al. 2009. Evolution of morphology, ontogeny and life cycles within the Crustacea Thecostraca. Arthro System Phylo, 67: 199-217.

Hoffman, D. 1972. The development of ovotestis and copulatory organs in a population of protrandric shrimp *Pandalus platyceros* Brandt from Lopez Sound, Washington. Biol Bull, 142: 251-270.

Hoffmann, G. E. and S. C. Hand. 1994. Global arrest of translation during invertebrate quiescence. Proc Natl Acad Sci USA, 91: 8492-8496.

Hopkins, C. C. E. 1977. The relationship between maternal body size and clutch size, development time and egg mortality in *Euchaeta norvegica* (Copepoda: Calanoida) from Loch Etive, Scotland. J Mar Biol Asso UK, 57: 723-733.

Hopkins, C. C. E. and D. Machin. 1977. Patterns of spermatophore distribution and placement in *Euchaeta norvegica* (Copepoda: Calanoida). J Mar Biol Ass U K, 57: 113-131.

Hopkins, P. M. 2012. The eyes have it: a brief history of crustacean neuroendocrinology. Gen Comp Endocrinol, 175: 357-366.

Hopkins, P. M., A. K. Chung and D. S. Durica. 1999. Limb regeneration in the fiddler crab, *Uca pugilator*: histological, physiological and molecular considerations. Am Zool, 39: 513-526.

Huang, D. J., S. Y. Wang and H. C. Chen. 2004. Effects of the endocrine disrupter chemicals chlordane and lindane on the male green neon shrimp (*Neocaridina denticulata*). Chemosphere, 57: 1621-1627.

Huberman, A. 2000. Shrimp endocrinology. A review. Aquaculture, 191: 191-208.

Hubschman, J. H. 1963. Development and function of neurosecretory sites in the eyestalks of larval *Palaemonetes* (Decapoda: Natantia). Biol Bull, 125: 96-113.

Hubschman, J. H. and A. C. Broad. 1974. The larval development of *Palaemonetes intermedius* Holthuis 1949 (Decapoda, palaemonidae) reared in the laboratory. Crustaceana, 26: 89-103.

Hui, J. H. L., S. S. Tobe and S. M. Chan. 2008. Characterization of the putative farnesoic acid O-methyltransferase (LvFAMeT) cDNA from white shrimp, *Litopenaeus vannamei*: Evidence for its role in molting. Peptides, 29: 252-260.

Hutchinson, T. H., N. A. Pounds, M. Hampel and T. D. Williams. 1999a. Impact of natural and synthetic steroids on the survival, development and reproduction of marine copepods (*Tisbe battagliai*). Sci Total Environ, 233: 167-179.

Hutchinson T. H., N. A. Pounds, M. Hampel and T. D. Williams. 1999b. Life cycle studies with marine copepods (*Tisbe battagliai*) exposed to 20-hydroxyecdyserone and diethylstilbestrol. Environ Toxicol Chem, 18: 2914-2920.

Hyne, R. V. 2011. Review of the reproductive biology of amphipods and their endocrine regulation: identification of mechanistic pathways for reproductive toxicants. Environ Toxicol Chem, 30: 2647-2657.

Ibarra, A. M., F. G. Arcos, T. R. Famula et al. 2005. Heritability of the categorical trait 'number of spawns' in Pacific white female shrimp *Penaeus (Litopenaeus) vannamei*. Aquaculture, 250: 95-101.

Ibarra, A. M., I. S. Racotta, F. G. Arcos and E. Palacios. 2007. Progress on the genetics of reproductive performance in penaeid shrimp. Aquaculture, 268: 23-43.

Ignace, D. D., S. I. Dodson and D. R. Kashian. 2011. Identification of the critical timing of sex determination in *Daphnia magna* (Crustacea, Branchiopoda) for use in toxicological studies. Hydrobiologia, 668: 117-123.

Ikhwanuddin, M., M. N. Azra, H. Siti-Aimuni and A. B. Abol-Munafi. 2012. Fecundity, embryonic and ovarian development of blue swimming crab, *Portunus pelagicus* (Linnaeus, 1758) in coastal water of Johor, Malaysia. Pak J Biol Sci, 15: 720-728.

Innes, D. J. 1997. Sexual reproduction of *Daphnia pulex* in a temporary habitat. Oecologia, 111: 53-60.

Innes, D. J. and P. D. Hebert. 1988. The origin and genetic basis of obligate parthenogenesis in *Daphnia pulex*. Evolution, 1024-1035.

Irigoien, X., B. Obermuller, R. N. Head et al. 2000. The effect of food on the determination of sex ratio in *Calanus* spp: evidence from experimental studies and field data. ICES J Mar Sci, 57: 1752-1763.

Isaeva, V. V. 2010. The diversity of ontogeny in animals with asexual reproduction and plasticity of early development. Rus J Mar Biol, 37: 209-216.

Isaeva, V. V. 2011. Pluripotent gametogenic stem cells of asexually reproducing invertebrates. In: *Embryonic Stem Cells – Basic Biology to Bioengineering*. (ed) M. S. Kallus. In-Tech Publishers, Rijeka, Crotaia, pp 449-478.

Isaeva, V. V., A. V. Akhmadieva, Ya. N. Alexsandrova and A. I. Shukalyuk. 2009. Morphofunctional organization of reserve stem cells providing for asexual and sexual reproduction of invertebrates. Rus J Dev Biol, 40: 57-68.

Isaeva, V. V., A. V. Akhmadieva, Y. N. Aleksandrova et al. 2011. Germinal granules in interstitial cells of the colonial hydroids *Obelia longissima* Pallas, 1766 and *Ectopleura crocea* Agassiz, 1862. Russ J Mar Biol, 37: 303-310.

Iyer, P. and J. Roughgarden. 2008. Dioecy as a specialization promoting sperm delivery. Evol Ecol Res, 10: 867-892.

Jacobson, T. and B. Sundelin. 2006. Reproductive effects of the endocrine disruptor fenarimol on a Baltic amphipod *Monoporeia affinis*. Environ Toxicol Chem, 25: 1126-1131.

Jaglarz, M. K. and K. R. Howard. 1995. The active migration of *Drosophila* primordial germ cells. Development, 121: 3495-3503.

Jarne, P. and J. R. Auld. 2006. Animals mix it up too: the distribution of self-fertilization among hermaphroditic animals. Evolution, 60: 1816-1824.

Jawinski, A. 2002. Day-to-day variability in abundance and size distribution of *Daphnia cucullata* and *Bosmina thersites* in the lake and in the diet of smelt (*Osmerus eperlanus*), a dominant offshore planktivore in Lake Mikolajskie. MSc. Thesis, University of Warsaw.

Jay, C. V. 1989. Prevalence, size and fecundity of the parasitic isopod *Argeia pugettensis* on its host shrimp *Crangon francisorum*. Am Midla Nat, 121: 68-77.

Jayaprakas, V. and C. Sambhu. 1998. Growth characteristics of white prawn *Penaeus indicus* (Decapoda-Crustacea) under dietary administration of protein hormones. Ind J Mar Sci, 27: 389-395.

Jerry, D. R., B. S. Evans, M. Kenway and K. Wilson. 2006. Development of a microsatellite DNA parentage marker suite for black tiger shrimp *Penaeus monodon*. Aquaculture, 255: 542-547.

Ji, Y. K., A. Wang, X. L. Lu et al. 2014. Mitochondrial genomes of two brachyuran crabs (Crustacea: Decapoda) and phylogenetic analysis. J Crust Biol, 34: 494-503.

Johnson, W. S., M. Stevens and L. Walting. 2001. Reproduction and development of marine peracaridans. Adv Mar Biol, 39: 105-260.

Joshi, P. C. and S. S. Khanna. 1984. Neurosecretory system of the thoracic ganglion and its relation to testicular maturation of the crab, *Potamon koolooense* (Rathbum). Jahrb Morphol Mikroskop Anat. 2. Z Mikroskop-anat Forsch, 98: 429-442.

Juanes, F. and L. D. Smith. 1995. The ecological consequences of limb damage and loss in decapod crustaceans: a review and prospectus. J Exp Mar Biol Ecol, 193: 197-223.

Kaczmarczyk, A. N and A. Kopp. 2011. Germline stem cell maintenance as a proximate mechanism of life-history trade-offs?. Bioessays, 33: 5-12.

Kagewade, P. V. 1981. The hermaphrodite prawn *Hippolysmata ensirostris* Kemp. Ind J Fish, 28: 189-194.

Kaiser, H., A. K. Gordon and T. G. Poulet. 2006. Review of the African distribution of the brine shrimp genus *Artemia*. Water SA, 32: 597-603.

Kalavathy, T., P. Mamatha and S.P. Reddy. 1999. Methyl farnesoate stimulates testicular growth in the freshwater crab *Oziotelphusa senex senex* Fabricius. Naturwissenschaften, 86: 394-395.

Kamiyama, T. 2011. Planktonic ciliates as a food source for the scyphozoan *Aurelia aurilia*: feeding activity and assimilation of the polyp stage. J Exp Mar Biol Ecol, 407: 207-215.

Kao, H. W. and E. S. Chang. 1996. Homeotic transformation of crab walking leg into claw by autotransplantation of claw tissue. Biol Bull, 190: 313-321.

Kao, H. W. and E. S. Chang. 1997. Limb regeneration in the eye sockets of crabs. Biol Bull, 193: 393-400.

Kao, H. W. and E. S. Chang. 1999. Limb regeneration following auto-or interspecies transplantation of crab limb tissues. Invert Reprod Dev, 35: 155-165.

Kasahara, S., S. Uye and T. Onbé. 1975. Calanoid copepod eggs in sea-bottom muds. II. Seasonal cycles of abundance in the populations of several species of copepods and their eggs in the Inland Sea of Japan. Mar Biol, 31: 25-29.

Kashian, D. R. and S. I. Dodson. 2004. Effects of vertebrate hormones on development and sex determination in *Daphnia magna*. Environ Toxicol Chem, 23: 1282-1288.

Katakura, Y. 1960. Transformation of ovary into testis following implantation of androgenous glands in *Armadillidium vulgare*, an isopod crustacean. Annot Zool Jpn: 33: 241-244.

Katakura, Y. 1989. Endocrine and genetic control of sex differentiation in the malacostracan Crustacea. Invert Reprod Dev, 16: 177-181.

Katakura, Y. and Y. Hasegawa. 1983. Masculinization of females of the isopod crustacean, *Armadillidium vulgare*, following injections of an active extract of the androgenic gland. Gen Comp Endocrinol, 49: 57-62.

Kato, Y., K. Kobayashi, S. Oda et al. 2008. Molecular cloning and sexually dimorphic expression of DM-domain genes in *Daphnia magna*. Genomics, 91: 94-101.

Kato, Y., K. Kobayashi, S. Oda et al. 2010. Sequence divergence and expression of a transformer gene in the branchiopod crustacean, *Daphnia magna*. Genomics, 95: 160-165.

Katre, S. 1977. The relation between body size and number of eggs in the freshwater prawn, *Macrobrachium lamarrei* (H. Milne Edwards) (Decapoda, Caridea). Crustaceana, 33: 17-22.

Kelman, D. and R. B. Emlet. 1999. Swimming and buoyancy in ontogenetic stages of the cushion star *Pteraster tesselatus* (Echinodermata: Asteroidea) and their implications for distribution and movement. Biol Bull, 197: 309-314.

Kerfoot, W. C. 1974. Egg-size cycle of a cladoceran. Ecology, 1259-1270.

Khalaila, I., T. Katz, U. Abdu et al. 2001. Effects of implantation of hypertrophied androgenic glands on sexual characters and physiology of the reproductive system in the female red claw crayfish, *Cherax quadricarinatus*. Gen Comp Endocrinol, 121: 242-249.

Kim, D. H., S. K. Kim, J. H. Choi et al. 2010. The effects of manipulating water temperature, photoperiod, and eyestalk ablation on gonad maturation of the swimming crab, *Portunus trituberculatus*. Crustaceana, 83: 129-141.

Kim, E., C. M. Ansell and J. L. Dudycha. 2014. Resveratrol and food effects on lifespan and reproduction in the model crustacean *Daphnia*. J Exp Zool Part A: Ecol Genet Physiol, 321: 48-56.

Kim, W. J., H. T. Jung, Y. Y. Chun et al. 2012. Genetic evidence for natural hybridization between red snow crab (*Chionoecetes japonicus*) and snow crab (*Chionoecetes opilio*) in Korea. J Shellfish Res, 31: 49-56.

Kinne, O. 1990. *Diseases of Marine Animals*. Biologische Anstalt Helgoland, Hamburg, Vol 3: p 696.

Kiorboe, T. 2006. Sex, sex ratios and the dynamics of pelagic copepod populations. Oecologia, 148: 40-50.

Kiorboe, T. and M. Sabitini. 1995. Scaling of fecundity, growth and development in marine planktonic copepods. Mar Ecol-Prog Ser, 120: 285-298.

Kirankumar, S. and T. J. Pandian. 2004. Interspecific androgenetic restoration of rosy barb using cadaveric sperm. Genome, 47: 66-73.

Kirankumar, S., V. Anathy and T. J. Pandian. 2003. Hormonal induction of supermale golden rosy barb and isolation of Y-chromosome specific markers. Gen Comp Endocrinol, 134: 62-71.

Klapper, W., K. Kühne, K. K. Singh et al. 1998. Longevity of lobsters is linked to ubiquitous telomerase expression. FEBS Lett, 439: 143-146.

Klaus, S., G. H. Goh, Y. Malkowsky et al. 2014. Seminal receptacle of the pill box crab *Limnopilos naiyanetri* Chuang and Ng, 1991 (Brachyura: Hymenosomatidae). J Crust Biol, 34: 407-411.

Knoblich, J. A. 2001. Asymmetric cell division during animal development. Nat Rev Mol Cell Biol, 2: 11-20.

Kochina, E. M. 1987. Cytotaxonomic study of cyclopoids of *Acanthocyclops "americanus-vernalis"* group (Crustacea, Copepoda). Vestnik Zoologii, Kiev, 3: 7-11.

Kolbasov, G. A. and G. B. Zevina. 1999. A new species of *Paralepas* (Cirripedia: Heteralepadidac) symbiotic with *Xenophora* (Mollusca: Gastropoda) with the first complemental male known for the family. Bull Mar Sci, 64: 391-398.

Koopman, H. N and Z. A. Siders. 2013. Variation in egg quality in blue crabs, *Callinectes sapidus*, from North Carolina: does female size matter?. J Crust Biol, 33: 481-487.

Krishnan, L. and P. A. John. 1974. Observations on the breeding biology of *Melita zeylanica* Stebbing, a brackish water amphipod. Hydrobiologia, 44: 413-430.

Kristensen, T., A. I. Nielsen, A. I. Jørgensen et al. 2012. The selective advantage of host feminization: a case study of the green crab *Carcinus maenas* and the parasitic barnacle *Sacculina carcini*. Mar Biol, 159: 2015-2023.

Krumm, J. L. 2013. Axial gynandromorphy and sex determination in *Branchinecta lindahli* (Branchiopoda: Anostraca). J Crust Biol, 33: 303-308.

Kulkarni, G., R. Nagabhushanam and P. Joshi. 1979. Effect of progesterone on ovarian maturation in a marine penaeid prawn *Parapenaeopsis hardwickii* (Miers, 1878). Ind J Exp Biol, 17: 986-987.

Kumar, S. R. and K. Altaff. 2003. Life span and egg production of *Macrothrix spinosa* King (Cladocera: Macrothoracidae) under laboratory conditions. J Aquat Biol, 18: 7-9.

Kumari, S. S. and T. J. Pandian. 1987. Effects of unilateral eyestalk ablation on moulting, growth, reproduction and energy budget of *Macrobrachium nobilii*. Asian Fisher Sci, 1: 1-17.

Kumari, S. and T. J. Pandian. 1991. Interaction of ration and unilateral eyestalk ablation on energetic of female *Macrobrachium nobilii*. Asian Fish Sci, 4: 227-244.

Kuo, C. M. and Y. H. Yang. 1999. Hyperglycemic responses to cold shock in the freshwater giant prawn *Macrobrachium rosenbergii*. J Comp Physiol, 169: 49-54.

Kuris, A. M., Z. Ra'anan, A. Sagi and D. Cohen. 1987. Morphotypic differentiation of male Malaysian giant prawns, *Macrobrachium rosenbergii*. J Crust Biol, 7: 219-237.

Kusk, K. O. and L. Wollenberger. 2007. Towards an internationally harmonized test method for reproductive and developmental effects of endocrine disrupters in marine copepods. Ecotoxicology, 16: 183-195.

Lacaze, E., O. Geffard, D. Goyet et al. 2011. Linking genotoxic responses in *Gammarus fossarum* germ cells with reproduction impairment, using the Comet assay. Environ Res, 111: 626-34.

Lacoste, A., S. A. Poulet, A. Cueff et al. 2001. New evidence of the copepod maternal food effects on reproduction. J Exp Mar Biol Ecol, 259: 85-107.

Lafferty, K. D. and A. M. Kuris. 2009. Parasitic castration: the evolution and ecology of body snatchers. Trends Parasitol, 25: 564-572.

Lai, J. H., J. C. del Alamo, J. Rodríguez-Rodríguez and J. C. Lasheras. 2010. The mechanics of the adhesive locomotion of terrestrial gastropods. J Exp Biol, 21: 3920-3933.

LaMontagne, J. M. and E. McCauley. 2001. Maternal effects in *Daphnia*: what mothers are telling their offspring and do they listen? Ecol Lett, 4: 64-71.

Lampert, W. 2011. *Daphnia : Development of a Model Organism in Ecology and Evolution.* (ed) O. Kinne. Internatl Ecol Inst, Germany. p 250.

Larsson, P. 1991. Intraspecific variability in response to stimuli for male and ephippia formation in *Daphnia pulex*. In: *Biology of Cladocera* (eds) V. Korinick and D. G. Frey Springer, Netherlands. pp 281-290.

Lasker, R., J. B. J. Wells and A. D. McIntyre. 1970. Growth, reproduction, respiration and carbon utilization of the sand-dwelling harpacticoid copepod, *Asellopsis intermedia*. J Mar Biol Asso UK, 50: 147-160.

Lass, S. and P. Spaak. 2003. Chemically induced anti-predator defences in plankton: a review. Hydrobiologia, 491: 221-239.

Laufer, H. 1992. U.S. Patent No. 5,161,481. Washington, DC: U.S. Patent and Trademark Office.

Laufer, H. and M. Landau. 1991. Endocrine control of reproduction in shrimp and other Crustacea. In: *Frontiers of Shrimp Research*. (eds) P. Dehrach and W. J. Davison. Elsevier Science Publication, Amsterdam. pp 65-81.

Laufer, H. and W. J. Biggers. 2001. Unifying concepts learnt from methyl farnesoate of invertebrate reproduction and post-embryonic development. Am Zool, 41: 442-457.

Laufer, H., D. Borst, F. C. Baker et al. 1987. Identification of a juvenile hormone-like compound in a crustacean. Science, 235: 202-205.

Laufer, H., J. S. Ahl and A. Sagi. 1993. The role of juvenile hormones in crustacean reproduction. Am Zool, 33: 365-374.

Laufer, H., P. Takac, J. S. B. Ahl and M. R. Laufer. 1997. Methyl farnesoate and the effect of eyestalk ablation on the morphogenesis of the juvenile female spider crab *Libinia emarginata*. Invert Reprod Dev, 31: 1-3.

Laufer, H., N. Demir and X. Pan. 2005. Methyl farnesoate controls adult male morphogenesis in the crayfish, *Procambarus clarkii*. J Insect Physiol, 51: 379-384.

Lavaniegos, B. E. 1995. Production of the euphausiid *Nyctiphanes simplex* in Vizcaino Bay, western Baja California. J Crust Biol, 15: 444-444.

Leelatanawit, R., K. Sittikankeaw, P. Yocawibun et al. 2009. Identification, characterization and expression of sex-related genes in testes of the giant tiger shrimp *Penaeus monodon*. Comp Biochem Physiol Part A: Mol Integ Physiol, 152: 66-76.

Le Grand-Hamelin, E. 1977. Obtention de néo-mâles fonctionnels et démonstration expérimentale de l'hétérogamétie femelle chez *ldotea balthica* (Crustacé Isopode). C R Sean Soc. Biol, 171: 176-180.

Le Grand, J. J., E. LeGrand-Hamelin and P. Juchault. 1987. Sex determination in Crustacea. Biol Rev, 62: 439-470.

Le Page, Y., N. Diotel and C. Vaillant. 2010. Aromatase, brain sexualization and plasticity: the fish paradigm. Euro J Neurosci, 32: 2105-2115.

Le Roux, A. 1963. Contribution a l'etude development larvaire d' *Hippolyte inermis* Leach (Crustacee, Decapode, Macroure). C R Hebd Sean Acad Sci Paris, 256: 3499-3501.

Le Roux, A. 1968. Description d' organes mandibulaires nouveaux chezles Crustaces Decapodes. C R Hebd Sean Acad Sci Paris, 266: 1414-1417.

Le Roux, A. 1980. Effect of de l'ablation des pedoncules oculaires et des quelques conditions d'elevage sur le development de *Pisidia longicornis* (Linne). Arch Zool Exp Gen, 121: 97-214.

Le Roux, A. 1984. Quelques effets de l'ablation des pédoncules oculaires sur les larves et les premiers stades juvéniles de *Palaemonetes varians* (Leach) (Decapoda, Palaemonidae). Bull Soc Zool France,109: 43-60.

Le Roux, M. L. 1931a. Castration parasitaire et caracteres sexuels secondaires chez les Gammariens. C R Hebd Sean Acad Sci Paris, 192: 889-891.

Le Roux, M. L. 1931b. La castration experimentale des femelles de Gammariens et la repercussion sur revolution des oostegites. C R Hebd Sean Acad Sci Paris, 193: 885-887.

Lehto, M. P. and C. R. Haag. 2010. Ecological differentiation between coexisting sexual and asexual strains of *Daphnia pulex*. J Anim Ecol, 79: 1241-1250.

Leone, I. C. and F. L. Mantelatto. 2015. Maternal investment in egg production: substrate- and population-specific effects on offspring performance of the symbiotic crab *Pachycheles monilifer* (Anomura: Porcellanidae). J Exp Mar Biol Ecol, 464: 18-25.

Lewis, C. and A. T. Ford. 2012. Infertility in male aquatic invertebrates: a review. Aquat Toxicol, 120: 79-89.

Lewis, R. W. 1970. The densities of three classes of marine lipids in relation to their possible role as hydrostatic agents. Lipids, 5: 151-153.

Li, F., J. Xiang, X. Zhang et al. 2003a. Gonad development characteristics and sex ratio in triploid Chinese shrimp (*Fenneropenaeus chinensis*). Mar Biotechnol, 5: 528-535.

Li, F., J. Xiang, X. Zhang et al. 2003b. Tetraploid induction by heat shocks in Chinese shrimp *Fenneropenaeus chinensis*. J Shellfish Res, 22: 541-545.

Li, F., C. Zhang, K. Yu et al. 2006. Larval metamorphosis and morphological characteristic analysis of triploid shrimp *Fenneropenaeus chinensis* (Osbeck, 1765). Aquac Res, 37: 1180-1186.

Li, H. and D. W. Borst. 1991. Characterization of a methyl farnesoate binding protein in hemolymph from *Libinia emarginata*. Gen Comp Endocrinol, 81: 335-342.

Li, H. Y., S. Y. Hong and Z. H. Jin. 2011. Fecundity and brood loss of sand shrimp, *Crangon uritai* (Decapoda: Crangonidae). J Crust Biol, 31: 34-40.

Li, Q., J. Xie, L. He et al. 2015. Identification of ADAM10 and ADAM17 with potential roles in the spermatogenesis of the Chinese mitten crab, *Eriocheir sinensis*. Gene, 562: 117-27.

Li, S., F. Li, B. Wang et al. 2010. Cloning and expression profiles of two isoforms of a CHH-like gene specifically expressed in male Chinese shrimp, *Fenneropenaeus chinensis*. Gen Comp Endocrinol, 167: 308-316.

Li, S., F. Li, R. Wen and J. Xiang. 2012. Identification and characterization of the sex-determiner transformer-2 homologue in Chinese shrimp, *Fenneropenaeus chinensis*. Sex Dev, 6: 267-278.

Li, S., F. Li, Y. Xie et al. 2013. Screening of genes specifically expressed in males of *Fenneropenaeus chinensis* and their potential as sex markers. J Mar Biol, 2013. doi. org/10.1155/2013/921067.

Liang, D. and S. I. Uye. 1996. Population dynamics and production of the planktonic copepods in a eutrophic inlet of the Inland Sea of Japan. 3. *Paracalanus* sp. Mar Biol, 127: 219-227.

Libertini, A., R. Trisolini and M. Rampin. 2008. Chromosome number, karyotype morphology, heterochromatin distribution and nuclear DNA content of some talitroidean amphipods (Crustacea: Gammaridea). Eur J Entomol, 105: 53-58.

Limburg, P. A. and L. J. Weider. 2002. 'Ancient' DNA in the resting egg bank of a microcrustacean can serve as a palaeolimnological database. Proc R Soc Lond, 269B: 281-287.

Little, G. 1969. The larval development of the shrimp, *Palaemon macrodactylus* Rathbun, reared in the laboratory, and the effect of eyestalk extirpation on development. Crustaceana, 17: 69-87.

Liu, A., L. Kong, M. Zhang et al. 2014. Cloning, expression and cellular localization of *Daphnia pulex* senescence-associated protein, DpSAP. Gene, 534: 424-430.

Liu, H-C and J. Lutzen. 2000. Asexual reproduction in *Sacculina plana* (Cirripedia: Rhizocephala), a parasite of six species of grapsid crabs from Taiwan. Zool Anz, 239: 277-287.

Longhurst, A. R. 1955a. A review of the Notostraca. Bull Br Mus Nat Hist (Zool), 3: 3-57.

Longhurst, A. R. 1955b. Evolution in Notostraca. Evolution, 9: 84-86.

Lopez, L. C. S., D. A. Gonçalves, A. Mantovani and R. I. Rios. 2002. Bromeliad ostracods pass through amphibian (*Scinaxax perpusillus*) and mammalian guts alive. Hydrobiologia, 485: 209-211.

Lovrich, G. A., D. Roccatagliata and L. Peresan. 2004. Hyperparasitism of the cryptoniscid isopod *Liriopsis pygmaea* on the lithodid *Paralomis granulosa* from the Beagle Channel, Argentina. Dis Aquat Org, 58: 71-77.

Lu, W., G. Wainwright, S. G. Webster et al. 2000. Clustering of mandibular organ-inhibiting hormone and moult inhibiting genes in the crab *Cancer pagurus* and implications for regulation of expression. Gene, 253: 197-207.

Lu, W., G. Wainwright, L. A. Olohan et al. 2001. Characterization of cDNA encoding molt-inhibiting hormone of the crab, *Cancer pagurus*; expression of MIH in non-X-organ tissues. Gene, 278: 149-159.

Lugo, J. M., Y. Morera, T. Rodriguez et al. 2006. Molecular cloning and characterization of the crustacean hyperglycemic hormone cDNA from *Litopenaeus schmitti*. J FEBS, 273: 5669-5677.

Lutzen, J. 1981. Field studies on regeneration in *Sacculina carcini* Thompson (Crustacea: Rhizocephala) in the Isefjord, Denmark. J Exp Mar Biol Ecol, 53: 241-249.

Lynch, M. and R. Ennis. 1983. Resource availability, maternal effects, and longevity. Exp Geront, 18: 147-165.

Machida, R. J., M. U. Miya, M. Nishida and S. Nishida. 2002. Complete mitochondrial DNA sequence of *Tigriopus japonicus* (Crustacea: Copepoda). Mar Biotechnol, 4: 406-417.

Maciel, F. E., M. A. Geihs, M. A. Vargas et al. 2008. Daily variation of melatonin content in the optic lobes of the crab *Neohelice granulata*. Com Biochem Physiol Part A: Mol Integ Physiol, 149: 162-166.

Madsen, A., J. S. Madin, C. H. Tan and A. H. Baird. 2014. The reproductive biology of the scleractinian coral *Plesiastrea versipora* in Sydney Harbour, Australia. Sex Early Dev Aquat Org, 1: 25-33.

Maiphae, S., W. Limbut, P. Choikaew and P. Peckhrat. 2010. The cladocera (Ctenopoda and Anomopoda) in rice fields during a crop cycle at Nakhon Si Thammarat province, southern Thailand. Crustaceana, 83: 1469-1482.

Mak, A. S. C., C. L. Choi, S. H. Tiu et al. 2005. Vitellogenesis in the red crab *Charybdis feriatus*: Hepatopancreas-specific expression and farnesoic acid stimulation of vitellogenin gene expression. Mol Reprod Dev, 70: 288-300.

Malati, E. F., B. Heidari and M. Zamani. 2013. The variations of vertebrate-type steroid hormones in the freshwater narrow-clawed crayfish *Astacus leptodactylus* (Eschscholtz, 1823) (Decapoda, Astacidae) during oocyte development. Crustaceana, 86: 129-138.

Malecha, S. R., P. A. Nevin, P. Ha et al. 1992. Sex-ratios and sex-determination in progeny from crosses of surgically sex-reversed freshwater prawns, *Macrobrachium rosenbergii*. Aquaculture, 105: 201-218.

Manning, R. B. and D. L. Felder. 1996. *Nannotheres moorei*, a new genus and species of minute pinnotherid crab from Belize, Caribbean Sea (Crustacea: Decapoda: Pinnotheridae). Proc Biol Soc Washington, 109: 311-317.

Manor, R., S. Weil, S. Oren et al. 2007. Insulin and gender: an insulin-like gene expressed exclusively in the androgenic gland of the male crayfish. Gen Comp Endocrinol, 150: 326-336.

Marazzo, A. and J. L. Valentin. 2003. *Penilia avirostris* (Crustacea: Ctenopoda) in a tropical Bay: variations in density and aspects of reproduction. Acta Oecol, 24: S251-S257.

Marcus, N. H. 1984. Recruitment of copepod nauplii into the plankton: importance of diapause eggs and benthic processes. Mar Ecol Prog Ser, 15: 47-54.

Marcus, N. H., R. Lutz, W. Burnett and P. Cable. 1994. Age, viability, and vertical distribution of zooplankton resting eggs from an anoxic basin: evidence of an egg bank. Limnol Oceanogr, 39: 154-158.

Margolis, L. and T. H. Butler. 1954. An unusual and heavy infection of a prawn, *Pandalus borealis* Kroyer by a nematode, *Contracaecum* sp. J Parasitol, 40: 649-655.

Mariappan, P. 2000. Studies on chela biology and behaviour of *Macrobrachium nobilii* with special reference to aquaculture. Ph.D Thesis, Bharathidasan University, Tiruchirapalli, India.

Mariappan, P., C. Balasundaram and B. Schmitz. 2000. Decapod crustacean chelipeds: an overview. J Biosci, 25: 301-313.

Marsden, G., D. Hewitt and E. Boglio. 2008. Methyl farnesoate inhibition of late stage ovarian development and fecundity reduction in the black tiger prawn, *Penaeus monodon*. Aquaculture, 280: 242-246.

Marshall, S. M. and A. P. Orr. 1952. On the biology of *Calanus finmarchicus*. VII. Factors affecting egg production. J Mar Biol Asso U K, 30: 527-549.

Martin, G., O. Sorokine, M. Moniatte et al. 1999. The structure of a glycosylated protein hormone responsible for sex determination in the isopod, *Armadillidium vulgare*. Euro J Biochem, 262: 727-736.

Martin, J. W. and G. E. Davis. 2001. An updated classification of the recent Crustacea. Nat Hist Mus, Los Angeles County, Sci Ser, 39: 1-124.

Martin, P. and G. Scholtz. 2012. A case of intersexuality in the parthenogenetic marmorkrebs (Decapoda: Astacida: Cambaridae). J Crust Biol, 32: 345-350.

Martin, P., K. Kohlmann and G. Scholtz. 2007. The parthenogenetic marmorkrebs (marbled crayfish) produces genetically uniform offspring. Naturwissenschaften, 94: 843-846.

Martins, J., K. Ribeiro, T. Rangel-Figueiredo and J. Coimbra. 2007. Reproductive cycle, ovarian development, and vertebrate-type steroids profile in the freshwater prawn *Macrobrachium rosenbergii*. J Crust Biol, 27: 220-228.

Martinez, M. 2009. Reproduction of the tadpole shrimp *Triops* (Notostraca) in Mexican waters. Curr Sci, 96: 91-97.

Martinez, M. M. 2001. Running in the surf: hydrodynamics of the shore crab *Grapsus tenuicrustatus*. J Exp Biol, 204: 3097-3112.

Martínez-Mayén, M. and R. Román-Contreras. 2011. Some reproductive aspects of *Latreutes fucorum* (Fabricius, 1798) (Decapoda, Hippolytidae) from Bahía De La Ascensión, Quintana Roo, Mexico. Crustaceana, 84: 1353-1365.

Martínez-Mayén, M. and R. Román-Contreras. 2013. Data on reproduction and fecundity of *Processa bermudensis* (Rankin, 1900) (Caridea, Processidae) from the southern coast of Quintana Roo, Mexico. Crustaceana, 86: 84-97.

Matsumoto, T., E. Ikuno, S. Itoi and H. Sugita. 2008. Chemical sensitivity of the male daphnid, *Daphnia magna*, induced by exposure to juvenile hormone and its analogs. Chemosphere, 72: 451-456.

Matzke-Karasz, R., R. Nagler and S. Hofmann. 2014. The ostracod springtail – camera recordings of a previously undescribed high speed escape jump in the genus *Tonycypris* (Ostracoda, Cypridoidea). Crustaceana, 87: 1072-1094.

Mazurová, E., K. Hilscherová, R. Triebskorn et al. 2008. Endocrine regulation of the reproduction in crustaceans: identification of potential targets for toxicants and environmental contaminants. Biologia, 63: 139-150.

McAllen, R. and E. Brennan. 2009. The effect of environmental variation on the reproductive development time and output of the high-shore rockpool copepod *Tigriopus brevicornis*. J Exp Mar Biol Ecol, 368: 75-80.

McConaugha, J. R. and J. D. Costlow. 1981. Ecdysone regulation of larval crustacean molting. Comp Biochem Physiol A, 68: 91-93.

McDermott, J. J. 2002. Relationships between the parasitic isopods *Stegias clibanarii* Richardson, 1904 and *Bopyrissa wolffi* Markham, 1978 (Bopyridae) and the intertidal hermit crab *Clibanarius tricolor* (Gibbes, 1850) (Anomura) in Bermuda. Ophelia, 56: 33-42.

McDermott, J. J. 2009. Hypersymbioses in the pinnotherid crabs (Decapoda: Brachyura: Pinnotheridae): a review. J Nat Hist, 43: 785-805.

McLain, D. K. and A. E. Pratt. 2011. Body and claw size at autotomy affect the morphology of regenerated claws of the sand fiddler crab, *Uca pugilator*. J Crust Biol, 31: 1-8.

Medina, M. and C. Barata. 2004. Static-renewal culture of *Acartia tonsa* (Copepoda: Calanoida) for ecotoxicological testing. Aquaculture, 229: 203-213.

Mejía-Ortíz, L. M., R. G. Hartnoll and M. López-Mejía. 2010. The abbreviated larval development of *Macrobrachium totonacum* Mejia, Alvarez and Hartnoll, 2003 (Decapoda, Palaemonidae), reared in the laboratory. Crustaceana, 83: 1-16.

Meladenov, P. V. and R. D. Burke. 1994. Echinodermata: asexual reproduction. In: *Reproductive Biology of Invertebrates*. (eds) K. G. Adiyodi and R. G. Adiyodi. Oxford and IBH Publishing, New Delhi, Vol 6, Part B. pp 339-383.

Melander, Y. 1950. Studies on the chromosomes of *Ulophysema öresundense*. Hereditas, 36: 233-255.

Mellors, W. K. 1975. Selective predation of ephippal *Daphnia* and the resistance of ephippal eggs to digestion. Ecology, 1975: 974-980.

Mergeay, J., J. Vanoverbecke, D. Verschuren and L. De Mester. 2007. Extinction and recolonization and dispersal through time in a planktonic crustacean. Ecology, 88: 3032-3043.

Mergeay, J., X. Aguilera, S. Declerck et al. 2008. The genetic legacy of polyploidy Bolivian *Daphnia*: the tropical Andes as a source for the North and South American *D. pulicaria* complex. Mol Ecol, 17: 1789-1800.

Mesa, L. M. and J. T. Eastman. 2012. Antarctic silverfish: life strategies of a key species in the high-Antarctic ecosystem. Fish Fisher, 13: 241-266.

Meyers, T. R. 1990. Diseases caused by protozoans and metazoans. In: *Diseases of Marine Animals*. (ed) O. Kinne. Biologische and Anstalt Helgoland, Hamburg, Vol 3. pp 350-389.

Michels, E., K. Cottenie, L. Neys and L. De Meester. 2001. Zooplankton on the move: first results on the quantification of dispersal of zooplankton in a set of interconnected ponds. Hydrobiologia, 442: 117-126.

Miller, C. B., J. A. Crain and N. H. Marcus. 2005. Seasonal variation of male type antennular setation in *Calanus finmarchicus*. Mar Ecol Prog Ser, 30: 217-219.

Miranda, I. and F. Mantelatto. 2010. Temporal dynamic of the relationship between the parasitic isopod *Aporobopyrus curtatus* (Crustacea: Isopoda: Bopyridae) and the anomuran crab *Petrolisthes armatus* (Crustacea: Decapoda: Porcellanidae) in southern Brazil. Lat Am J Aquat Res, 38: 210-217.

Mittenthal, J. E. 1980. On the form and size of crayfish legs regenerated after grafting. Biol Bull, 159: 700-713.

Mittenthal, J. E. 1981. Intercalary regeneration in legs of crayfish: distal segments. Dev Biol, 88: 1-14.

Miyashita, L. K., M. Pompeu, S. A. Gaeta and R. M. Loper. 2010. Seasonal contrasts in abundance and reproductive parameters of *Penilia avirostris* (Cladocera, Ctenopoda) in a coastal subtropical area. Mar Biol, 157: 2511-2519.

Mlinarec, J., M. Mčžić, M. Pavlica et al. 2011. Comparative karyotype investigations in the European crayfish *Astacus astacus* and *A. leptodactylus* (Decapoda, Astacidae). Crustaceana, 84: 1497-1510.

Modlin, R. F. and P. A. Harris. 1989. Observations on the natural history and experiments on the reproductive strategy of *Hargeria rapax* (Tanaidacea). J Crust Biol, 9: 578-586.

Mohamed, K. S. and A. D. Diwan. 1991. Neuroendocrine regulation of ovarian maturation in the Indian white prawn *Penaeus indicus* H. Milne Edwards. Aquaculture, 98: 381-393.

Mohammed, D. S., S. D. Salman and M. H. Ali. 2010. A morphological and molecular study on *Artemia franciscana* (Branchiopoda, Anostraca) from Basrah, Iraq. Crustaceana, 83: 941-956.

Montgomery, E. M. and A. R. Palmer. 2012. Effects of body size and shape on locomotion in the bat star (*Patiria miniata*). Biol Bull, 222: 222-232.

Morelli, M. and Aquacop. 2003. Effects of heat-shock on cell division and microtubule organization in zygotes of the shrimp *Penaeus indicus* (Crustacea, Decapoda) observed with confocal microscopy. Aquaculture, 216: 39-53.

Moritz, C. 1987. A note on the hatching and viability of *Ceriodaphnia* ephippia collected from lake sediment. In: *Cladocera*. (eds) L. Forró and D. G. Frey. Springer Verlag, Netherlands. pp 309-314.

Morritt, D. and J. I. Spicer. 1996. Developmental ecophysiology of the beachflea *Orchestia gammarellus* (Pallas) (Crustacea: Amphipoda). I. Female control of the embryonic environment. J Exp Mar Biol Ecol, 207: 191-203.

Mosco, A., S. Pegoraro, P. G. Giulianini and P. Edomi. 2013. Cloning of the crustacean hyperglycemic hormone gene promoter of *Astacus leptodactylus*. J Crust Biol, 33: 56-61.

Mozley, A. 1932. A biological study of a temporary pond in western Canada. Am Nat, 235-249.

Mu, X. and G. A. LeBlanc. 2002. Developmental toxicity of testosterone in the crustacean *Daphnia magna* involves anti-ecdysteroidal activity. Gen Comp Endocrinol, 129: 127-133.

Mu, X. and G. A. LeBlanc. 2004. Cross communication between signaling pathways: juvenoid hormones modulate ecdysteroid activity in a crustacean. J Exp Zool Part A: Comp Exp Biol, 301: 793-801.

Muñoz, J. and F. Pacios. 2010. Global biodiversity and geographical distribution of diapausing aquatic invertebrates: the case of the cosmopolitan brine shrimp, *Artemia* (Branchiopoda, Anostraca). Crustaceana, 83: 465-480.

Munuswamy, N. 1982. Aspects of reproductive biology of the fairy shrimp *Streptocephalus dichotomus* Baird (Cratacea: Anostraca), Ph.D Thesis, University of Madras, Chennai.

Munuswamy, N. 2005. Fairy shrimp as live food in aquaculture. Aquafeeds, 2: 10-12.

Munuswamy, N. and T. Subramoniam. 1985a. Oogenesis and shell gland activity in a freshwater fairy shrimp *Streptocephalus dichotomus* Baird (Crustacea: Anostraca). Cytobios, 44: 137-147.

Munuswamy, N. and T. Subramoniam. 1985b. Studies on the oviductal secretion and ovulation in *Streptocephalus dichotomus* Baird, 1860 (Anostraca). Crustaceana, 49: 113-118.

Munuswamy, N. and T. Subramoniam. 1985c. Influence of mating on ovarian and shell gland activity in a freshwater fairy shrimp *Streptocephalus dichotomus* (Anostraca). Crustaceana, 49: 225-232.

Munuswamy, N., S. Satyanarayanan and A. Priyadharshini. 2009. Embroynic development and occurrence of p 26 and artemin-like protein in the cryptobiotic cysts of freshwater fairy shrimp *Streptocephalus dichotomus* Baird. Curr Sci, 96: 103-110.

Murakami, T., T. H. Lee, K. Suzuki and F. Yamazaki. 2004. An unidentified gland-like tissue near the androgenic gland of red swamp crayfish, *Procambarus clarkii*. Fisher Sci, 70: 561-568.

Murugan, G., A. M. Maeda-Martínez, G. Criel and H. J. Dumont. 1996. Unfertilized oocytes in streptocephalids: Resorbed or released?. J Crust Biol, 16: 54-60.

Murugan, N. 1975a. The biology of *Ceriodaphnia cornuta* Sars (Cladocera: Daphnidae). J Inland Fish Soc Ind, 7: 8-87.

Murugan, N. 1975b. Egg production, development and growth in *Moina micrura* Kurz (1874) (Cladocera: Moinidae). Freshwat Biol, 5: 245-250.

Murugesan, P., T. Balasubramanian and T. J. Pandian. 2010. Does haemocoelom exclude embryonic stem cells and asexual reproduction in invertebrates. Curr Sci, 98: 768-771.

Muthukrishnan, J. 1994. Arthropoda-Insecta. In: *Reproduction Biology of Invertebrates*. (eds) K. G. Adiyodi and R. G. Adiyodi. Oxford and IBH Publishers, New Delhi Vol 6, Part B. pp 167-292.

Muthupriya, P. and K. Altaff. 2004. Fecundity of *Mesocyclops thermocyclopoids* (Cyclopoidae: Coepoda) in relation to different inoculum rates and composition of algal food. Ind Hydro Biol, 7: 1-5.

Muthuvelu, S., P. Murugesan, M. Muniasamy et al. 2013. Changes in benthic macrofaunal assemblages in relation to bottom trawling in Cuddalore and Parangipettai coastal waters, Southeast coast of India. Ocean Sci J, 48: 1-13.

Nagamine, C., A. W. Knight, A. Maggenti and G. Paxman. 1980a. Effects of androgenic gland ablation on male primary and secondary sexual characteristics in the Malaysian prawn, *Macrobrachium rosenbergii* (de Man) (Decapoda, Palaemonidae), with first evidence of induced feminization in a non-hermaphroditic decapod. Gen Comp Endocrinol, 41: 423-441.

Nagamine, C., A. W. Knight, A. Maggenti and G. Paxman. 1980b. Masculinization of female *Macrobrachium rosenbergii* (de Man) (Decapoda, Palaemonidae) by androgenic gland implantation. Gen Comp Endocrinol, 41: 442-457.

Nagaraju, G. P. C. 2007. Is methyl farnesoate a crustacean hormone? Aquaculture, 272: 39-54.

Nagaraju, G. P. C. 2011. Reproductive regulators in decapod crustaceans: an overview. J Exp Biol, 214: 3-16.

Nagaraju, G. P. C. and D. W. Borst. 2008. Methyl farnesoate couples environmental changes to testicular development in the crab *Carcinus maenas*. J Exp Biol, 211: 2773-2778.

Nagaraju, G. P. C., N. J. Suraj and P. S. Reddy. 2003. Methyl farnesoate stimulates gonad development in *Macrobrachium malcolmsonii* Milne Edwards. Crustaceana, 76: 1171-1178.

Nagaraju, G. P. C., P. R. Reddy and P. S. Reddy. 2004. Mandibular organ: it's relation to body weight, sex, molt and reproduction in the crab, *Oziotelphusa senex senex* Fabricius (1791). Aquaculture, 232: 603-612.

Nagaraju, G. P., G. L. Prasad and P. S. Reddy. 2005. Isolation and characterization of mandibular organ-inhibiting hormone from the eyestalks of freshwater crab *Oziotelphusa senex senex*. Int J Appl Sci Eng, 3: 61-68.

Nagaraju, G. P. C., P. R. Reddy and P. S. Reddy. 2006. *In vitro* methyl farnesoate secretion by mandibular organs isolated from different molt and reproductive stages of the crab *Oziotelphusa senex senex*. Fish Sci, 72: 410-414.

Nair, K. B. 1939. The reproduction, oogenesis and development of *Mesopodopsis orientalis* Tatt. Proc: Plant Sci, 9: 175-223.

Nair, K. K. C. and K. Anger. 1979a. Life cycle of *Corophium insidiosum* (Crustacea, Amphipoda) in laboratory culture. Helgoländer Wissens Meeresunters, 32: 279-294.

Nair, K. K. C. and K. Anger. 1979b. Experimental studies on the life cycle of *Jassa falcata* (Crustacea, Amphipoda). Helgoländer Wissens Meeresunters, 32: 444-452.

Nakanauchi, M. and T. Takeshita. 1983. Ascidian one-half embryo can develop into functional adults. J Exp Zool, 227: 155-158.

Nakatani, I. 1999. An albino of the crayfish *Procambarus clarkii* (Decapoda: Cambaridae) and its offspring. J Crust Biol, 19: 380-383.

Nakatani, I. 2000. Reciprocal transplantation of leg tissue between albino and wild crayfish *Procambarus clarkii* (Decapoda: Cambaridae). J Crust Biol, 20: 453-459.

Nakayama, C., S. Peixoto, S. Lopes et al. 2008. Methodos de extrusao manual e electrica dos espermatoforos of *Farfantepenaeus paulensis* (Decapoda: Penaeidae). Gencia Rural, 38: 2018-2022.

Nandini, S. and S. S. S. Sarma. 2007. Effect of algal and animal diets on life history of the freshwater copepod *Eucyclops serrulatus* (Fischer, 1851). Aquat Ecol, 41: 75-84.

Nandini, S., A. R. N. Ortiz and S. S. S. Sarma. 2011. *Elaphoidella grandidieri* (Harpacticoida: Copepoda). Demographic characteristics and possible use as live prey in aquaculture. J Env Biol, 32: 505-511.

Nemoto, T., K. Kamada and K. Hara. 1972. Fecundity of a euphausiid crustacean *Nematoscelis difficilis* in the North Pacific Ocean. Mar Biol, 14: 41-47.

Nemoto, T., J. Mauchline and K. Kamada. 1976. Brood size and chemical composition of *Pareuchaeta norvegica* (Crustacea: Copepoda) in Loch Etive, Scotland. Mar Biol, 36: 151-157.

Nevalainen, L. and K. Sarmaja-Korjonen. 2008. Timing of sexual reproduction in chydorid cladocerans (Anomopoda, Chydoridae) from nine lakes in southern Finland. Eston J Ecol, 57: 21-36.

Nicol, S. and Y. Endo. 1999. Krill fisheries: development, management and ecosystem implications. Aquat Liv Resour, 12: 105-120.

Niehoff, B. 2004. The effect of food limitation on gonad development and egg production of the planktonic copepod *Calanus finmarchicus*. J Exp Mar Biol Ecol, 307: 237-259.

Nielson, C. 1990. Bryozoa Entoprocta. In: *Reproductive Biology of Invertebrates*. (eds) K. G. Adiyodi and R. G. Adiyodi. Oxford and IBH Publishers, New Delhi, Vol 4, Part B. pp 201-254.

Niiyama, H. 1959. A comparative study of the chromosomes in decapods, isopods and amphipods, with some remarks on cytotaxonomy and sex-determination in the Crustacea. Mem Facul Fisher Hokkaido Univ, 7: 1-60.

Nikitin, N. S. 1977. Experimental morphological study of morphogenetic potential of homogenous conglomerates of different cell types from the freshwater sponge *Ephydatia fluviatilis*. Sov J Dev Biol, 8: 460-467.

Nithya, M. and N. Munuswamy. 2002. Immunocytochemical identification of crustacean hyperglycemic hormone-producing cells in the brain of a freshwater fairy shrimp, *Streptocephalus dichotomus* Baird (Crustacea: Anostraca). Hydrobiologia, 486: 325-333.

Normark, B. B., O. P. Judson and N. A. Moran. 2003. Genomic signatures of ancient asexual lineages. Biol J Linn Soc, 79: 69-84.

Norris, B. J., F. E. Coman, M. J. Sellars and N. P. Preston. 2005. Triploid induction in *Penaeus japonicus* (Bate) with 6-dimethylaminopurine. Aquac Res, 36: 202-206.

O'Brien, J. 1984. Precocious maturity of the majid crab, *Pugettia producta*, parasitized by the rhizocephalan barnacle, *Heterosaccus californicus*. Biol Bull, 166: 384-395.

Ocampo, E. H., J. D. Nuñez, M. Cledón and J. A. Baeza. 2014. Parasitic castration in slipper limpets infested by the symbiotic crab *Calyptraeotheres garthi*. Mar Biol, 161: 2107-2120.

Oda, S., N. Tatarazako, H. Watanabe et al. 2005. Production of male neonates in *Daphnia magna* (Cladocera, Crustacea) exposed to juvenile hormones and their analogs. Chemosphere, 61: 1168-1174.

Oh, C. W. and J. N. Kim. 2008. Reproductive biology of *Exopalaemon carinicauda* (Decapoda, Palaemonidae) in the Hampyong Bay of Korea. Crustaceana, 81: 949-962.

Ojima, Y. 1958. A cytological study on the development and maturation of the parthenogenetic and sexual eggs of *Daphnia pulex* (Crustacea, Cladocera). Kwansei Gakuen Univ Ann Stud, 6: 123-176.

Okumura, T. and M. Hara. 2004. Androgenic gland cell structure and spermatogenesis during the molt cycle and correlation to morphotypic differentiation in the giant freshwater prawn, *Macrobrachium rosenbergii*. Zool Sci, 21: 621-628.

Okumura, T. and K. Sakiyama. 2004. Hemolymph levels of vertebrate-type steroid hormones in female kuruma prawn *Marsupenaeus japonicus* (Crustacea: Decapoda: Penaeida) during natural reproductive cycle and induced ovarian development by eyestalk ablation. Fisher Sci, 70: 372-380.

Okumura, T., C. H. Han, Y. Suzuki et al. 1992. Changes in hemolymph vitellogenin and ecdysteroid levels during the reproductive and nonreproductive molt cycles in the freshwater prawn *Macrobrachium nipponense*. Zool Sci, 9: 37-45.

Okuno, A., Y. Hasegawa and H. Nagasawa. 1997. Purification and properties of androgenic gland hormone from the terrestrial isopod *Armadillidium vulgare*. Zool Sci, 14: 837-842.

Okuno, A., Y. Hasegawa, T. Ohira et al. 1999. Characterization and cDNA cloning of androgenic gland hormone of the terrestrial isopod *Armadillidium vulgare*. Biochem Biophy Res Comm, 264: 419-423.

Oliveira, R. F. 2006. Neuroendocrine mechanism of alternative reproductive tactics in fish. In: *Fish Physiology,Behaviour and Fish Physiology*. (eds) K. Sloman, S. Batshine and R. Wilson. Elsevier, Amsterdam, Vol. 24. pp 297-357.

Olmstead, A. W. and G. A. LeBlanc. 2002. Juvenoid hormone methyl farnesoate is a sex determinant in the crustacean *Daphnia magna*. J Exp Zool, 293: 736-739.

Olmstead, A. W. and G. A. LeBlanc. 2007. The environmental endocrine basis of gynandromorphism (intersex) in a crustacean. Int J Biol Sci, 3: 77-84.

Olsen, L. C., R. Aasland and A. Fjose. 1997. A vasa-like gene in zebrafish identifies putative primordial germ cells. Mech Dev, 66: 95-105.

Onbe, T. 1999. Ctenopoda and Onycopoda (= Cladocera). In: *South Atlantic Zooplankton*. (ed) D. Boltoveskoy. Backhuys Publishers, Leiden. pp 797-813.

Otani, S., S. Maegawa, K. Inoue et al. 2002. The germ cell lineage identified by vas-mRNA during the embryogenesis in goldfish. Zool Sci, 19: 519-526.

Overstreet, R. M. 1978. Marine melodies? worms, germs and other symbionts. From the Northern Gulf of Mexico. MASGP78-021 Mississippi-Alabama Sea Grant Consortium. Ocean Springs, Mississippi.

Palma, P., V. L. Palma, R. M. Fernandes et al. 2009. Endosulfan sulphate interferes with reproduction, embryonic development and sex differentiation in *Daphnia magna*. Ecotoxicol Environ Saf, 72: 344-350.

Pandian, T. J. 1967. Changes in chemical composition and caloric content of developing eggs of the shrimp *Crangon crangon*. Helgoländer wissens Meeresunters, 16: 216-224.

Pandian, T. J. 1970a. Ecophysiological studies on the developing eggs and embryos of the European lobster *Homarus gammarus*. Mar Biol, 5: 154-167.

Pandian, T. J. 1970b. Yolk utilization and hatching time in the Canadian lobster *Homarus americanus*. Mar Biol, 7: 249-254.

Pandian, T. J. 1972. Egg incubation and yolk utilization in the isopod *Ligia oceanica*. Proc Ind Natl Sci Acad, 38: 430-441.

Pandian, T. J. 1975. Mechanisms of heterotrophy. In: *Marine Ecology*. (ed) O. Kinne. John Wiley, London, Vol 2, Part 1. pp 61-249.

Pandian, T. J. 1987. Fish. In: *Animal Energetics: Bivalvia through Reptilia*. (eds) T. J. Pandian and F. J. Vernberg. Academic Press, San Diego, Vol 2. pp 357-465.

Pandian, T. J. 1994. Arthropoda-Crustacea. In: *Reproduction Biology of Invertebrates*. (eds) K. G. Adiyodi and R. G. Adiyodi. Oxford and IBH Publishers, New Delhi, Vol 6, Part B. pp 39-166.

Pandian, T. J. 2010. *Sexuality in Fishes*. Science Publishers/CRC Press, USA. p 208.

Pandian, T. J. 2011. *Sex Determination in Fish*. Science Publishers/CRC Press, USA. p 270.

Pandian, T. J. 2012. *Genetic Sex Differentiation in Fish*. CRC Press, USA. p 214.

Pandian, T. J. 2013. *Endocrine Sex Differentiation in Fish*. CRC Press, USA. p 303.

Pandian, T. J. 2014. *Environmental Sex Differentiation in Fish*. CRC Press, USA. p 299.

Pandian, T. J. and K. H. Schumann. 1967. Chemical composition and caloric content of egg and zoea of the hermit crab *Eupagurus bernhardus*. Helgoländer wissens Meeresunters, 16: 225-230.

Pandian, T. J. and J. Flüchter. 1968. Rate and efficiency of yolk utilization in developing eggs of the sole *Solea solea*. Helgoländer wissens Meeresunters, 18: 53-60.

Pandian, T. J. and S. Katre. 1972. Effect of hatching time on larval mortality and survival of the prawn *Macrobrachium idae*. Mar Biol, 13: 330-337.

Pandian, T. J. and C. Balasundaram. 1980. Contribution to the reproductive biology and aquaculture of *Macrobrachium nobilli*, Proc Symp Invert Reprod, Madras University, Madras, 1: 183-193.

Pandian, T. J. and C. Balasundaram. 1982. Moulting and spawning cycles in *Macrobrachium nobilii* (Henderson and Mathai). Int J Invert Reprod, 5: 21-30.

Panouse, J. B. 1943. Influence de l'ablation du peduncule oculaire sur la croissance de kovaire chez la crevette *Leander serratus*. C R Hepd Acad Sci Paris, T 217: 553-555.

Paolucci, M., C. Di Cristo and A. Di Cosmo. 2002. Immunological evidence for progesterone and estradiol receptors in the freshwater crayfish *Austropotamobius pallipes*. Mol Reprod Dev, 63: 55-62.

Parnes, S., I. Khalaila, G. Hulata and A. Sagi. 2003. Sex determination in crayfish: are intersex *Cherax quadricarinatus* (Decapoda, Parastacidae) genetically females? Genet Res, 82: 107-116.

Parrish, K. K. and D. F. Wilson. 1978. Fecundity studies on *Acartia tonsa* (Copepoda: Calanoida) in standardized culture. Mar Biol, 46: 65-81.

Patton, W. K. 2014. On the natural history and functional morphology of the clam shrimp, *Lynceus brachyurus* Müller (Branchiopoda: Laevicaudata). J Crust Biol, 34: 677-703.

Pauwels, K., R. Stoks, A. Verbiest and L. De Meester. 2007. Biochemical adaptation for dormancy in subitaneous and dormant eggs of *Daphnia magna*. Hydrobiologia, 594: 91-96.

Pawlos, D., K. Formicki, A. Korzelecka-Orkisz and A. Werinicki. 2010. Hatching process in the signal crayfish *Pacifastacus leniusculus* (Dane 1852) (Decapoda, Astacidae). Crustaceana, 83: 1167-1180.

Peeters, L. 1996. Genetic manipulation performs on warm water fish. Center Universities of the Polynesian Francaise, Tahiti.

Peixoto, S., W. Wasielesky and R. O. Cavalli. 2011. Broodstock maturation and reproduction of the indigenous pink shrimp *Farfantepenaeus paulensis* in Brazil: an updated review on research and development. Aquaculture, 315: 9-15.

Pennafirma, S. and A. Soares-Gomes. 2009. Population biology and reproduction of *Kallapseudes schubartii* Mane-Garzon, 1949 (Peracarida, Tanaidacea) in a tropical coastal lagoon, Itaipu, Southeastern, Brazil. Crustaceana, 82: 1509-1526.

Pérez-Martínez, C., J. Barea-Arco, J. M. Conde-Porcuna and R. Morales-Baquero. 2007. Reproduction strategies of *Daphnia pulicaria* population in a high mountain lake of Southern Spain. Hydrobiologia, 594: 75-82.

Peterson, J. K., D. R. Kashian and S. I. Dodson. 2001. Methoprene and 20–OH-ecdysone affect male production in *Daphnia pulex*. Environ Toxicol Chem, 20: 582-588.

Peterson, W. T. 1986. Development, growth and survivorship of the copepod *Calanus marshallae* in the laboratory. Mar Ecol Prog Ser, 111: 79-86.

Petrusek, A., M. Černý and E. Audenaert. 2004. Large intercontinental differentiation of *Moina micrura* (Crustacea: Anomopoda): one less cosmopolitan cladoceran?. Hydrobiologia, 526: 73-81.

Phang, V. P. E. 1975. Studies on *Thompsonia* species a parasite of the edible swimming crab *Portunus pelagicus*. Malay Nat J, 29: 90-98.

Plodsomboon, S., A. M. Maeda-Martínez, H. Obregón-Barboza and L. O. Sanoamuang. 2012. Reproductive cycle and genitalia of the fairy shrimp *Branchinella thailandensis* (Branchiopoda: Anostraca). J Crust Biol, 32: 711-726.

Pongtippatee, P., R. Luppanakane, P. Thaweethamsewee et al. 2009. Delay of the egg activation process in the black tiger shrimp *Penaeus monodon* by manipulation of magnesium levels in spawning water. Aquac Res, 41: 227-232.

Pongtippatee, P., K. Laburee, P. Thaweethamsewee et al. 2012. Triploid *Penaeus monodon*: sex ratio and growth rate. Aquaculture, 356-357: 7-13.

Pongtippatee-Taweepreda, P., J. Chavadej, P. Plodpai et al. 2005. Egg activation in the black tiger shrimp *Penaeus monodon*. Aquaculture, 234: 183-198.

Poulet, S. A., A. Ianora, A. Miralto and L. Meijer. 1994. Do diatoms arrest embryonic development of copepods? Mar Ecol Prog Ser, 111: 79-86.

Preechaphol, R., R. Leelatanawit, K. Sittikankeaw et al. 2007. Expressed sequence tag analysis for identification and characterization of sex-related genes in the giant tiger shrimp *Penaeus monodon*. J Biochem Mol Biol, 40: 501-510.

Preetha, P. E. and K. Altaff. 1996. Fecundity in relation to different types of feed composition in *Sinodiaptomus (Rhinediaptomus) indicus*. (Copepoda: *Calanoida*). Proc Ind Natl Acad, 62B: 191-198.

Prestwich, G. D., M. J., Bruce, I. Ujváry and E. S. Chang. 1990. Binding proteins for methyl farnesoate in lobster tissues: detection by photo affinity labeling. Gen Comp Endocrinol, 80: 232-237.

Qian, Y. Q., L. Dai, J. S. Yang et al. 2009. CHH family peptides from an 'eyeless' deep-sea hydrothermal vent shrimp, *Rimicaris kairei*: characterization and sequence analysis. Comp Biochem Physiol B. Biochem Mol Biol, 154: 37-47.

Quinitio, E. T., A. Hara, K. Yamauchi and S. Nakao. 1994. Changes in the steroid hormone and vitellogenin levels during the gametogenic cycle of the giant tiger shrimp, *Penaeus monodon*. Comp Biochem Physiol, 109: 21-26.

Ra'anan, Z. and A. Sagi. 1985. Alternative mating strategies in male morphotypes of the freshwater prawn *Macrobrachium rosenbergii* (De Man). Biol Bull, 169: 592-601.

Ra'anan, Z. and D. Cohen. 1985. Ontogeny of social structure and population dynamics of giant freshwater prawn *Macrobrachium rosenbergii* (De Man). In: *Crustacea Issue 3 Crustacean Growth Factors in Adult Growth*. (ed) A. Wenner. A. A. Balkema Publishers, Amsterdam. pp 277-311.

Rabet, N., D. Montero and S. Lacau. 2014. The effects of soils and soil stay on the egg morphology of Neotropical *Eulimnadia* (Branchiopoda: Limnadiidae). J Limnol, 73(1). Doi: 10.4081/j limnol.2014.707.

Radhika, M., A. A. Nazar, N. Munuswamy and K. Nellaiappan. 1998. Sex-linked differences in phenol oxidase in the fairy shrimp *Streptocephalus dichotomus* Baird and their possible role (Crustacea: Anostraca). Hydrobiologia, 377: 161-164.

Raghukumar, C., S. Raghukumar, G. Sheelu et al. 2004. Buried in time: culturable fungi in a deep-sea sediment core from the Chagos Trench, Indian Ocean. Deep Sea Res Part I: Oceanograp Res Papers, 51: 1759-1768.

Raikova, E. V. 1973. Life cycle and systematic position of *Polypodium hydriforme* Ussa (Coelenterata), a cnidarian parasite of the eggs of Acipenseridae. Proc Second Int Symp Cnidaria. Seto Mar Biol Lab, Shirahama. pp 165-174.

Ram, J. L., X. Fei, S. M. Danaher et al. 2008. Finding females: pheromone-guided reproductive tracking behavior by male *Nereis succinea* in the marine environment. J Exp Biol, 211: 757-765.

Ramasubramanian, V. and N. Munuswamy. 1993. Scanning electron microscopic studies on the hatching of *Artemia parthenogenetica* cysts. Cytobios, 76: 75-80.

Raymond, C. S., M. W. Murphy, M. G. O'Sullivan et al. 2000. *Dmrt1*, a gene related to worm and fly sexual regulators, is required for mammalian testis differentiation. Genes Dev, 14: 2587-2595.

Reddy, P. R., G. P. C. Nagaraju and P. S. Reddy. 2004. Involvement of methyl farnesoate in the regulation of molting and reproduction in the freshwater crab *Oziotelphusa senex senex*. J Crust Biol, 24: 511-515.

Reddy, P. R., P. Kiranmayi, K. T. Kumar and P. S. Reddy. 2006. 17α-Hydroxyprogesterone induced ovarian growth and vitellogenesis in the freshwater rice field crab *Oziotelphusa senex senex*. Aquaculture, 254: 768-775.

Reddy, P. S. and R. Ramamurthi. 1998. Methyl farnesoate stimulates ovarian maturation in the freshwater crab *Oziotelphusa senex senex* Fabricius. Curr Sci, 74: 68-70.

Reeve, M. R. 1963. The filter-feeding in *Artemia*: 1. In pure cultures of plant cells. J Exp Biol, 40: 195-205.

Reinhard, E. G. 1942. Studies on the life history and host-parasite relationship of *Peltogaster paguri*. Biol Bull, 83: 401-415.

Reinhard, E. G. 1949. Experiments on the determination and differentiation of sex in the bopyrid *Stegophryxus hyptius* Thompson. Biol Bull, 96: 17-31.

Reinhard, E. G. and T. von Brand. 1944. The fat content of *Pagurus* parasitized by *Peltogaster* and its relation to theories of sacculinization. Physiol Zool, 17: 31-41.

Revathi, P. and N. Munuswamy. 2010. Effect of tributyltin on the early embryonic development in the freshwater prawn *Macrobrachium rosenbergii* (De Man). Chemosphere, 79: 922-927.

Reverberi, G. and M. Pitotti. 1942. Il ciclo biologico e la determinazione fenotipica del sesso di *Ione thoracica* Montagu, Bopiride parassita di *Callianassa laticauda* Otto. Pubbli Staz Zool Napoli, 19: 111-184.

Rice, M. E. and J. F. Pilger. 1993. Sipuncula. In: *Reproduction Biology of Invertebrates*. (eds) K. G. Adiyodi and R. G. Adiyodi. Oxford and IBH Publishers, New Delhi, Vol 6, Part A. pp 297-310.

Richman, S. 1958. The transformation of energy by *Daphnia pulex*. Ecol Monogr, 273-291.

Rigaud, T. and P. Juchault. 1998. Sterile intersexuality in an isopod induced by the interaction between a bacterium (*Wolbachia*) and the environment. Can J Zool, 76: 493-499.

Rinkevich, B. 2009. Stem cells: autonomy interactors that emerge as causal agents and legitimate units of selection. In: *Stem Cells in Marine Organisms*. (eds) B. Rinkevich and V. Matranga. Springer Verlag, Dordrecht. pp 1-19.

Rinkevich, B. and V. Matranga. 2009. *Stem Cells in Marine Organisms*. Springer, Dordrecht. p 371.

Rinkevich, Y., V. Matranga and B. Rinkevich. 2009. Stem cells in aquatic invertebrates: common premises and emerging unique themes. In: *Stem Cells in Marine Organisms*. (eds) B. Rinkevich and V. Matranga. Springer Verlag, Dordrecht. pp 61-104.

Ritchie, L. E. and J. T. Høeg. 1981. The life history of *Lernaeodiscus porcellanae* (Cirripedia: Rhizocephala) and co-evolution with its porcellanid host. J Crust Biol, 1: 334-347.

Roberts, S. and C. R. Kennedy. 1993. Platyhelminthes-Eucestoda. In: *Reproduction Biology of Invertebrates*. (eds) K. G. Adiyodi and R. G. Adiyodi. Oxford and IBH Publishers, New Delhi, Vol 6, Part A. pp 197-218.

Robey, J. and J. C. Groeneveld. 2014. Fecundity of the langoustine *Metanephrops mozambicus* MacPherson, 1990 (Decapoda, Nephropidae) in eastern South Africa. Crustaceana, 87: 814-826.

Rocha, S. S., S. L. D. S. Bueno, R. M. Shimizu and F. L. Mantelatto. 2013. Reproductive biology and population structure of *Potimirim brasiliana* Villalobos, 1959 (Decapoda, Atyidae) from a littoral fast-flowing stream, Sao Paulo state, Brazil. Crustaceana, 86: 67-83.

Rodriguez, E. M., L. S. L. Greco, D. A. Medesani et al. 2002a. Effect of methyl farnesoate, alone and in combination with other hormones, on ovarian growth of the red swamp crayfish, *Procambarus clarkii*, during vitellogenesis. Gen Comp Endocrinol, 125: 34-40.

Rodriguez, E. M., D. A. Medesani, L. S. L. Greco and M. Fingerman 2002b. Effects of some steroids and other compounds on ovarian growth of the red swamp crayfish, *Procambarus clarkii*, during early vitellogenesis. J Exp Zool, 292: 82-87.

Rodriguez, E. M., D. A. Medesani and M. Fingerman. 2007. Endocrine disruption in crustaceans due to pollutants: a review. Comp Biochem Physiol, 146 A: 661-671.

Roessler, E. W. 1995. Review of Colombian Conchostraca (Crustacea)—ecological aspects and life cycles—family Cyclestheriidae. Hydrobiologia, 298: 113-124.

Rogers, D. C. 2014. Larger hatching fractions in ovarian dispersed anostraca eggs (Branchiopoda). J Crust Biol, 34: 135-143.

Romero-Rodríguez, J. and R. Román-Contreras. 2008. Aspects of the reproduction of *Bopyrinella thorii* (Richardson, 1904) (Isopoda, Bopyridae), a branchial parasite of *Thor floridanus* Kingsley, 1878 (Decapoda, Hippolytidae) in Bahía de la Ascensión, Mexican Caribbean. Crustaceana, 81: 1201-1210.

Rompolas, P., J. Azimzadeh, W. F. Marshall and S. M. King. 2013. Analysis of ciliary assembly and function in planaria. Methods Enzymol, 525: 245-264.

Rosen, O., R. Manor, S. Weil et al. 2010. A sexual shift induced by silencing of a single insulin-like gene in crayfish: ovarian up regulation and testicular degeneration. PLoS ONE, 5: e15281-e15281.

Ross, R. M. and L. B. Quetin. 2000. Reproduction in euphausiacea. In: *Krill: Biology, Ecology and Fisheries*. (ed) I. Everson. Fish Aquat Res Ser, Vol 6. pp 150-181.

Rossi, V. and P. Menozzi. 2012. Inbreeding and out-breeding depression in geographical parthenogens *Heterocypris incongruens* and *Eucypris virens* (Crustacea: Ostracoda). Itl J Zool, doi org/10.1080/11250003.2012.718375: 1-9.

Rozzi, M. C., V. Rossi, G. Benassi and P. Menozzi. 1991. Effetto della temperatura e del fotoperiodo sull'attivazione di uova durature di dloni elettroforetici di *Heterocypris incongruens* (Ostracoda). Società Italieana di Ecologia Atti, 12: 591-594.

Rubiliani, C. 1985. Response by two species of crabs to a rhizocephalan extract. J Invert Pathol, 45: 304-310.

Rubiliani, C., M. Rubiliani-Durozoi and G. G. Payen. 1980. Effets de la Sacculine sur les gonades, les glandes androgenes et filesysteme nerveux central des crabes *Carcinus maenas* (L.) et *C. mediterraneus* Czerniavsky. Bull Zool Soc France, 105: 95-100.

Rudoph, E. H. 1999. Intersexuality in freshwater crayfish *Somastacus spinifrons* (Philipp 1982) (Decapoda, Parastacidae). Crustaceana, 72: 325-337.

Rudoph, E., A. Verdi and J. Tapia. 2001. Intersexuality in the burrowing crayfish *Parastacus varicosus* Faxon, 1898 (Decapoda,Parastacidae). Crustaceana, 74: 27-37.

Runham, N. W. 1993. Mollusca. In: *Reproduction Biology of Invertebrates.* (eds) K. G. Adiyodi and R. G. Adiyodi. Oxford and IBH Publishers, New Delhi, Vol 6, Part A. pp 310-383.

Sabbadin, A. 1978. Genetics of the colonial ascidian *Botryllus schlosseri.* In: *Marine Organisms.* (ed) B. Beardmore. Plenum, New York. pp 185-209.

Saboor, A. and K. Altaff. 2012. Decmocraphic characteristics of *Onychocamptus bengalensis* (Copepoda: Harpecticoida)-A potential live feed for aquaculture. Revel Sci, 2: 51-59.

Sadovy, Y. and D. Y. Shapiro. 1987. Criteria for the diagnosis of hermaphroditism in fishes. Copeia,1987: 136-156.

Saffman, E. E. and P. Lasko. 1999. Germline development in vertebrates and invertebrates. Cell Mol Life Sci, 55: 1141-1163.

Sagawa, K., H. Yamagata and Y. Shiga. 2005. Exploring embryonic germ line development in the water flea, *Daphnia magna,* by zinc-finger-containing VASA as a marker. Gene Expres Patterns, 5: 669-678.

Sagi, A. and I. Khalaila. 2001. The crustacean androgen: a hormone in an isopod and androgenic activity in decapods. Am Zool, 41: 477-484.

Sagi, A., D. Cohen and Y. Milner. 1990. Effect of androgenic gland ablation on morphotypic differentiation and sexual characteristics of male freshwater prawns, *Macrobrachium rosenbergii.* Gen Comp Endocrinol, 77: 15-22.

Sagi, A., J. S. B. Ahl, H. Danaee and H. Laufer. 1992. Methyl farnesoate and reproductive behaviour in male morphotype of the spider crab *Libinia emarginata.* Am Zool, 31: 87A.

Sagi, A., I. Khalaila, A. Barki, G. Hulata and I. Karplus. 1996. Intersex red claw crayfish, *Cherax quadricarinatus* (von Martens): functional males with pre-vitellogenic ovaries. Biol Bull, 190: 16-23.

Sagi, A., E. Snir and I. Khalaila. 1997. Sexual differentiation in decapod crustaceans: role of the androgenic gland. Invert Reprod Dev, 31: 55-61.

Sagi A., R. Manor, C. Segall et al. 2002. On intersexuality in the crayfish *Cherax quadricarinatus*: an inducible sexual plasticity model. Invert Reprod Dev, 41: 27-33.

Sagi, A., R. Manor and T. Ventura. 2013. Gene silencing in crustaceans: from basic research to biotechnologies. Genes, 4: 620-645.

Sainath, S. B. 2009. Elucidation of effect of dopamine, serotonin and melatonin in the regulation of growth, reproduction and metabolism in the crab *Oziopelphusa senex senex.* S. V. University, Thirupati, India.

Sainath, S. B. and P. S. Reddy. 2011. Effect of selected biogenic amines on reproduction in the freshwater edible crab, *Oziotelphusa senex senex.* Aquaculture, 313: 144-148.

Sakwińska, O. 1998. Plasticity of *Daphnia magna* life history traits in response to temperature and information about a predator. Freshwat Biol, 39: 681-687.

Sam, S. T. and S. Krishnaswamy. 1979. Effect of osmolarity of the medium upon hatching of undried eggs of *Streptocephalus dichotomus* Baird (Anostraca, Crustacea). Archiv Hydrobiol, 86: 125-130.

Sandoz, M. and R. Rogers. 1944. The effect of environmental factors on hatching, moulting, and survival of zoea larvae of the blue crab *Callinectes sapidus* Rathbun. Ecology, 25: 216-228.

Santhanam, P. and P. Perumal. 2012. Feeding, survival, egg production and hatching rate of the marine copepod *Oithona rigida* Giesbrecht (Copepoda: Cyclopoida) under experimental conditions. J Mar Biol Ass India, 54: 38-44.

Santhanam, P. and P. Perumal. 2013. Developmental biology of brackishwater copepod *Oithona rigida* Giesbrecht: A laboratory investigation. Indian J Geo-Mar Sci, 42: 236-243.

Santhanam, P., N. Jeyaraj and K. Jothiraj. 2013. Effect of temperature and algal food on egg production and hatching of copepod, *Paracalanus parvus*. J Env Biol, 34: 243-246.

Santhanam, P., K. Jothiraj, N. Jeyaraj et al. 2015. Effect of mono algal diet on growth, survival and egg production in *Nannocalanus minor* (Copepoda, Calanoida). Ind J Mar Sci, (in press).

Sarma, S. S. S., S. Nandini and R. D. Gulati. 2005. Life history strategies of cladocerans: comparisons of tropical and temperate taxa. Hydrobiologia, 542: 315-333.

Sarojini, R., R. Nagabhushanam and M. Fingerman. 1995. *In vivo* inhibition by dopamine of 5-hydroxytryptamine-stimulated ovarian maturation in the red swamp crayfish, *Procambarus clarkii*. Experientia, 51: 156-158.

Sarojini, R., R. Nagabhushanam and M. Fingerman. 1997. An *in vitro* study of the inhibitory action of methionine enkephalin on ovarian maturation in the red swamp crayfish, *Procambarus clarkii*. Comp Biochem Physiol Part C: Pharmacol, Toxicol Endocrinol, 117: 207-210.

Sarvala, J. 1979. A parthenogenetic life cycle in a population of *Canthocamptus staphylinus* (Copepoda, Harpacticoida). Hydrobiologia, 62: 113-129.

Sassaman, C. 1995. Sex determination and evolution of unisexuality in the conchostracan shrimp *Eulimnada texana*. Hydrobiologia, 298: 45-65.

Sassaman, C. M. and M. Fugate. 1997. Gynandromorphism in Anostraca: multiple mechanisms of origin?. Hydrobiologia, 359:163-169.

Sassaman, C., M. A. Simovich and M. Fugate. 1997. Reproductive isolation and genetic differentiation in North American species of *Triops* (Crustacea: Branchiopoda: Notostraca). Hydrobiologia, 359: 125-147.

Scanabissi, F., E. Eder and M. Cesari. 2005. Male occurrence in Austrian populations of *Triops cancariformis* (Branchiopoda, Notostraca) and ultrastructural observations of the male gonad. Invert Biol, 124: 57-65.

Scanabissi, F., M. Cesari, S. K. Reed and S. C. Weeks. 2006. Ultrastructure of the male gonad and male gametogenesis in the clam shrimp *Eulimnadia texana* (Crustacea, Branchiopoda, Spinicaudata). Invert Biol, 125: 117-124.

Schatte, J. and R. Saborowski. 2006. Change of external sexual characteristics during consecutive moults in *Crangon crangon* L. Helgol Mar Res, 60: 70–73.

Schneider, S. D., J. S. Boates and M. Forbes. 1994. Sex ratios of *Corophium volutator* (Pallas) (Crustacea: Amphipoda) in Bay of Fundy populations. Can J Zool, 72: 1915-1921.

Scholtz, G., A. Abzhanov, F. Alwes et al. 2009. Development, genes and decapod evolution. Decapod Crust Phylogen, Decapod Crustacean Phylogenetics. (eds) J. W. Martin, K. A. Crandall, D. L. Felder, CRC Press, Boca Ratan, 18: 31-46.

Schön, I., R. K. Butlin, H. I. Griffiths and K. Martens. 1998. Slow molecular evolution in an ancient asexual ostracod. Proc R Soc Lond, 265B: 235-242.

Schroeder, P. C. 1989. Annelida-Polychaeta. In: *Reproduction Biology of Invertebrates.* (eds) K. G. Adiyodi and R. G. Adiyodi. Oxford and IBH Publishers, New Delhi, Vol 4, Part A. pp 383-442.

Schwartz, S. S. and P. D. N. Hebert. 1987. Breeding system of *Daphniopsis ephemeralis*: adaptations to a transient environment. Hydrobiologia, 145: 195-200.

Sellars, M. J., F. E. Coman, B. M. Degnan and N. P. Preston. 2006. The effectiveness of heat, cold and 6-dimethylaminopurine shocks for inducing tetraploidy in the kuruma shrimp, *Marsupenaeus japonicus* (Bate). J Shellfish Res, 25: 631-637.

Sellars, M. J., A. T. Wood, T. J. Dixon et al. 2009. A comparison of heterozygosity, sex ratio and production traits in two classes of triploid *Penaeus (Marsupenaeus) japonicus* (kuruma shrimp): Polar Body I vs II triploids. Aquaculture, 296: 207-212.

Sellars, M. J., S. M. Arce and P. L. Hertzler. 2012a. Triploidy induction in the Pacific white shrimp *Litopenaeus vannamei*: an assessment of induction agents and parameters, embryo viability, and early larval survival. Mar Biotechnol, 14: 740-751.

Sellars, M. J., A. Wood, B. Murphy et al. 2012b. Triploid black tiger shrimp (*Penaeus monodon*) performance from egg to harvest age. Aquaculture, 324: 242-249.

Seo, Y. S., H. M. Park and C. W. Oh. 2012. Reproductive biology of *Argis lar* from the East Sea of Korea (Decapoda, Natantia). Crustaceana, 85: 551-569.

Shane, B. S. 1994. *Introduction to Ecotoxicology.* CRC Press, Boca Raton, FL.

Sheader, M. 1978. Distribution and reproductive biology of *Corophium insidiosum* (Amphipoda) on the north-east coast of England. J Mar Biol Asso U K, 58: 585-596.

Sheela, S. G., T. J. Pandian and S. Mathavan. 1999. Electroporatic transfer, stable integration, expression and transmission of pZpBypGH and pZpBrtGH in Indian catfish, *Heteropnuestes fossilis*. Aquac Res, 30: 233-248.

Shields, J. D. 1994. The parasitic dinoflagellates of marine crustaceans. Annu Rev Fish Dis, 4: 241-271.

Shields, J. D. and F. E. Wood. 1993. Impact of parasites on the reproduction and fecundity of the blue sand crab *Portunus pelagicus* from Moreton Bay, Australia. Mar Ecol Prog Ser, 92: 159-159.

Shih, J. 1997. Sex steroid-like substances in the ovaries, hepatopancreases, and body fluid of female *Mictyris brevidactylus*. Zool Stud, 36: 136-145.

Shillaker, R. O. and P. G. Moore. 1987. The biology of brooding in the amphipods *Lembos websteri* Bate and *Corophium bonnellii* Milne Edwards. J Exp Mar Biol Ecol, 110: 113-132.

Shostak, S. 1993. Cnidaria. In: *Reproduction Biology of Invertebrates.* (eds) K. G. Adiyodi and R. G. Adiyodi. Oxford and IBH Publishers, New Delhi, Vol 6, Part A. pp 44-105.

Shukalyuk, A., V. Isaeva, E. Kizilova and S. Baiborodin. 2005. Stem cells in the reproductive strategy of colonial rhizocephalan crustaceans (Crustacea: Cirripedia: Rhizocephala). Invert Reprod Dev, 48: 41-53.

Shukalyuk, A. I., K. A. Golovnima, S. I. Baiborodin et al. 2007. Vasa-related genes and their expression in stem cells of colonial parasitic rhizocephalan barnacle *Polyascus polygenea* (Arthropoda, Crustacea: Crripedia: Rhizocephala). Cell Biol Int, 31: 97-108.

Shunmugasundaram, G. K., K. Rani and N. Munuswamy. 1996. Studies on the role of antioxidants in the cryptobiotic cysts of the brine shrimp, *Artemia parthenogenetica*. Biomed Lett, 53: 17-22.

Shuster, S. M. 1987. Alternative reproductive behaviors: three discrete male morphs in *Paracerceis sculpta,* an intertidal isopod from the northern Gulf of California. J Crust Biol, 7: 318-327.

Shuster, S. M. 1991. Changes in female anatomy associated with the reproductive moult in *Paracerceis sculpta*, a semelparous isopod crustacean. J Zool Lond, 225: 365-379.

Shuster, S. M. and C. Sassaman. 1997. Genetic interaction between male mating strategy and sex ratio in a marine isopod. Nature, 388: 373-377.

Shuster, S. M., S. J. Embry, C. R. Hargis and A. Nimer. 2014. The inheritance of autosomal and sex-linked cuticular pigmentation patterns in the marine isopod, *Paracerceis sculpta* Holmes, 1904 (Isopoda: Sphaeromatidae). J Crust Biol, 34: 460-466.

Simon, J. C., F. Delmotte, C. Rispe and T. Crease. 2003. Phylogenetic relationships between parthenogens and their sexual relatives: the possible routes to parthenogenesis in animals. Biol J Linn Soc, 79: 151-163.

Simovich, M. A. and S. A. Hathaway. 1997. Diversified bet-hedging as a reproductive strategy of some ephemeral pool anostracans (Branchiopoda). J Crust Biol, 17: 38-44.

Sivagnanam, S. 2005. Studies on the characterization and biochemical composition of different strains of *Artemia* (crustacea: Anostraca) in South India, Ph.D Thesis, University of Madras, Chennai.

Sivagnanam, S., V. Krishnakumar, S. Kulasekarapandian and N. Munuswamy. 2011. Present status of the native parthenogenetic strain of *Artemia* sp. in the salterns of Tamil Nadu. Ind J Fish, 58: 61-65.

Sivagnanam, S., V. Krishnakumar and N. Munuswamy. 2013. Morphology and Ultrastructure of cysts in different species of the brine shrimp, *Artemia* from Southern India. Int J Aquat Biol, 1: 266-272.

Skold, H. N., M. Olist, M. Skold and B. Akesson. 2009. Stem cells in asexual reproduction of marine invertebrates. In: *Stem cells in Marine Organisms*. (eds) B. Rinkevich and V. Matranga. Springer Verlag, Dordrecht. pp 205-138.

Slusarczyk, M. 2001. Food threshold for diapause in *Daphnia* under the threat of fish predation. Ecology, 82: 1089-1096.

Ślusarczyk, M. 2004. Environmental plasticity of fish avoidance diapause response in *Daphnia magna*. J Limnol, 63: 70-74.

Ślusarczyk, M. and E. Rygielska. 2004. Fish faeces as the primary source of chemical cues inducing fish avoidance diapause in *Daphnia magna*. Hydrobiologia, 526: 231-234.

Slusarczyk, M. and B. Rybicka. 2011. Role of temperature in diapause response to fish kairomones in crustacean *Daphnia*. J Insect Physiol, 57: 676-680.

Soffker, M. and C.R. Tyler. 2012. Endocrine disrupting chemicals and sexual behaviors in fish: a critical review on effects and possible consequences. Crit Rev Toxicol, 42: 653-668.

Spano, N., E. M. D. Porporato and S. Ragonese. 2013. Spatial distribution of Decapoda in the Strait of Sicily (Central Mediterranean Sea) based on a trawl survey. Crustaceana, 86: 139-157.

Spiegel, M. and E. Spiegel. 1986. Cell-Cell interaction during sea urchin morphogenesis. In: *Development Biology: A Comprehensive Synthesis*: (ed) L. W. Browder Plenum Press, New York, Vol 2. pp 195-240.

Sprague, V. and J. Couch. 1971. An annotated list of protozoan parasites, hyperparasites, and commensals of decapod Crustacea. J Protozool, 18: 526-537.

Spremberg, U., J. T. Høeg, L. Buhl-Mortensen and Y. Yusa. 2012. Cypris settlement and dwarf male formation in the barnacle *Scalpellum scalpellum*: a model for an androdioecious reproductive system. J Exp Mar Biol Ecol, 422: 39-47.

Sreekumar, S. and R. G. Adiyodi. 1983. Role of eyestalk principle (s) in spermatogenesis of the freshwater shrimp, *Macrobrachium idella*. In: Proc 1st Int. Biennial Conf Warm Water Aquaculture, 9-11.

Staelens, J., D. Rombaut, I. Vercauteren et al. 2008. High-density linkage maps and sex-linked markers for the black tiger shrimp (*Penaeus monodon*). Genetics, 179: 917-925.

Steel, C. G. H. and X. Vafopoulou. 1998. Ecdysteroid titres in haemolymph and other tissues during moulting and reproduction in the terrestrial isopod, *Oniscus asellus* (L.). Invert Reprod Dev, 34: 187-194.

Steele, V. J. and D. H. Steele. 1970. The biology of *Gammarus* (Crustacea, Amphipoda) in the northwestern Atlantic. II. *Gammarus setosus* Dementieva. Can J Zool, 48: 659-671.

Stephens, G. C. and R. A. Schinske. 1961. Uptake of amino acids by marine invertebrates. Limnol Oceanogr, 6: 175-181.

Stirnadel, H. A. and D. Ebert. 1997. Prevalence, host specificity and impact on host fecundity of microparasites and epibionts in three sympatric *Daphnia* species. J Ani Ecol, 66: 212-222.

Stross, R. G. and J. C. Hill. 1969. Photoperiod control of winter diapause in the freshwater crustacean *Daphnia*: Biol Bull 134: 176-198.

Subramoniam, T. 1981. Protandric hermaphrodities in a mole crab *Emerita asiatica* (Decapoda: Anomura). Biol Bull, 160: 161-174.

Subramoniam, T. 2000. Crustacean ecdysteriods in reproduction and embryogenesis. Comp Biochem Physiol Part C: Pharmacol Toxicol Endocrinol, 125: 135-156.

Sudha, K. and G. Anilkumar. 1996. Seasonal growth and reproduction in a highly fecund brachyuran crab, *Metopograpsus messor* (Forskal) (Grapsidae). Hydrobiologia, 319: 15-21.

Sugumar, V. and N. Munuswamy. 2006. Ultrastructure of cyst shell and underlying membranes of three strains of the brine shrimp *Artemia* (Branchiopoda: Anostraca) from South India. Microsc Res Tech, 69: 957-963.

Sugumaran, J., M. S. Naveed and K. Altaff. 2009. Ecology of meiofaunal harpacticoid copepods of Chennai coast, Tamil Nadu, India. J Aquat Biol, 24: 35-39.

Sun, P. S., T. M. Weatherby, M. F. Dunlap et al. 2000. Developmental changes in structure and polypeptide profile of the androgenic gland of the freshwater prawn *Macrobrachium rosenbergii*. Aquac Int, 8: 327-334.

Sutcliffe, D. W. 2010. Reproduction in *Gammarus* (Crustacea, Amphipoda): basic processes. Freshwat Forum, 2: 102-128.

Suvaparp, R., V. Chawengsri, N. Tayaputch et al. 2001. Dissipation of endosulfan in the rice field. *In*: Proc Rice Res Inst, Dept of Agricult, Bangkok. pp 129-139.

Suzuki, S. 1999. Androgenic gland hormone is a sex-reversing factor but cannot be a sex-determining factor in the female crustacean isopods *Armadillidium vulgare*. Gen Comp Endocrinol, 115: 370-378.

Svane, I. 1986. Sex determination in *Scalpellum scalpellum* (Cirripedia: Thoracica: Lepadomorpha), a hermaphroditic goose barnacle with dwarf males. Mar Biol, 90: 249-253.

Svensen, C. and K. Tande. 1999. Sex change and female dimorphism in *Calanus finmarchicus*. Mar Ecol Prog Ser, 176: 93-102.

Sweeting, R. A. 1981. Hermaphrodite roach in the River Lee. Thames Water, Lea Division.

Swetha, C. H., S. B. Sainath, P. R. Reddy and P. S. Reddy. 2011. Reproductive endocrinology of female crustaceans: perspective and prospective. J Mar Sci Res Dev S, 3, 2. Doi.org/10.4172/2155-9910.s3-001.

Taborsky, M. 1998. Sperm competition in fish:bourgeois males and parasitic spawning. Trends Ecol Evol, 13: 222-227.

Taborsky, M. 2001. The evolution of bourgeois and cooperative reproductive behaviours in fish. J Hered, 72: 100-110.

Takac, P., H. Laufer and G. D. Prestwich. 1993. Characterization of methyl farnesoate (MF) binding proteins and the metabolism of MF by some tissues of the spider crab *Libinia emarginata*. Am Zool, 33: 10A.

Takahashi, T. and J. Lützen. 1998. A sexual reproduction as part of the life cycle in *Sacculina polygenea* (Cirripedia: Rhizocephala: Sacculinidae). J Crust Biol, 18: 321-331.

Taketomi, T., M. Motono and M. Miyawaki. 1989. On the function of the mandibular gland of decapod crustacean. Cell Biol Int Reprod, 13: 463-469.

Taketomi, Y. and S. Nishikawa. 1996. Implantation of androgenic glands into immature female crayfish, *Procambarus clarkii*, with masculinization of sexual characteristics. J Crust Biol, 16: 232-239.

Takeuchi, I. and J. M. Guerra-Garcîa. 2002. *Paraprotella saltatrix*, a new species of the Caprellidea (Crustacea, Amphipoda) from Phuket Island, Thailand. Phuket Mar Biol Center Res Bull, 23: 273-280.

Tamone, S. L. and E. S. Chang. 1993. Methyl farnesoate stimulates ecdysteroid secretion from crab Y-organs *in vitro*. Gen Comp Endocrinol, 89: 425-432.

Tamone, S. L., M. M. Adams and J. M. Dutton. 2005. Effect of eyestalk-ablation on circulating ecdysteroids in hemolymph of snow crabs, *Chionecetes opilio*: Physiological evidence for a terminal molt. Int Comp Biol, 45: 166-171.

Tatarazako, N., S. Oda, H. Watanabe et al. 2003. Juvenile hormone agonists affect the occurrence of male *Daphnia*. Chemosphere, 53: 827-833.

Taylor, D. J., T. J. Crease and W. M. Brown. 1999. Phylogenetic evidence for a single long-lived clade of crustacean cyclic parthenogens and its implications for the evolution of sex. Proc R Soc Lond, 266 B: 791-797.

Techa, S. and J. S. Chung. 2013. Ecdysone and retinoid-X receptors of the blue crab, *Callinectes sapidus*: cloning and their expression patterns in eyestalks and Y-organs during the molt cycle. Gene, 527: 139-153.

Thibaut, R. and C. Porte. 2004. Effects of endocrine disrupters on sex steroid synthesis and metabolism pathways in fish. J Ster Biochem Mol Biol, 92: 485-494.

Thomas, F., F. Renaud and F. Cezilly. 1996a. Assortative pairing by parasitic prevalence in *Gammarus insensibilis* (Amphipoda): patterns and processes. Anim Behav, 52: 683-690.

Thomas, F., O. Verneau, F. Santalla et al. 1996b. The influence of intensity of infection by a trematode parasite on the reproductive biology of *Gammarus insensibilis* (Amphipoda). Int J Parasitol, 26: 1205-1209.

Thomson, M. 2014. Ovoviviparous reproduction in Australian specimens of the intertidal isopod *Cirolana harfordi*. Invert Reprod Dev, 58: 218-225.

Thorp, J. H. and A. P. Covich. 2009. *Ecology and Classification of North American Freshwater Invertebrates*. Academic Press, New York.

Tighe-Ford, D. J. 1977. Effects of juvenile hormone analogues on larval metamorphosis in the barnacle *Elminius modestus* Darwin (Crustacea: Cirripedia). J Exp Mar Biol Ecol, 26: 163-176.

Tilden, A., L. McGann, J. Schwartz et al. 2001. Effect of melatonin on hemolymph glucose and lactate levels in the fiddler crab *Uca pugilator*. J Exp Zool, 290: 379-383.

Timms, B. V. 2009. Biodiversity of large branchiopods of Australia. Curr Sci, 96: 74-80.

Tinikul, Y., A. J. Mercier, N. Soonklang and P. Sobhon. 2008. Changes in the levels of serotonin and dopamine in the central nervous system and ovary, and their possible roles in the ovarian development in the giant freshwater prawn, *Macrobrachium rosenbergii*. Gen Comp Endocrinol, 158: 250-258.

Tiu, S. H. K. and S. M. Chan. 2007. The use of recombinant protein and RNA interference approaches to study the reproductive functions of a gonad-stimulating hormone from the shrimp *Metapenaeus ensis*. J FEBS, 274: 4385-4395.

Toledo, A., C. Cruz, G. F. Ragoso et al. 1997. *In vitro* culture of *Taenia crassiceps* larval cells and cyst regeneration after injection into mice. J Parasitol, 83: 189-193.

Touir, A. 1977. Données nouvelles concernant l'endocrinologie sexuelle des Crustacés Décapodes Natantia hermaphrodites et gonochoriques. III. Mise en évidence d'un contrôle neurohormonal du maintien de l'appareil génital mâle et des glandes androgènes exercé par le protocérébron médian. Bull Soc Zool Fr, 102: 375-400.

Toumi, H., M. Boumaiza, M. Millet et al. 2015. Investigation of differences in sensitivity between 3 strains of *Daphnia magna* (Crustacea Cladocera) exposed to malathion (organophosphorous pesticide). J Environ Sci Health, Part B, 50: 34-44.

Treerattrakool, S., S. Panyim, S. M. Chan et al. 2008. Molecular characterization of gonad-inhibiting hormone of *Penaeus monodon* and elucidation of its inhibitory role in vitellogenin expression by RNA interference. J FEBS, 275: 970-980.

Treerattrakool, S., S. Panyim and A. Udomkit. 2011. Induction of ovarian maturation and spawning in *Penaeus monodon* broodstock by double-stranded RNA. Mar Biotechnol, 13: 163-169.

Trivers, R. 1972. Parental investment and sexual selection. In: *Sexual Selection and Descent of Man*. (ed) B. Campbell. Aldine, Chicago. pp 136-179.

Tropea, C., M. Arias, N. S. Calvo and L. S. L. Greco. 2012. Influence of female size on offspring quality of the freshwater crayfish *Cherax quadricarinatus* (Parastacidae: Decapoda). J Crust Biol, 32: 883-890.

Tsai, M. L. and C. F. Dai. 2001. Life history plasticity and reproductive strategy enabling the invasion of *Ligia exotica* (Crustacea: Isopoda) from the littoral zone to an inland creek. Mar Ecol Prog Ser, 210: 175-184.

Tsai, M. L., J. J. Li and C. F. Dai. 1999. Why selection favors protandrous sex change for the parasitic isopod, *Ichthyoxenus fushanensis* (Isopoda: Cymothoidae). Evol Ecol, 13: 327-338.

Tsukimura, B. 2001. Crustacean vitellogenesis: Its role in oocyte development. Am Zool, 41: 465-476.

Tsukimura, B., W. K. Nelson and C. J. Linder. 2006. Inhibition of ovarian development by methyl farnesoate in the tadpole shrimp, *Triops longicaudatus*. Comp Biochem Physiol Part A: Mol Integ Physiol, 144: 135-144.

Tsutsui, N., Y. K. Kim, S. Jasmani et al. 2005. The dynamics of vitellogenin gene expression differs between intact and eyestalk ablated kuruma prawn *Penaeus (Marsupeaneus) japonicus*. Fisher Sci, 71: 249-256.

Tucker, B. W. 1930. Memoirs: On the effects of an epicaridan parasite, *Gyge branchialis*, on *Upogebia littoralis*. Q J Microsc Sci, 2: 1-118.

Turgeon, J. and P. D. N. Herbert. 1994. Evolutionary interactions between sexual and all female taxa of *Cyprinotus* (Ostracoda, Cyprididae). Evolution, 48: 1855-1856.

Turgeon, J. and P. D. N. Herbert. 1995. Genetic characterization of breeding systems, ploidy levels and species boundaries in *Cypricercus* (Ostracoda). Heredity 75: 561-570.

Turner, C. L. 1935. The abbarent secondary sexual characters of the genus *Cambarus*. Am Midland Nat, 16: 863-872.

Turra, A. 2004. Intersexuality in hermit crabs: reproductive role and fate of gonopores in intersex individuals. J Mar Biol Assoc UK, 84: 757-759.

Urano, S., S. Yamaguchi, S. Yamato and Y. Yusa. 2009. Evolution of dwarf males and a variety of sexual modes in barnacles: an ESS approach. Evol Ecol Res, 11: 713-729.

Utinomi, H. 1961. A revision of the nomenclature of the family Nephtheidae (Octocorallia: Alcyonacea). II. The boreal genera *Gersemia, Duva, Drifa* and *Pseudodrifa* (n. g.). Publ Seto Mar Biol Lab 9: 229-246.

Uye, S. I. 1985. Resting egg production as a life history strategy of marine planktonic copepods. Bull Mar Sci, 37: 440-449.

Van der Spoel, S. 1979. Strobilization in pteropod (Gastropoda: Opisthobranchiata). Malacologia, 18: 27-30.

Ventura, T., R. Manor and D. Eliahu et al. 2009. Temporal silencing of an androgenic gland specific insulin-like gene affecting phenotypical gender differences and spermatogenesis. Endocrinology 150: 1278-1286.

Ventura, T., E. D. Aflalo, S. Weil et al. 2011a. Isolation and characterization of a female-specific DNA marker in the giant freshwater prawn *Macrobrachium rosenbergii*. Heredity, 107: 456-461.

Ventura, T., R. Manor, E. D. Aflalo et al. 2011b. Expression of an androgenic gland specific to insulin-like peptide during the course of sexual and morphotypic differentiation. ISRN Endocrinology, 2011 ID 476283.

Vergilino, R., C. Belzile and F. Dufresne. 2009. Genome size evolution and polyploidy in the *Daphnia pulex* complex (Cladocera: Daphniidae). Biol J Linn Soc, 97: 68-79.

Verslycke, T., K. De Wasch, H.F. De Brabander and C.R. Janssen. 2002. Testosterone metabolism in the estuarine mysid *Neomysis integer* (Crustacea; Mysidacea): identification of testosterone metabolites and endogenous vertebrate-type steroids. Gen Comp Endocrinol, 126: 190–199.

Verslycke, T., S. Poelmans, K. De Wasch et al. 2004. Testosterone and energy metabolism in the estuarine mysid *Neomysis integer* (Crustacea: Mysidacea) following exposure to endocrine disruptors. Environ Toxicol Chem, 23: 1289-1296.

Verslycke, T., A. Ghekiere, S. Raimondo and C. Janssen. 2007. Mysid crustaceans as test models for the screening and testing of endocrine-disrupting chemicals. Ecotoxicology, 16: 205-219.

Vikas, P. A., N. K. Sajeshkumar, P. C. Thomas et al. 2012. Aquaculture related invasion of the exotic *Artemia franciscana* and displacement of the autochthonous *Artemia* populations from the hypersaline habitats of India. Hydrobiologia, 684: 129-142.

Villalobos-Rojas, F. and I. S. Wehrtmann. 2014. Secondary sexual characters and spermatophores of *Solenocera agassizii* (Decapoda: Solenoceridae), including a comparison with other solenocerid shrimp. Sex Early Dev Aquat Org, 1: 45-55.

Vogt, G. 2002. Functional anatomy. In: *Biology of Freshwater Crayfish*. (ed) D. M. Holdich. Blackwell Science, Oxford. pp 53-151.

Vogt, G. 2007. Exposure of the eggs to 17α-methyl testosterone reduced hatching success and growth and elicited teratogenic effects in postembryonic life stages of crayfish. Aquat Toxicol, 85: 291-296.

Vogt, G. 2008. Investigation of hatching and early post-embryonic life of freshwater crayfish by *in vitro* culture, behavioral analysis, and light and electron microscopy. J Morphol, 269: 790-811.

Vogt, G. 2010. Suitability of the clonal marbled crayfish for biogerontological research: a review and perspective, with remarks on some further crustaceans. Biogerontology, 11: 643-669.

Vogt, G. 2011. Marmorkrebs: natural crayfish clone as emerging model for various biological disciplines. J Biosci, 36: 377-382.

Vogt, G., L. Tolley and G. Scholtz. 2004. Life stages and reproductive components of the marmorkrebs (marbled crayfish), the first parthenogenetic decapod crustacean. J Morphol, 261: 286-311.

Vogt, G., M. Huber, M. Thiemann et al. 2008. Production of different phenotypes from the same genotype in the same environment by developmental variation. J Exp Biol, 211: 510-523.

Voordouw, M. J. and B. R. Anholt. 2002. Environmental sex determination in a splash pool copepod. Biol J Linn Soc, 76: 511-520.

Voordouw, M. J., H. E. Robinson and B. R. Anholt. 2005. Paternal inheritance of the primary sex ratio in a copepod. J Evol Biol, 18: 1304-1314.

Waddy, S. L., D. E. Aiken and D. P. V. de Kleijn. 1995. Control of growth and reproduction. In: *Biology of the Lobster Homarus americanus*. (ed) J. R. Factor. Academic Press, San Diego. pp 217-266.

Wainwright, G., S. G. Webster, M. C. Wilkinson et al. 1996. Structure and significance of mandibular organ-inhibiting hormone in the crab, *Cancer pagurus*. Involvement in multihormonal regulation of growth and reproduction. J Biol Chem, 271: 12749-12754.

Walker, G. 2004. Swimming speeds of the larval stages of the parasitic barnacle, *Heterosaccus lunatus* (Crustacea: Cirripedia: Rhizocephala). J Mar Biol Asso UK, 84: 737-742.

Walker, G. 2005. Sex determination in the larvae of the parasitic barnacle *Heterosaccus lunatus*: an experimental approach. J Exp Mar Biol Ecol, 318: 31-38.

Walker, I. 1979. Mechanisms of density-dependent population regulation in the marine copepod *Amphiascoides* sp.(Harpacticoida). Mar Ecol Prog Ser, 1: 209-221.

Walker, S. P. 1977. *Probopyrus pandalicola*: discontinuous ingestion of shrimp hemolymph. Exp Parasitol, 41: 198-205.

Walley, L. J. 1965. The development and function of the oviducal gland in *Balanus balanoides*. J Mar Biol Ass U K, 45: 115-128.

Wang, C. C., J. Y. Liu and L. S. Chou. 2014. Egg bank spatial structure and functional size of three sympatric branchiopods (Branchiopoda) in Siangtian Pond, Taiwan. J Crust Biol, 34: 412-421.

Warburg, M. R. 2013. Intra-and inter-specific variability in some aspects of the reproduction of oniscid isopods. Crustaceana, 86: 98-109.

Warrier, S. R., R. Tirumalai and T. Subramoniam. 2001. Occurrence of vertebrate steroids, estradiol 17β and progesterone in the reproducing females of the mud crab *Scylla serrata*. Comp Biochem Physiol A Mol Integr Physiol, 130: 283-294.

Wear, R. G. 1974. Incubation in British decapod Crustacea, and the effects of temperature on the rate and success of embryonic development. J Mar Biol Asso UK, 54: 745-762.

Webster, S. G., R. Keller and H. Dircksen. 2012. The CHH-superfamily of multifunctional peptide hormones controlling crustacean metabolism, osmoregulation, moulting, and reproduction. Gen Comp Endocrinol, 175: 217-233.

Weeks, A. R., F. Marec and J. A. Breeuwer. 2001. A new species that consists of entirely haploid females. Science, 292: 2479-2482.

Weeks, S. C., V. Marcus and S. Alvarez. 1997. Notes on the life history of the clam shrimp, *Eulimnadia texana*. Hydrobiologia, 359: 191-197.

Weeks, S. C., T. F. Sanderson, S. K. Reed et al. 2006a. Ancient androdioecy in the freshwater crustacean *Eulimnadia*. Proc R Soc Lond, 273 B: 725-734.

Weeks, S. C., M. Zofolkova and B. Knott. 2006b. Limnadiid clam shrimp biogeography in Australia (Crustacea: Branchiopoda: Spinicaudata). J R Soc West Aust, 89: 155-161.

Weeks, S. C., C. Benvenuto and S. K. Reed. 2006c. When males and hermaphrodites coexist: a review of androdioecy in animals. Integ Comp Biol, 46: 449-464.

Weeks, S. C., T. F. Sanderson, M. Zofkova and B. Knott. 2008. Breeding systems in the clam shrimp family Limnadiidae (Branchiopoda, Spinicaudata). Invert Biol, 127: 336-349.

Weeks, S. C., E. G. Chapman, D. C. Rogers et al. 2009. Evolutionary transitions among dioecy, androdioecy and hermaphroditism in limnadiid clam shrimp (Branchiopoda: Spinicaudata). J Evol Biol, 22: 1781-1799.

Weeks, S. C., C. Benvenuto, T. F. Sanderson and R. J. Duff. 2010. Sex chromosome evolution in the clam shrimp, *Eulimnadia texana*. J Evol Biol, 23: 1100-1106.

Weeks, S. C., J. S. Brantner, T. I. Astrop et al. 2014a. The evolution of hermaphroditism from dioecy in crustaceans: selfing hermaphroditism described in a fourth spinicaudatan genus. Evol Biol, 41: 251-261.

Weeks, S. C., C. Benvenuto, S. K. Reed et al. 2014b. A field test of a model for the stability of androdioecy in the freshwater shrimp, *Eulimnadia texana*. J Evol Biol, 27: 2080-2095.

Weider, L. J., M. J. Beaton and P. D. N. Hebert. 1987. Clonal diversity in high-arctic populations of *Daphnia pulex*, a polyploid apomictic complex. Evolution, 41: 1335-1346.

Weider, L. J., W. Lampert, M. Wessels et al. 1997. Long–term genetic shifts in a microcrustacean egg bank associated with anthropogenic changes in the Lake Constance ecosystem. Proc R Soc Lond. 264 B: 1613-1618.

Weider, L. J., A. Hobaek, J. K. Colbourne et al. 1999. Holarctic phylogeography of an asexual species complex I. Mitochondrial DNA variation in Arctic *Daphnia*. Evolution, 53: 777-792.

Weismann, A. 1883. Die Entstehung der sexualzellen bei den Hydramedusen: Zugleichcin Beitrug zur Kenntniss des Baues und der Lebenserscheinungen dieser Gruppe. Fischer vertae, Jena, p 295.

Weissberger, E. J., P. A. Jumars, L. M. Mayer and L. L. Schick. 2008. Structure of a northwest Atlantic Shelf macrofaunal assemblage with respect to seasonal variation in sediment nutritional quality. J Sea Res, 60: 164-175.

Weng, H. T. 1987. The parasite barnacle, *Sacculina granifera* Boschma, affecting the commercial sand crab, *Portunus pelagicus* (L.), in populations from two different environments in Queensland. J Fish Dis, 10: 221-227.

Wenner, E. L. and N. T. Windsor. 1979. Parasitism of Galatheid crustaceans from the Norfolk Canyon and Middle Atlantic Bight by bopyrid isopods 1, 2). Crustaceana, 37: 293-303.

Whitefield, P. J. and N. A. Evans. 1983. Parthenogenesis and asexual multiplication among parasitic platyhelminths. Parasitology, 86: 121-160.

Wickham, D. E. 1980. Aspects of the life history of *Carcinonemertes errans* (Nemertea: Carcinonemertidae), an egg predator of the crab *Cancer magister*. Biol Bull, 159: 247-257.

Wilber, D. H. 1989. Reproductive biology and distribution of stone crabs (Xanthidae, *Menippe*) in the hybrid zone on the northeastern Gulf of Mexico. Mar Ecol Prog Ser, 52: 235-244.

Williams, J. D. and C. B. Boyko. 2012. The global diversity of parasitic isopods associated with crustacean hosts (Isopoda: Bopyroidea and Cryptoniscoidea). PloS ONE, 7(4): e35350.

Williamson, C. E. and J. W. Reid. 2009. Copepoda. In: *Ecology and Classification of North American Freshwater Invertebrates.* (eds) J. H. Thorp and A. P. Covich. Academic Press, San Diego, 2nd edn. pp 915-954.

Wilson, D. F. and K. K. Parrish. 1971. Remating in a planktonic marine calanoid copepod. Mar Biol, 9: 202-204.

Wilson, H. W. 1907. On some phenomenon of coalescence and regeneration in sponges. J Exp Zool, 5: 250-255.

Wilson, K., V. Cahill, E. Ballment and J. Benzie. 2000. The complete sequence of the mitochondrial genome of the crustacean *Penaeus monodon*: are malacostracan crustaceans more closely related to insects than to branchiopods? Mol Biol Evol, 17: 863-874.

Wittmann, K. J. 1978. Adoption, replacement, and identification of young in marine Mysidacea (Crustacea). J Exp Mar Biol Ecol, 32: 259-274.

Wojtasik, B. and M. Bryłka-Wołk. 2010. Reproduction and genetic structure of a freshwater crustacean *Lepidurus arcticus* from Spitsbergen. Polish Polar Res, 31: 33-44.

Wood, A. T., G. J. Coman, A. R. Foote and M. J. Sellars. 2011. Triploid induction of black tiger shrimp, *Penaeus monodon* (Fabricius) using cold shock. Aquac Res, 42: 1741-1744.

Wowor, D., V. Muthu, R. Meier et al. 2009. Evolution of life history traits in Asian freshwater prawns of the genus *Macrobrachium* (Crustacea: Decapoda: Palaemonidae) based on multilocus molecular phylogenetic analysis. Mol Phylogenet Evol, 52: 340-350.

Wu, L. T. and K. H. Chu. 2010. Characterization of an ovary-specific glutathione peroxidase from the shrimp *Metapenaeus ensis* and its role in crustacean reproduction. Comp Biochem Physiol Part B: Biochem Mol Biol, 155: 26-33.

Wu, P., D. Qi and L. Chen et al. 2009. Gene discovery from an ovary cDNA library of oriental river prawn *Macrobrachium nipponense* by ESTs annotation. Comp Biochem Physiol Part D: Genomics Proteomics, 4: 111-120.

Wyban, J. and J. N. Sweeney. 1991. Intensive Shrimp Production Technology. The Oceanic Institute Shrimp Manual. The Ocean Institute, Honolulu

Wylie, C. 2000. Germ cells. Cell 96: 165-174.

Wyngaard, G. A., B. E. Taylor and D. L. Mahoney. 1991. Emergence and dynamics of cyclopoid copepods in an unpredictable environment. Freshwat Biol, 25: 219-232.

Xiang, J., F. Li, C. Zhang et al. 2006. Evaluation of induced triploid shrimp *Penaeus (Fenneropenaeus) chinensis* cultured under laboratory conditions. Aquaculture, 259: 108-115.

Xie, T. and A. C. Spradling. 2000. A niche maintaining germ line stem cells in the *Drosophila* ovary. Science, 290: 328-330.

Xie, Y., F. Li, B. Wang et al. 2010. Screening of genes related to ovary development in Chinese shrimp *Fenneropenaeus chinensis* by suppression subtractive hybridization. Comp Biochem Physiol Part D: Genomics Proteomics, 5: 98-104.

Xu, E. Y., F. L. Moore and R. A. R. Pera. 2001. A gene family required for human germ cell development evolved from an ancient meiotic gene conserved in metazoans. Proc Natl Acad Sci USA, 98: 7414-7419.

Xu, X., S. Song, Q. Wang et al. 2009. Analysis and comparison of a set of expressed sequence tags of the parthenogenetic water flea *Daphnia carinata*. Mol Genet Genomics, 282: 197-203.

Yamaguchi, S. and K. Endo. 2003. Molecular phylogeny of Ostracoda (Crustacea) inferred from 185 ribosomal DNA sequences: implications for its origin and diversification. Mar Biol, 143: 23-38.

Yamamoto, H., T. Okino, E. Yoshimura et al. 1997. Methyl farnesoate induces larval metamorphosis of the barnacle, *Balanus amphitrite* via protein kinase C activation. J Exp Zool, 278: 349-355.

Yamauchi, M. M., M. U. Miya and M. Nishida. 2003. Complete mitochondrial DNA sequence of the swimming crab, *Portunus trituberculatus* (Crustacea: Decapoda: Brachyura). Gene, 311: 129-135.

Yang, C. P., H. X. Li, L. Li et al. 2014. Population structure, morphometric analysis and reproductive biology of *Portunus sanguinolentus* (Herbst, 1783) (Decapoda: Brachyura: Portunidae) in Honghai Bay, South China Sea. J Crust Biol, 34: 722-730.

Yang, W. J. and K. R. Rao. 2001. Cloning of precursors for two MIH/VIH-related peptides in the prawn, *Macrobrachium rosenbergii*. Biochem Biophys Res Commun, 289: 407-413.

Yang, W. X., I. Mirabdullayev, H. U. Dahms and J. S. Hwang. 2009. Karyology of *Acanthocyclops* Kiefer, 1927 (Copepoda, Cyclopoida), including the first karyological reports on *Acanthocyclops trajani* Mirabdullayev and Defaye, 2002 and *A. einslei* Mirabdullayev and Defaye, 2004. Crustaceana, 82: 487-492.

Yano, I. 1987. Maturation of kuruma prawns *Penaeus japonicus* cultured in earthern ponds. NOAA Tech Rep NMFS 47: 3-7.

Yano, I. and J. A. Wyban. 1992. Induced ovarian maturation of *Penaeus vannamei* by injection of lobster [*Homarus americanus*] brain extract. Bull Natl Res Inst Aquac, 21: 1-7.

Yano, I. and R. Hoshino. 2006. Effects of 17 β-estradiol on the vitellogenin synthesis and oocyte development in the ovary of kuruma prawn (*Marsupenaeus japonicus*). Comp Biochem Physiol Part A: Mol Integ Physiol, 144: 18-23.

Yano, I., B. Tsukimura, J. N. Sweeney and J. A. Wyban. 1988. Induced ovarian maturation of *Penaeus vannamei* by implantation of lobster ganglion. J World Aquacult Soc, 19: 204-209.

Yano, I., M. Fingerman and R. Nagabhushanam. 2000. Endocrine control of reproductive maturation in penaeid shrimp. In: *Recent Advances in Marine Biotechnology Aquaculture, Part A: Seaweeds and Invertebrates.* (eds) M. Fingerman and R. Nagabhushanam. Science Publishers, Enfield. pp 161-176.

Yao, J. J., W. Luo, Y. L. Zhao et al. 2010. Morphogenesis of the eyestalk and expression of moult-inhibiting hormone during embryonic development of the freshwater prawn, *Macrobrachium rosenbergii* (Decapoda, Palaemonidae). Crustaceana, 83: 903-913.

Yue, G. H., G. L. Wang, B. Q. Zhu et al. 2008. Discovery of four natural clones in a crayfish species *Procambarus clarkii*. Int J Biol Sci, 4: 279.

Yusa, Y., S. Yamato and M. Marunura. 2001. Ecology of parasitic barnacle *Koleolepas avis*; relationships to the hosts, distribution, left-right asymmetry and reproduction. J Mar Biol Ass UK, 81: 781-788.

Yusa, Y., M. Yoshikawa, J. Kitaura et al. 2012. Adaptive evolution of sexual systems in pedunculate barnacles. Proc R Soc Lond, 279 B: 959-966.

Zaleski, M. A. and S. L. Tamone. 2014. Relationship of molting, gonadosomatic index, and methyl farnesoate in male snow crab (*Chionoecetes opilio*) from the eastern Bering Sea. J Crust Biol, 34: 764-772.

Zarkower, D. 2006. Somatic sex determination. In: *WormBook.* (eds) The *C. elegance* Research Community WormBook Doi/10.1995/wormbook.184.1.

Zeidberg, L. D. 2004. Allometry measurements from in situ video recordings can determine the size and swimming speeds of juvenile and adult squid *Loligo opalescens* (Cephalopoda: Myopsida). J Exp Biol, 207: 4195-4203.

Zeleny, C. 1905. Compensatory regulation. J Exp Zool, 2: 1-102.

Zhang, L. and C. E. King. 1993. Life history divergence of sympatric diploid and polyploidy populations of the brine shrimp *Artemia parthenogenetica*. Oecologia, 93: 177-183.

Zhang, L., C. Yang, Y. Zhang et al. 2007. A genetic linkage map of Pacific white shrimp (*Litopenaeus vannamei*): sex-linked microsatellite markers and high recombination rates. Genetica, 131: 37-49.

Zhao, G. F., G. L. Li and C. H. Zhu. 2009. Preliminary study on sex differentiation of *Litopenaeus vannamei*. J Guangdong Ocean Univ, 3: 1-5.

Zhao, R., Y. Xuan, X. Li and R. Xi. 2008. Age-related changes of germline stem cell activity, niche signaling activity and egg production in *Drosophila*. Aging Cell, 7: 344-354.

Zhu, D. F., Z. H. Hu and J. M. Shen. 2011. Molt-inhibiting hormone from the swimming crab *Portunus trituberculatus* (Miers, 1876): PCR cloning, tissue distribution and expression of recombinant protein in *Esherichia coli* (Migula, 1895). Crustaceana, 84: 1481-1496.

Zmora, N., J. Trant, Y. Zohar and J. S. Chung. 2009a. Molt-inhibiting hormone stimulates vitellogenesis at advanced ovarian developmental stages in the female blue crab, *Callinectes sapidus* 1: an ovarian stage dependent involvement. Saline Systems, 5: 7.

Zmora, N., A. Sagi, Y. Zohar and J. S. Chung. 2009b. Molt-inhibiting hormone stimulates vitellogenesis at advanced ovarian developmental stages in the female blue crab, *Callinectes sapidus* 2: novel specific binding sites in hepatopancreas and cAM P as a second messenger. Saline Systems, 5(6). doi:10.1186/1746-1448-5-6.

Zoja, R. 1895. Sullo sviluppo dei blastomeri isolati delle nova di alcune medusa. Arch Entwicklungsmech Org, 1: 578-595, 2: 1-37.

Zou, E. and M. Fingerman. 1997a. Synthetic estrogenic agents do not interfere with sex differentiation but do inhibit molting of the cladoceran *Daphnia magna*. Bull Environ Contam Toxicol, 58: 596-602.

Zou, E. and M. Fingerman. 1997b. Effects of estrogenic xenobiotics on molting of the water flea, *Daphnia magna*. Ecotoxicol Environ Saf, 38: 281-285.

Zou, E. and M. Fingerman. 2003. Endocrine disruption of sexual development, reproduction and growth in crustaceans by environmental organic contaminants: current perspectives. Curr Top Pharmacol, 7: 69-80.

Zucker, N., B. Stafki and S. C. Weeks. 2001. Maintenance of androdioecy in the freshwater clam shrimp *Eulimnadia texana*: longevity of males relative to hermaphrodites. Can J Zool, 79: 393-401.

Żukowska-Arendarczyk, M. 1981. Effect of hypophyseal gonadotropins (FSH and LH) on the ovaries of the sand shrimp *Crangon crangon* (Crustacea: Decapoda). Mar Biol, 63: 241-247.

Zúñiga-Panduro, de J. M., R. Casillas-Hernández, R. Garza-Torres et al. 2014. Abnormalities and possible mosaicism during embryonic cell division after cold shock in zygotes of the Pacific white shrimp, *Litopenaeus vannamei*, related to failure of induction of tetraploidy and triploidy. J Crust Biol, 34: 367-376.

Zupo, V. 1994. Strategies of sexual inversion in *Hippolyte inermis* Leach (Crustacea, Decapoda) from a Mediterranean sea grass meadow. J Exp Mar Biol Ecol, 178: 131-145.

Zupo, V. 2000. Effect of microalgal food on the sex reversal of *Hippolyte inermis* (Crustacea, Decapoda). Mar Ecol Prog Ser, 201: 251-259.

Zupo, V. 2001. Influence of diet on sex differentiation of *Hippolyte inermis* (Crustacea: Natantia) in the field. Hydrobiologia, 449: 131-140.

Zupo, V. and P. Messina. 2007. How do dietary diatoms cause the sex reversal of the shrimp *Hippolyte inermis* Leach (Crustacea, Decapoda). Mar Biol, 151: 907-917.

Zurlini, G., I. Ferrari and A. Nassogne. 1978. Reproduction and growth of *Euterpina acutifrons* (Copepoda: Harpacticoida) under experimental conditions. Mar Biol, 46: 59-64.

Author Index

Belanger, R. M., 125
Bell, G., 90
Benazzi, M., 3
Benazzi-Lentati, G., see Benazzi, M.
Bengtsson, B. E., see Breitholtz, M.
Benvenuto, C., 31, 156
Benzie, J. A. H., 166
Bernice, R., 153
Biggers, W. J., see Laufer, H.
Billinghurst, Z., 193-194
Biswas, A., 34
Blower, S. M., 200, 208, 210, 213
Blueweiss, L., 46,
Boddeke, R., 110,
Bode, S. N. S., 92
Bookhout, C. G., see Costlow, J. D.
Boore, J. L., 158
Borok, Z., 19
Borowsky, B., 65, 195
Borst, D. W., 188-189, 196
Borst, D. W., see Li, H., Nagaraju, G. P. C.,
     Vogel, J. M.
Bosch, T. C. G., 20
Bottrell, H. H., 34
Bouchon. D., 208, 213
Bourdon, R., 202-203
Bowman, T. E., 202
Boxshall, G. A., 32, 198-201, 203, 224
Boyko, C. B., see Williams, J. D.
Brantner, J. S., 70-71, 107-108
Breitholtz, M., 193-195
Brendonck, L., 132-133, 140, 148, 153-154
Brennan, E., see McAllen, R.
Broad, A. C., see Hubschman, J. H.
Brock, J. A., 44, 57-58, 201
Brook, H. J., 35, 72, 76, 105, 109
Brooks, B. W., 195
Brown, J. J., see Ahl, J. S. B.
Brown, L. A., see Banta, A. M.
Brown, R. J., 195
Brown, W. M., see Boore, J. L.
Browne, R. A., 36, 91, 142-143
Bryłka-Wołk, M., see Wojtasik, B.
Buhl-Mortensen, L., 16, 31, 53, 76, 109
Buikema Jr, A. L., 47
Bulnheim, H. P., 108, 213-214
Buřič, M., 94, 116
Burke, R. D., see Meladenov, P. V.
Burnett, A. L., 20
Butler, T. H., 110
Butler, T. H., see Margolis, L.
Butlin, R., 3, 22

C

Cáceres, C. E., 138-139, 144-146, 149-150,
     153-155
Calado, R., 69, 72, 200, 240-208, 210, 213
Callaghan, T. R., 162
Callan, H. G., 208-209
Campos, E. O., 12
Campos-Ramos, R., 31, 91-92, 169, 176, 219
Carbonell, A., 48, 75, 116 ,121
Carius, H. J., 44
Carlisle, D. B., 181
Carotenuto, Y., 214
Carr, S. D., 13,
Cartes, J. E., 31
Cattley, J. G., 215
Chakraborthy, C., 17, 19
Chamberlain, G. W., 182
Chan, S. M., 185
Chan, S. M., see Tiu, S. H. K.
Chandran, M. R., see Altaff, K.
Chang, E. S., 184
Chang, E. S., see Kao, H. W., Tamone, S. L.
Chaparro, O. R., 212
Chaplin, J. A., 92
Chapman, A. D., 45
Charmantier, G., 185-186
Charniaux-Cotton, H., 174-175, 177, 180-181
Charniaux Cotton, H., see Ginburger-Vogel, T.
Charnov, E. L., 13, 111
Chaves, A. R., 189
Chen, T., 152
Chittleborough, R. G., 74
Choy, S. C., 181
Christy, J. H., 63
Chu, K. H., 190, 195
Chu, K. H., see Wu, L. T.
Chung, J. S., see Techa, S.
Clark, M. S., 144, 152, 226
Clegg, J. S., 141, 144, 149-152, 226
CMFRI., 13, 35-37, 55, 58, 76-78, 87, 200, 225
Cobos, V., 216
Coccia, E., 191
Cohen, A. C., 48, 53, 91, 115, 222
Cohen, D., see Ra'anan, Z.
Cohen, F. P., 37
Colbourne, J. K., 95, 98, 160
Coman, F. E., 166, 169
Conklin, E. G., 19
Cordaux, R., 208, 213
Corley, L. S., 9
Correa, C., 114
Costa-Souza, A. C., 78, 83, 114-115

Species Index

Subject Index

Printed and bound by CPI Group (UK) Ltd, Croydon, CR0 4YY

01/11/2024

01782622-0006